D0065451

The Facts On File

DICTIONARY
of
CHEMISTRY

The Facts On File

DICTIONARY
of
CHEMISTRY

Revised and Expanded Edition

**Edited by
John Daintith**

Facts On File Publications
New York, New York ● Oxford, England

The Facts On File Dictionary of Chemistry
Revised and Expanded Edition

Copyright © 1981 and 1988 by Charles Letts & Co., Ltd.

Library of Congress Catalog Card Number 88-045477

ISBN 0-8160-1866-9

CIP data available on request

British CIP data available on request

Printed in the United States of America

10 9 8 7 6 5 4 3 2 1

FOREWORD

This dictionary is one of a series designed for use in schools. It is intended for students of Chemistry, but we hope that it will also be helpful to other science students, and to anyone interested in the life sciences. The other books in the series are the Facts On File dictionaries of *Biology, Physics,* and *Mathematics.*

This book contains concise definitions of over 2000 headwords. Originally published in 1980, this new edition has been extensively revised and expanded. In particular, we have included terms for new materials, new techniques, and for topics affecting the environment.

We would like to thank all the people who have cooperated in producing this book. A list of writers and editors is given on the acknowledgments page. We are also grateful to the many people who have given additional help and advice.

ACKNOWLEDGMENTS

Contributors
Julia Brailsford B.Sc.
John Connor B.Sc.
Derek Cooper Ph.D. F.R.I.C.
Rich Cutler B.Sc.
D.E. Edwards B.Sc. M.Sc.
Richard Hadfield B.Sc.
Valerie Illingworth B.Sc. M.Phil.
Alan Isaacs B.Sc. Ph.D.
Elizabeth Martin M.A.
P.R. Mercer B.A. T.C.
R.S. Smith B.Sc. C.Chem. M.R.I.C.
Derek Stefaniw B.Sc. Ph.D.
Elizabeth Tootil M.Sc.
J. Truman B.Sc.
David Eric Ward B.Sc. M.Sc. Ph.D.

Note

Note that the melting points and boiling points given here are at standard pressure unless otherwise stated. Relative densities of liquids are at standard pressure with the liquid at 20°C relative to water at 4°C. Relative densities of gases are relative to air, both gases being at standard temperature and pressure. Abbreviations used in text are:

p.n.	proton number (atomic number)
r.a.m.	relative atomic mass (atomic weight)
m.p.	melting point
b.p.	boiling point
r.d.	relative density

A

absolute alcohol Pure alcohol (ethanol).

absolute temperature Symbol: T A temperature defined by the relationship:
$$T = \theta + 273.15$$
where θ is the Celsius temperature. The absolute scale of temperature was a fundamental scale based on Charles' law applied to an ideal gas:
$$V = V_0(1 + \alpha\theta)$$
where V is the volume at temperature θ, V_0 the volume at 0, and α the thermal expansivity of the gas. At low pressures (when real gases show ideal behaviour) α has the value $1/273.15$. Therefore, at $\theta = -273.15$ the volume of the gas theoretically becomes zero. In practice, of course, substances become solids at these temperatures. However, the extrapolation can be used for a scale of temperature on which $-273.15°C$ corresponds to $0°$ (absolute zero). The scale is also known as the *ideal-gas scale*; on it temperature intervals were called *degrees absolute* (°A) or *degrees Kelvin* (°K), and were equal to the Celsius degree. It can be shown that the absolute temperature scale is identical to the thermodynamic temperature scale (on which the unit is the kelvin).

absolute zero The zero value of thermodynamic temperature; 0 kelvin or $-273.15°C$.

absorption A process in which a gas is taken up by a liquid or solid, or in which a liquid is taken up by a solid. In absorption, the substance absorbed goes into the bulk of the material. Solids that absorb gases or liquids often have a porous structure. The absorption of gases in solids is sometimes called *sorption*. *Compare* adsorption.

absorption indicator (adsorption indicator) An indicator used for titrations that involve a precipitation reaction. The method depends upon the fact that at the equivalence point there is a change in the nature of the ions absorbed by the precipitate particles. Fluorescein – a fluorescent compound – is commonly used. For example, in the titration of sodium chloride solution with added silver nitrate, silver chloride is precipitated. Sodium ions and chloride ions are absorbed in the precipitate. At the end point, silver ions and nitrate ions are in slight excess and silver ions are then absorbed. If fluorescein is present, negative fluorescein ions absorb in preference to nitrate ions, producing a pink complex.

absorption spectrum *See* spectrum.

abundance 1. The relative amount of a given element amongst others; for example, the abundance of oxygen in the Earth's crust is approximately 50% by mass.
2. The amount of a nuclide (stable or radioactive) relative to other nuclides of the same element in a sample. The *natural abundance* is the abundance of a nuclide as it occurs naturally. For instance, chlorine has two stable isotopes of masses 35 and 37. The abundance of ^{35}Cl is 75.5%; that of ^{37}Cl is 24.5%. For some elements the abundance of a particular nuclide depends on the source.

accelerator A catalyst added to increase the rate of cross-linking reactions in polymers.

acceptor The atom or group to which a pair of electrons is donated in a coordinate bond. Pi-acceptors accept electrons into pi, p or d orbitals.

accumulator (secondary cell, storage battery) An electric cell or battery that can be charged by passing an electric current through it. The chemical reaction in the cell is reversible. When the cell begins to run down, current in the opposite direction will convert the reaction products back in-acetal

to their original forms. The most common example is the lead-acid accumulator, used in vehicle batteries.

acetal A type of organic compound formed by addition of an alcohol to an aldehyde. Addition of one alcohol molecule gives a *hemiacetal*. Further addition yields the full acetal. Similar reactions occur with ketones to produce hemiketals and ketals.

acetaldehyde *See* ethanal.

acetamide *See* ethanamide.

acetate *See* ethanoate.

acetic acid *See* ethanoic acid.

acetone *See* propanone.

acetylation *See* acylation.

acetyl chloride *See* ethanoyl chloride.

acetylene *See* ethyne.

acetyl group *See* ethanoyl group.

Acheson process *See* carbon.

achiral Describing a molecule that does not exhibit optical activity.

acid A substance that gives rise to hydrogen ions (or H_3O^+) when dissolved in water. An acid in solution will have a pH below 7. This definition does not take into account the competitive behaviour of acids in solvents and it refers only to aqueous systems. The *Lowry-Brønsted theory* defines an acid as a substance that exhibits a tendency to release a proton, and a base as a substance that tends to accept a proton. Thus, when an acid releases a proton, the ion formed is the *conjugate base* of the acid. Strong acids (e.g. HNO_3) react completely with water to give H_3O^+, i.e. HNO_3 is stronger than H_3O^+ and the conjugate base NO_3^- is weak. Weak acids (e.g. CH_3COOH and C_6H_5COOH) are only partly dissociated because H_3O^+ is a stronger acid than the free acids and the

ions CH_3COO^- and $C_6H_5COO^-$ are moderately strong bases. *See also* Lewis acid.

acid anhydride A type of organic compound of general formula RCOOCOR',

Acid anhydride

where R and R' are alkyl or aryl groups. They are prepared by reaction of an acyl halide with the sodium salt of a carboxylic acid, e.g.:
$RCOCl + R'COO^-Na^+ \rightarrow RCOOCOR' + NaCl$
Like the acyl halides, they are very reactive acylating agents. Hydrolysis is to carboxylic acids:
$RCOOCOR' + H_2O \rightarrow RCOOH + R'COOH$
See also acylation.

acid–base indicator An indicator that is either a weak base or a weak acid and whose dissociated and undissociated forms differ markedly in colour. The colour change must occur within a narrow pH range. Examples are methyl orange and phenolphthalein.

acid dyes The sodium salts of organic acids used in the dyeing of silks and wool. They are so called because they are applied from a bath acidified with dilute sulphuric or acetic acid.

acid halide *See* acyl halide.

acidic Having a tendency to release a proton or to accept an electron pair from a donor. In aqueous solutions the pH is a measure of the acidity, i.e. an acidic solution is one in which the concentration of H_3O^+ exceeds that in pure water at the same temperature; i.e. the pH is lower than 7.

2

acidic hydrogen A hydrogen atom in a molecule that enters into a dissociation equilibrium when the molecule is dissolved in a solvent. For example, in acetic acid (CH_3COOH) the acidic hydrogen is the one on the carboxyl group, $-COOH$.

acidimetry Volumetric analysis or acid-base titration in which a standard solution of an acid is added to the unknown (base) solution plus the indicator. *Alkalimetry* is the converse, i.e. the base is in the burette.

acidity constant See dissociation constant.

acid rain See pollution.

acid salt (acidic salt) A salt in which there is only partial replacement of the acidic hydrogen of an acid by metal or other cations. For polybasic acids the formulae are of the type $NaHSO_4$ (sodium hydrogen sulphate) and $Na_3H(CO_3)_2$.$2H_2O$ (sodium sesquicarbonate). For monobasic acids such as HF the acid salts are of the form KHF_2 (potassium hydrogen fluoride). Although the latter were at one time formulated as a normal salt plus excess acid (i.e. KF.HF) it is preferable to treat these as hydrogen-bonded systems of the type K^+ $(F-H-F)^-$.

acid value A measure of the free acid present in fats, oils, resins, plasticizers, and solvents, defined as the number of milligrams of potassium hydroxide required to neutralize the free acids in one gram of the substance.

actinic radiation Radiation that can cause a chemical reaction; for example, ultraviolet radiation is actinic.

actinides See actinoids.

actinium A toxic radioactive metallic element that is the first member of the actinoid series. It occurs in minute quantities in uranium ores, but is usually obtained by neutron bombardment of radium, and has been used as a source of alpha particles.

Symbol: Ac; m.p. 1050°C; b.p. 3200°C (calc.); r.d. 10.07 (calc.); p.n. 89; stablest isotope ^{227}Ac (half-life 21.6 years).

actinoids (actinides) A group of 15 radioactive elements whose electronic configurations display filling of the 5f level. As with the lanthanoids, the first member, actinium, has no f electrons (Ac $[Rn]6d^17s^2$) but other members also show deviations from the smooth trend of f electron filling expected from simple considerations, e.g. thorium Th $[Rn]6d^27s^2$, berkelium Bk $[Rn]5f^86d^17s^2$. The actinoids are all radioactive and their chemistry is often extremely difficult to study. In general, artificial methods using high-energy bombardment are used to generate them. *See also* transuranic elements.

activated charcoal See charcoal.

activated complex The partially bonded system of atoms in the transition state of a chemical reaction.

activation energy Symbol: E_a The minimum energy a particle, molecule, etc., must acquire before it can react; i.e. the energy required to *initiate* a reaction regardless of whether the reaction is exothermic or endothermic. Activation energy is often represented as an energy barrier that must be overcome if a reaction is to take place. *See* Arrhenius equation.

active mass See mass action, law of.

activity 1. Symbol: a Certain thermodynamic properties of a solvated substance are dependent on its concentration (e.g. its tendency to react with other substances). Real substances show departures from ideal behaviour and a corrective concentration term – the activity – has to be introduced into equations describing real solvated systems.
2. Symbol: A The average number of atoms disintegrating per unit time in a radioactive substance.

3

activity coefficient Symbol: f A measure of the degree of deviation from ideality of a solvated substance, defined as:

$$a = fc$$

where a is the activity and c the concentration. For an ideal solute $f = 1$; for real systems f can be less or greater than unity.

acyclic Describing a compound that is not cyclic (does not contain a ring in its molecules).

acylation A reaction that introduces the acyl group (RCO–). Acylating agents are acyl halides (R.CO.X) and acid anhydrides (R.CO.O.CO.R), which react with such nucleophiles as H_2O, ROH, NH_3, and RNH_2. In these compounds a hydrogen atom of a hydroxyl or amine group is replaced by the RCO– group. In *acetylation* the acetyl group (CH_3CO–) is used. In *benzoylation* the benzoyl group (C_6H_5CO–) is used. Acylation is used to prepare crystalline derivatives of organic compounds to identify them (e.g. by melting point) and to protect –OH groups in synthetic reactions.

acyl anhydride See acid anhydride.

acyl group The group of atoms RCO–.

Acyl halide

acyl halide (acid halide) A type of organic compound of the general formula RCOX, where X is a halogen (acyl chloride, acyl bromide, etc.).
Acyl halides can be prepared by the reaction of carboxylic acid with a halogenating agent. Commonly, phosphorus halides are used (e.g. PCl_5) or a sulphur dihalide oxide (e.g. $SOCl_2$):
$RCOOH + PCl_5 \rightarrow RCOCl + POCl_3 + HCl$

$RCOOH + SOCl_2 \rightarrow RCOCl + SO_2 + HCl$
The acyl halides have irritating vapours and fume in moist air. They are very reactive to the hydrogen atom of compounds containing hydroxyl (–OH) or amine (–NH_2) groups. See acylation.

addition reaction A reaction in which additional atoms or groups of atoms are introduced into an unsaturated compound, such as an alkene or ketone. A simple example is the addition of bromine across the double bond in ethene:
$H_2C:CH_2 + Br_2 \rightarrow BrH_2CCH_2Br$
Addition reactions can be induced either by electrophiles or by nucleophiles. See also electrophilic addition, nucleophilic addition.

addition polymerization See polymerization.

adduct See coordinate bond.

adiabatic change A change during which no energy enters or leaves the system.
In an adiabatic expansion of a gas, mechanical work is done by the gas as its volume increases and the gas temperature falls. For an ideal gas undergoing a reversible adiabatic change it can be shown that
$$pV^\gamma = K_1$$
$$T^\gamma p^{1-\gamma} = K_2$$
$$\text{and } TV^{\gamma-1} = K_3$$
where K_1, K_2, and K_3 are constants and γ is the ratio of the principal specific heat capacities.
Compare isothermal change.

adipic acid See hexanedioic acid.

adsorption A process in which a layer of atoms or molecules of one substance forms on the surface of a solid or liquid. All solid surfaces take up layers of gas from the surrounding atmosphere. The adsorbed layer may be held by chemical bonds (*chemisorption*) or by weaker van der Waals forces (*physisorption*).
Compare absorption.

adsorption indicator *See* absorption indicator.

aerosol *See* sol.

afterdamp *See* firedamp.

air The mixture of gases that surrounds the Earth. The composition of dry air, by volume, is
nitrogen 78.08%
oxygen 20.95%
argon 0.93%
carbon dioxide 0.03%
neon 0.0018%
helium 0.0005%
krypton 0.0001%
xenon 0.00001%
Air also contains a variable amount of water vapour, as well as particulate matter (e.g. dust and pollen), and small amounts of other gases.

air gas *See* producer gas.

alabaster A mineral form of gypsum ($CaSO_4.2H_2O$).

alanine *See* amino acids.

alcohol A type of organic compound of the general formula ROH, where R is a hydrocarbon group. Examples of simple alcohols are methanol (CH_3OH) and ethanol (C_2H_5OH).
Alcohols have the −OH group attached to a carbon atom that is not part of an aromatic ring: C_6H_5OH, in which the −OH group is attached to the ring, is thus a phenol. Phenylmethanol ($C_6H_5CH_2OH$) does have the characteristic properties of alcohols.
Alcohols can have more than one −OH group; those containing two, three or more such groups are described as *dihydric*, *trihydric*, and *polyhydric* respectively (as opposed to those containing one −OH group, which is *monohydric*). Examples are glycol ($HOCH_2.CH_2OH$) and glycerol ($CH_2OH.CH(OH).CH_2OH$).
Alcohols are further classified according to the environment of the −C−OH grouping. If the carbon atom is attached to two hy-

primary

secondary

tertiary

Alcohols

drogen atoms, the compound is a *primary alcohol*. If the carbon atom is attached to one hydrogen atom and two other groups, it is a *secondary alcohol*. If the carbon atom is attached to three other groups, it is a *tertiary alcohol*.
Alcohols can be prepared by:
(1) Hydrolysis of haloalkanes using aqueous potassium hydroxide:

$$RI + OH^- \rightarrow ROH + I^-$$

(2) Reduction of aldehydes by nascent hydrogen (from sodium amalgam in water):

$$RCHO + 2[H] \rightarrow RCH_2OH$$

The main reactions are:
(1) Oxidation by potassium dichromate(VI) in sulphuric acid. Primary alcohols give aldehydes, which are further oxidized to carboxylic acids:
$[RCH_2OH \rightarrow RCHO \rightarrow RCOOH]$

aldehyde

Secondary alcohols are oxidized to ketones.

(2) Formation of esters with acids. The reaction, which is reversible, is catalysed by H^+ ions:

[ROH + R'COOH = R'COOR]

(3) Dehydration over hot pumice (400°C) to alkenes:

$$RCH_2CH_2OH - H_2O \rightarrow RCH{:}CH_2$$

(4) Reaction with sulphuric acid. Two types of reaction are possible. With excess acid at 160°C dehdyration occurs to give an alkene:

$$RCH_2CH_2OH + H_2SO_4 \rightarrow H_2O + RCH_2CH_2.HSO_4$$
$$RCH_2CH_2.HSO_4 \rightarrow RCH{:}CH_2 + H_2SO_4$$

With excess alcohol at 140°C an ether is formed:

$$2ROH \rightarrow ROR + H_2O$$

See also acylation, haloalkane.

Aldehyde

aldehyde A type of organic compound with the general formula (RCHO, where the $-$CHO group (the aldehyde group) consists of a carbonyl group attached to a hydrogen atom. Simple examples of aldehydes are methanal (formaldehyde, HCHO) and ethanal (acetaldehyde, CH_3CHO).

Aldehydes are formed by oxidizing a primary alcohol. In the laboratory potassium dichromate(VI) is used in sulphuric acid. They can be further oxidized to carboxylic acids. Reduction (using a catalyst or nascent hydrogen from sodium amalgam in water) produces the parent alcohol.

Aldehydes undergo a number of reactions:

(1) They act as reducing agents, being oxidized to carboxylic acids in the process. These reactions are used as tests for aldehydes using such reagents as Fehling's solution and Tollen's reagent (silver-mirror test).

(2) They form addition compounds with hydrogen cyanide to give 'cyanohydrins'.

For example, propanal gives 2-hydroxybutanonitrile:

$$C_2H_5CHO + HCN \rightarrow C_2H_5CH(OH)CN$$

(3) They form addition compounds with the hydrogensulphate(IV) ion (HSO_3^-):

$$RCHO + HSO_3^- \rightarrow RCH(OH)(HSO_3)$$

(4) They undergo condensation reactions with such compounds as hydrazine, hydroxylamine, and their derivatives.

(5) With alcohols they form hemiacetals and acetals.

(6) They polymerize readily. Polymethanal or methanal trimer can be formed from methanal depending on the conditions. Ethanal gives ethanal trimer or ethanal tetramer.

See also acetal, Cannizzaro reaction, condensation reaction, ketone.

aldohexose An aldose sugar with six carbon atoms. *See* sugar.

aldol *See* aldol reaction.

aldol reaction A reaction in which two molecules of aldehyde combine to give an *aldol* — i.e. a compound containing both aldehyde and alcohol functional groups. The reaction is base-catalysed; the reaction of ethanal refluxed with sodium hydroxide gives:

$$2CH_3CHO \rightarrow CH_3CH(OH)CH_2CHO$$

The mechanism is similar to that of the Claisen condensation: the first step is removal of a proton to give a carbanion, which subsequently attacks the carbon of the carbonyl group on the other molecule:

$$CH_3CHO + OH^- \rightarrow {}^-CH_2CHO + H_2O$$

See also Claisen condensation.

aldopentose An aldose sugar with five carbon atoms. *See* sugar.

aldose A sugar containing an aldehyde (CHO) or potential aldehyde group. *See* sugar.

alicyclic compound An aliphatic cyclic compound, such as cyclohexane.

aliphatic compound An organic compound with properties similar to those of the alkanes, alkenes, and alkynes and their derivatives. Most aliphatic compounds have an open chain structure but some, such as cyclohexane and sucrose, have rings. The term is used in distinction to aromatic compounds, which are similar to benzene. *Compare* aromatic compound.

alkali A water-soluble strong base. Strictly the term refers to the hydroxides of the alkali metals (group I) only, but in common usage it refers to *any* soluble base. Thus borax solution may be described as mildly alkaline.

alkali metals (group I elements) A group of soft reactive metals, each representing the start of a new period in the periodic table and having an electronic configuration consisting of a noble-gas structure plus one outer electron. The alkali metals are lithium (Li), sodium (Na), potassium (K), rubidium (Rb), caesium (Cs), and francium (Fr).

The elements all easily form positive ions M^+ and consequently are highly reactive (particularly with any substrate that is oxidizing). As the group is descended there is a gradual decrease in ionization potential and an increase in the size of the atoms; the group shows several smooth trends which follow from this. For example, lithium reacts in a fairly controlled way with water, sodium ignites, and potassium explodes. There is a general decrease in the following: melting points, heats of sublimation, lattice energy of salts, hydration energy of M^+, ease of decomposition of nitrates and carbonates, and heat of formation of the "-ide" compounds (fluoride, hydride, oxide, carbide, chloride).

Lithium has the smallest ion and therefore the highest charge/size ratio and is polarizing with a tendency towards covalent character in its bonding; the remaining elements form typical ionic compounds in which ionization, M^+X^-, is regarded as complete. The slightly anomalous position of lithium is illustrated by the similarity of its chemistry to that of magnesium. For example, lithium hydroxide is much less soluble than the hydroxides of the other group I elements; lithium perchlorate is soluble in several organic solvents. Be-

cause of the higher lattice energies associated with smaller ions lithium hydride and nitride are fairly stable compared to NaH which decomposes at 345°C. Na_2N, K_3N etc., are not obtained pure and decompose below room temperature.

The oxides also display the trend in properties as lithium forms M_2O with only traces of M_2O_2, sodium forms M_2O_2 and at high temperatures and pressures MO_2, potassium, rubidium, and caesium form M_2O_2 if oxygen is restricted but MO_2 if burnt in air. Hydrolysis of the oxides or direct reaction of the metal with water leads to the formation of the hydroxide ion.

Salts of the bases MOH are known for all acids and these are generally white crystalline solids. The ions M^+ are hydrated in water and remain unchanged in most reactions of alkali metal salts.

Because of the ease of formation of the ions M^+ there are very few coordination compounds of the type ML_n^+ apart from solvated species of very low correlation times. The group I elements form a variety of organometallic compounds; the bonding in lithium alkyls and aryls is essentially covalent but the heavier elements form ionic compounds. Organo-alkali metal compounds – particularly the lithium compounds –are widely used in synthetic organic chemistry.

Francium is formed only by radioactive decay and in nuclear reactions; all the isotopes of francium have short half-lives the longest of which is 21 minutes (francium-223). The few chemical studies which have been carried out indicate that it would have similar properties to those of the other metals.

alkalimetry *See* acidimetry.

alkaline-earth metals (group II elements) A group of moderately reactive metals, harder and less volatile than the alkali metals. The term *alkaline earth* strictly refers to the oxides, but is often used loosely for the elements themselves. The electronic configurations are all those of a noble-gas structure with an additional two electrons in the outer s orbital. The elements are beryllium (Be), magnesium (Mg), calcium (Ca), strontium (Sr), barium (Ba), and radium (Ra). The group shows an increasing tendency to ionize to the divalent state M^{2+}. The first member, beryllium has a much higher ionization potential than the others and the smallest atomic radius. Thus it has a high charge/size ratio and consequently the bonding in beryllium compounds is largely covalent. The chemistry of the heavier members of the group is largely that of divalent ions.

The group displays a typical trend towards metallic character as the group is descended. For example, beryllium hydroxide is amphoteric; magnesium hydroxide is almost insoluble in water and is slightly basic; calcium hydroxide is sparingly soluble and distinctly basic; and strontium and barium hydroxides are increasingly soluble in water and strongly basic. The group also displays a smooth trend in the solubilities of the sulphates ($MgSO_4$ is soluble, $CaSO_4$ sparingly soluble, and $BaSO_4$ very insoluble). The trend to increasing metallic character is also shown by the increase in thermal stabilities of the carbonates and nitrates with increasing relative atomic mass.

The elements all burn in air (beryllium must be finely powdered) to give the oxide MO (covalent in the case of beryllium) and for barium the peroxide, BaO_2 in addition to BaO. The heavier oxides, CaO, SrO, and BaO, react with water to form hydroxides, $M(OH)_2$; magnesium oxide reacts only at high temperatures and beryllium oxide not at all. The metals Ca, Sr, and Ba all react readily with water to give the hydroxide:

$$M + 2H_2O \rightarrow M^{2+} + 2OH^- + H_2$$

In contrast, magnesium requires dilute acids in order to react (to the salt plus hydrogen), and beryllium is resistant to acid attack. A similar trend is seen in the direct reaction of hydrogen: under mild conditions calcium, strontium, and barium give ionic hydrides, high pressures are required to form magnesium hydride, and beryllium hydride can not be prepared by direct combination.

Because of its higher polarizing power, beryllium forms a range of complexes in which the beryllium atom should be treated as an electron acceptor (i.e. the vacant p orbitals are being used). Complexes such as etherates, acetylethanoates, and the tetrafluoride (BeF_4^{2-}) are formed, all of which are tetrahedral. In contrast Mg^{2+}, Ca^{2+}, Sr^{2+}, and Ba^{2+} have poor acceptor properties and form only weak complexes, even with donors such as ammonia or edta. Magnesium forms Grignard reagents (RMgX), which are important in organic synthesis, and related compounds R_2Mg. MgX_2 and R_2Mg are known. The few organic compounds of Ca, Sr, and Ba are ionic. All isotopes of radium are radioactive and radium was once widely used for radiotherapy. The half-life of ^{226}Ra (formed by decay of ^{238}U) is 1600 years.

alkaloid One of a group of natural organic compounds found in plants. They contain oxygen and nitrogen atoms; most are poisonous. However, they include a number of important drugs with characteristic physiological effects, e.g. morphine, codeine, caffeine, cocaine, and nicotine.

alkane A type of hydrocarbon with general formula C_nH_{2n+2}. Alkanes are saturated compounds, containing no double or triple bonds. Methane (CH_4) and ethane (C_2H_6) are typical examples. The alkanes are fairly unreactive (their former name, the *paraffins*, means 'small affinity'). In ultraviolet radiation they react with chlorine to give a mixture of substitution products. There are a number of ways of preparing alkanes:

(1) From a sodium salt of a carboxylic acid treated with sodalime:

$$RCOO^-Na^+ + NaOH \rightarrow RH + Na_2CO_3$$

(2) By reduction of a haloalkane with nascent hydrogen from the action of ethanol on a zinc-copper couple:

$$RX + 2[H] \rightarrow RH + HX$$

(3) By the Wurtz reaction – i.e. sodium in dry ether on a haloalkane:

$$2RX + 2Na \rightarrow 2NaX + RR$$

(4) By the Kolbé electrolytic method:

$$RCOO^- \rightarrow RR$$

(5) By refluxing a haloalkane with magnesium in dry ether to form a Grignard reagent:

$$RI + Mg \rightarrow RMgI$$

With acid this gives the alkane:

$$RMgI + H \rightarrow RH$$

The main source of lower molecular weight alkanes is natural gas (for methane) and crude oil.

alkene A type of hydrocarbon with the formula C_nH_{2n}. The alkenes (formerly called *olefins*) are unsaturated compounds containing double carbon-carbon bonds. They can be obtained from crude oil by cracking alkanes. Important examples are ethene (C_2H_4) and propene (C_3H_6), both of which are used in plastics production and as starting materials for the manufacture of other organic chemicals.

The methods of synthesizing alkenes are:
(1) The elimination of HBr from a haloalkane using an alcoholic solution of potassium hydroxide:

$$RCH_2CH_2Br + KOH \rightarrow KBr + H_2O + RCH:CH_2$$

(2) The dehydration of an alcohol by passing the vapour over hot pumice (400°C):

$$RCH_2CH_2OH \rightarrow RCH:CH_2 + H_2O$$

The reactions of alkenes include:
(1) Hydrogenation using a catalyst (usually nickel at about 150°C):

$$RCH:CH_2 + H_2 \rightarrow RCH_2CH_3$$

(2) Addition reactions with halogen acids to give haloalkanes:

$$RCH:CH_2 + HX \rightarrow RCH_2CH_2X$$

The addition follows Markovnikoff's rule.
(3) Addition reactions with halogens, e.g.

$$RCH:CH_2 + Br_2 \rightarrow RCHBrCH_2Br$$

(4) Hydration using concentrated sulphuric acid, followed by dilution and warming:

$$RCH:CH_2 + H_2O \rightarrow RCH(OH)CH_3$$

(5) Oxidation by cold potassium permanganate solutions to give diols:

$$RCH:CH_2 + H_2O + [O] \rightarrow RCH(OH)CH_2OH$$

Ethene can be oxidized in air using a silver catalyst to the cyclic compound epoxy ethane (C_2H_4O).
(6) Polymerization to polyethene (by the Ziegler or Phillips process).
See also oxo reaction, ozonolysis.

alkoxide An organic compound containing an ion of the type RO^-, where R is an alkyl group. Alkoxides are made by the reaction of metallic sodium on an alcohol. Sodium ethoxide ($C_2H_5O^-Na^+$) is an example.

alkylbenzene A type of organic hydrocarbon containing one or more alkyl groups substituted onto a benzene ring. Methylbenzene ($C_6H_5CH_3$) and 1,3-dimethylbenzene are simple examples. Alkylbenzenes can be made by a Friedel-Crafts reaction or by the Fittig reaction. Industrially, large quantities of methylbenzene are made by the hydroforming of crude oil.

Substitution of alkylbenzenes can occur at the benzene ring. The alkyl group directs the substituent into the 2- or 4-position. Substitution of hydrogen atoms on the alkyl group can also occur.

alkyne A type of hydrocarbon with the general formula C_nH_{2n-2}. The alkynes are unsaturated compounds containing triple carbon-carbon bonds. The simplest member of the series is ethyne (C_2H_2), which can be prepared by the action of water on calcium dicarbide. The alkynes were formerly called the *acetylenes*.

$$CaC_2 + 2H_2O \rightarrow Ca(OH)_2 + C_2H_2$$

In general, alkynes can be made by the cracking of alkanes or by the action of a hot alcoholic solution of potassium hydroxide on a dibromoalkane, for example:

$$BrCH_2CH_2Br + KOH \rightarrow KBr + CH_2:CHBr + H_2O$$
$$CH_2:CHBr + KOH \rightarrow CHCH + KBr + H_2O$$

The main reactions of the alkynes are:
(1) Hydrogenation with a catalyst (usually nickel at about 150°C):

$$C_2H_2 + H_2 \rightarrow C_2H_4$$
$$C_2H_4 + H_2 \rightarrow C_2H_6$$

(2) Addition reactions with halogen acids:

$$C_2H_2 + HI \rightarrow H_2C:CHI$$
$$H_2C:CHI + HI \rightarrow CH_3CHI_2$$

(3) Addition of halogens; for example, with bromine in carbon tetrachloride:

$$C_2H_2 + Br_2 \rightarrow BrHC:CHBr$$
$$BrHC:CHBr + Br_2 \rightarrow Br_2HCCHBr_2$$

(4) With dilute sulphuric acid at 60–80°C and mercury(II) catalyst, ethyne forms ethanal:

$$C_2H_2 + H_2O \rightarrow H_2C:C(OH)H$$

This enol form converts to the aldehyde:

$$CH_3COH$$

(5) Ethyne polymerizes if passed through a hot tube to produce some benzene:

$$3C_2H_2 \rightarrow C_6H_6$$

(6) Ethyne forms unstable dicarbides (acetylides) with ammoniacle solutions of copper(I) and silver(I) chlorides.

alkyl group A group obtained by removing a hydrogen atom from an alkane or other aliphatic hydrocarbon.

alkyl halide See haloalkane.

allotropy The ability of certain elements to exist in more than one physical form. Sulphur and phosphorus are the most common examples. Allotropy is more common in groups IV, V, and VI of the periodic table than in other groups. See also enantiotropy, monotropy.

alloy A mixture of two or more metals (e.g. bronze or brass) or a metal with small amounts of non-metals (e.g. steel). Alloys may be completely homogeneous mixtures or may contain small particles of one phase in the other phase.

Alnico (Trademark) Any of a group of very hard brittle alloys used to make powerful permanent magnets. They contain nickel, aluminium, cobalt, and copper in various proportions. Iron, titanium, and niobium can also be present. They have a high remanence and coercive force.

alpha particle A He^{2+} ion emitted at a high velocity by a radioactive substance. Alpha particles are used to cause nuclear disintegration reactions.

alum A type of double salt. Alums are double sulphates obtained by crystallizing mixtures in the correct proportions. They have the general formulae:

$$M_2SO_4.M'_2(SO_4)_3.24H_2O$$

Where M is a univalent metal or ion, and M' is a trivalent metal. Thus, aluminium potassium sulphate (called potash alum, or simply alum) is

$$K_2SO_4.Al_2(SO_4)_3.24H_2O$$

Aluminium ammonium sulphate (called ammonium alum) is

$$(NH_4)_2SO_4.Al_2(SO_4)_3.24H_2O$$

The name 'alum' originally came from the presence of Al^{3+} as the trivalent ion, but is also applied to other salts containing trivalent ions, thus, chromium(III) potassium sulphate (chrome alum) is

$$K_2SO_4.Cr_2(SO_4)_3.24H_2O$$

alumina See aluminium oxide.

aluminate See aluminium hydroxide.

aluminium A soft moderately reactive metal; the second element in group III of the periodic table. It has the electronic structure of neon plus three additional outer electrons. There are numerous minerals of aluminium; it is the most common metallic element in the Earth's crust (8.1% by weight) and the third in order of abundance. Commercially important minerals are bauxite (hydrated Al_2O_3), corundum (anhydrous Al_2O_3), cryolite (Na_3AlF_6), and clays and mica (aluminosilicates). The metal is produced on a massive scale by the Hall-Heroult method in which alumina, a non-electrolyte, is dissolved in molten cryolite and electrolysed. The bauxite contains iron, which would contaminate the product, so the bauxite is dissolved in hot alkali, the iron oxide is removed by filtration, and the pure alumina then precipitated by acidification. Molten aluminium is tapped off from the base of the cell and oxygen evolved at the anode. The aluminium atom is much bigger than boron and its ionization potential is not particularly high. Consequently aluminium forms positive ions Al^{3+}. However, it also has nonmetallic chemical properties. Thus, it is amphoteric and also has a number of covalently bonded compounds.

Unlike boron, aluminium does not form a vast range of hydrides – AlH_3 and Al_2H_6 may exist at low pressures, and the only stable hydride, $(AlH_3)_n$, must be prepared by reduction of aluminium trichloride. The ion AlH_4^- is widely used in the form of $LiAlH_4$ as a vigorous reducing agent.

The reaction of aluminium metal with oxygen is very exothermic but at ordinary temperatures an impervious film of the oxide protects the metal from further attack. This oxide film also protects aluminium from oxidizing acids. There is only one oxide, Al_2O_3 (alumina), but a variety of polymorphs and hydrates are known. It is relatively inert and has a high melting point, and for this reason is widely used as a furnace lining and for general refractory brick. Aluminium metal will react with alkalis releasing hydrogen and producing initially $Al(OH)_3$ then $Al(OH)_4^-$.

Aluminium reacts readily with the halogens; in the case of chlorine thin sheets will enflame. The fluoride has a high melting point (1290°C) and is ionic. The other halides are dimers in the vapour phase (two halogen bridges). Aluminium also forms a sulphide (Al_2S_3), nitride (AlN), and carbide (Al_4C), the latter two at extremely high temperatures.

Because of aluminium's ability to expand its coordination number and tendency towards covalence it forms a variety of complexes such as AlF_6^{2-} and $AlCl_4^-$. A number of very reactive aluminium alkyls are also known, some of which are important as polymerization catalysts. Symbol: Al; m.p. 659.8°C; b.p. 2450°C; r. d. 2.702; p.n. 13; r.a.m. 26.97.

aluminium acetate (aluminium ethanoate, $Al(OOCCH_3)_3$) A white solid soluble in water. It is usually obtained as the dibasic salt, *basic aluminium ethanoate* $(Al(OH)(C_2H_2O_2)_2)$. It is prepared by dissolving aluminium hydroxide in ethanoic acid and is used extensively as a mordant in dyeing and as a size for paper and cardboard products. The solution is hydrolysed and contains various complex aluminium-hydroxyl species and colloidal aluminium hydroxide.

aluminium bromide $(AlBr_3)$ A white solid soluble in water and many organic solvents.

aluminium chloride $(AlCl_3)$ A white covalent solid that fumes in moist air:
$$AlCl_3 + 3H_2O \rightarrow Al(OH)_3 + 3HCl$$
It is prepared by heating aluminium in dry chlorine or dry hydrogen chloride. Vapour-density measurements show that its structure is a dimer; it consists of Al_2Cl_6 molecules in the vapour. The $AlCl_3$ structure would be electron deficient. Aluminium chloride is used in Friedel-Crafts reactions in organic preparations.

aluminium fluoride (AlF_3) A white crystalline solid that is slightly soluble in water but insoluble in most organic solvents. Its primary use is as an additive to the cryolite (Na_3AlF_6) electrolyte in the production of aluminium.

aluminium hydroxide $(Al(OH)_3)$ A white powder prepared as a colourless gelatinous precipitate by adding ammonia solution or a small amount of sodium hydroxide solution to a solution of an aluminium salt. It is an amphoteric hydroxide. It is used as a foaming agent in fire extinguishers and as a mordant in dyeing.

Its amphoteric nature causes it to dissolve in excess sodium hydroxide solution to form the *aluminate* ion.
$$Al(H_2O)_3(OH)_3 + OH^- \rightarrow$$
$$Al(H_2O)_2(OH)_4^- + H_2O$$
No aluminium carbonate is produced since solutions of carbonates are sufficiently alkaline to precipitate aluminium hydroxide from solutions of aluminium salts. When precipitating, aluminium hydroxide readily absorbs coloured matter from dyes to form lakes.

aluminium nitrate $(Al(NO_3)_3.9H_2O)$ A hydrated white crystalline solid prepared by dissolving freshly prepared aluminium hydroxide in nitric acid. It cannot be prepared by the action of dilute nitric acid on aluminium since the metal is rendered passive.

aluminium oxide (alumina, Al_2O_3) A white powder that is almost insoluble in water. Because of its amphoteric nature it will react with both acids and alkalis. Aluminium oxide occurs naturally as bauxite,

11

corundum, and white sapphire; it is manufactured by heating aluminium hydroxide. It is used in the extraction by electrolysis of aluminium, as an abrasive (corundum), in furnace linings (because of its refractory properties), and as a catalyst (e.g. in the dehydration of alcohols).

aluminium potassium sulphate (potash alum, $Al_2(SO_4)_3.K_2SO_4.24H_2O$, $AlK(SO_4)_2.12H_2O$) A white solid, soluble in water but insoluble in alcohol, prepared by mixing equimolecular quantities of solutions of ammonium and aluminium sulphate followed by crystallization. It is used as a mordant for dyes, as a waterproofing agent, and as a tanning additive.

aluminium sulphate ($Al_2(SO_4)_3.$ $18H_2O$) A white crystalline solid. It is used as a size for paper, a precipitating agent in sewage treatment, a foaming agent in fire control, and as a fireproofing agent. Its solutions are acidic by hydrolysis, containing such species as $Al(H_2O)_5(OH)^{2+}$.

aluminium trimethyl See trimethylaluminium.

amalgam An alloy of mercury with one or more other metals. Amalgams may be liquid or solid. An amalgam of sodium (Na/Hg) with water is used as a source of nascent hydrogen.

americium A highly toxic radioactive silvery element of the actinoid series of metals. A transuranic element, it is not found naturally on Earth but is synthesized from plutonium. ^{241}Am has been used in gamma-ray radiography.
Symbol: Am; m.p. 994°C; b.p. 2607°C; r.d. 13.67; p.n. 95; stablest isotope ^{243}Am (half-life 7.4×10^3 years).

amide 1. A type of organic compound of general formulae $RCONH_2$ (primary), $(RCO)_2NH$ (secondary), and $(RCO)_3N$ (tertiary). Amides are white, crystalline solids and are basic in nature, some being soluble in water. Amides can be formed by reaction of ammonia with acid chlorides or anhydrides:

$$(RCO)_2O + 2NH_3 \rightarrow RCONH_2 + RCOO^-NH_4^+$$

Reactions of amides include:
(1) reaction with hot acids to give carboxylic acids:
$$RCONH_2 + HCl + H_2O \rightarrow RCOOH + NH_4Cl$$
(2) Reaction with nitrous acid to give carboxylic acids and nitrogen:
$$RCONH_2 + HNO_2 \rightarrow RCOOH + N_2 + H_2O$$
(3) Dehydration by phosphorus(V) oxide to give a nitrile:
$$RCONH_2 - H_2O \rightarrow RCN$$
See also Hofmann degradation.
2. An inorganic salt containing the NH_2^-

primary

secondary

tertiary

Amines

ion. They are formed by the reaction of ammonia with certain metals (such as sodium and potassium).

amine A compound containing a nitrogen atom bound to hydrogen atoms or hydrocarbon groups. They have the general formula R_3N, where R can be hydrogen. Amines can be prepared by reduction of amides or nitro compounds.

An amine is classified according to the number of organic groups bonded to the nitrogen atom: one, primary; two, secondary; three, tertiary. Since amines are basic in nature they can form the quaternium ion, R_3NH^+. All three classes, plus a quaternium salt, can be produced by the Hofmann reaction (which occurs in a sealed vessel at $100°C$):

$$RX + NH_3 \rightarrow RNH_3{}^+ X^-$$
$$RNH_3{}^+ X^- + NH_3 \rightleftharpoons RNH_2 + NH_4X$$
$$RNH_2 + RX \rightarrow R_2NH_2{}^+ X^-$$
$$R_2NH_2{}^+ X^- + NH_3 \rightleftharpoons R_2NH + NH_4X$$
$$R_2NH + RX \rightarrow R_3NH^+ X^-$$
$$R_3NH^+ X^- + NH_3 \rightleftharpoons R_3N + NH_4X$$
$$R_3N + RX \rightarrow R_4N^+X^-$$

Reactions of amines include:

(1) Reaction with acids to form salts:
$$R_3N + HX \rightarrow R_3NH^+X^-$$
(2) Reaction with acid chlorides to give N-substituted acid amides (primary and secondary amines only):
$$RNH_2 + R'COCl \rightarrow R'CONHR + HX$$

amino acids Derivatives of carboxylic acids in which a hydrogen atom in an aliphatic acid has been replaced by an amino group. Thus, from ethanoic acid, the amino acid 2-aminoethanoic acid (glycine) is formed. All are white, crystalline, soluble in water (but not in alcohol), and with the sole exception of the simplest member, all are optically active.

In the body the various proteins are assembled from the necessary amino acids and it is important therefore that all the amino acids should be present in sufficient quantities. In man, twelve of the twenty amino acids can be synthesized by the body itself. Since these are not required in the diet they are known as *non-essential amino acids*. The remaining eight cannot be synthesized by the body and have to be

un-ionized

zwitterion

Amino acid

supplied in the diet. They are known as *essential* amino acids.

The amino acids that occur in proteins all have the $-NH_2$ group and the $-COOH$ group attached to the same carbon atom. They are thus *alpha amino acids*, the carbon atom being the alpha carbon. They have complex formulae and are usually referred to by their common names, rather than systematic names:

alanine $CH_3CH(NH_2)COOH$
arginine
$NH_2C(NH)NH(CH_2)_3CH(NH_2)COOH$
asparagine $NH_2COCH_2CH(NH_2)COOH$
aspartic acid $COOHCH_2CH(NH_2)COOH$
cysteine $SHCH_2CH(NH_2)COOH$
cystine $[HOOCCH(NH_2)CH_2S]_2$
glutamic acid
$COOH(CH_2)_2CH(NH_2)COOH$
glutamine $NH_2CH(CH_2)_2(CONH_2)COOH$
glycine $CH_2(NH_2)COOH$
histidine $C_3H_3N_2CH_2CH(NH_2)COOH$
isoleucine
$(CH_3)CH_2CH(CH_3)CH(NH_2)COOH$
leucine $(CH_3)_2CHCH_2CH(NH_2)COOH$
lysine $NH_2(CH_2)_4CH(NH_2)COOH$
methionine $CH_3S(CH_2)_2CH(NH_2)COOH$
phenylalanine $C_6H_5CH_2CH(NH_2)COOH$
proline $NH(CH_2)_3CHCOOH$
serine $CH_2OHCH(NH_2)COOH$
threonine $CH_3CHOHCH(NH_2)COOH$

13

tryptophan
$C_6H_4NHC_2HCH_2CH(NH_2)COOH$
tyrosine $C_6H_4OHCH_2CH(NH_2)COOH$
valine $(CH_3)_2CHCH(NH_2)COOH$
Note that proline is in fact a cyclic imino acid, with the nitrogen atom bonded to the alpha carbon.
See also optical activity, Phillips synthesis.

aminobenzene *See* aniline.

aminoethane *See* ethylamine.

amino group The group $-NH_2$.

ammine A complex in which ammonia molecules are coordinated to a metal ion; e.g. $[Cu(NH_3)_4]^{2+}$.

ammonia (NH_3) A colourless gas with a characteristic pungent odour. On cooling and compression it forms a colourless liquid, which becomes a white solid on further cooling. Ammonia is very soluble in water (a saturated solution at 0°C contains 36.9% of ammonia): the aqueous solution is alkaline and contains a proportion of free ammonia. Ammonia is also soluble in ethanol. It occurs naturally to a small extent in the atmosphere, and is usually produced in the laboratory by heating an ammonium salt with a strong alkali. Ammonia is synthesized industrially from hydrogen and atmospheric nitrogen by the Haber process.

The compound does not burn readily in air but ignites, giving a yellowish-brown flame, in oxygen. It will react with atmospheric oxygen in the presence of platinum or a heavy metal catalyst – a reaction used as the basis of the commercial manufacture of nitric acid, which involves the oxidation of ammonia to nitrogen monoxide and then to nitrogen dioxide. Ammonia coordinates readily to form ammines and reacts with sodium or potassium to form inorganic amides and with acids to form ammonium salts; for example, it reacts with hydrogen chloride to form ammonium chloride:

$$NH_3(g) + HCl(g) \rightarrow NH_4Cl(g)$$

Ammonia is also used commercially in the manufacture of fertilizers, mainly ammonium nitrate, urea, and ammonium sulphate. It is used to a smaller extent in the refrigeration industry. Liquid ammonia is an excellent solvent for certain substances, which ionize in the solutions to give ionic reactions similar to those occurring in aqueous solutions. Ammonia is marketed as the liquid, compressed in cylinders ('anhydrous ammonia'), or as aqueous solutions of various strengths. The saturated solution is known commercially as '880 ammonia'.

ammoniacal Describing a solution in aqueous ammonia.

ammonium carbonate ($(NH_4)_2CO_3$) A white solid that crystallizes as plates or prisms. It is very soluble in water and readily decomposes on heating to ammonia, carbon dioxide, and water. The white solid sold commercially as ammonium carbonate is actually a double salt of ammonium hydrogencarbonate (NH_4HCO_3) and ammonium aminomethanoate (NH_2-CO_2NH_4). This salt is manufactured from ammonium chloride and calcium carbonate. It decomposes on exposure to air into ammonium hydrogencarbonate and ammonia, and it reacts with ammonia to give the true ammonium carbonate. Commercial ammonium carbonate is used in baking powders, smelling salts, and in the dyeing and wool-scouring industries.

ammonium chloride (sal ammoniac, NH_4Cl) A white crystalline solid with a characteristic saline taste. It is very soluble in water (37 g per 100 g of water at 20°C). Ammonium chloride can be manufactured by the action of ammonia on hydrochloric acid. It sublimes on heating because of the equilibrium:

$$NH_4Cl(s) = NH_3(g) + HCl(g)$$

Ammonium chloride is used in galvanizing, as a flux for soldering, in dyeing and calico printing, and in the manufacture of Leclanché and 'dry' cells.

ammonium ion The ion NH_4^+, formed by coordination of NH_3 to H^+. *See also* quaternary ammonium compound.

ammonium nitrate (NH_4NO_3) A colourless crystalline solid that is very soluble in water (871 g per 100 g of water at 100°C). It is usually manufactured by the action of ammonia on nitric acid. It is used in the manufacture of explosives and, because of its high nitrogen content, as a fertilizer.

ammonium sulphate (($NH_4)_2SO_4$) A colourless crystalline solid that is soluble in water. When heated carefully it gives ammonium hydrogen sulphate, which on stronger heating yields nitrogen, ammonia, sulphur dioxide, and water. Ammonium sulphate is manufactured by the action of ammonia on sulphuric acid. It is the most important ammonium salt because of its widespread use as a fertilizer. Its only drawback as a fertilizer is that it tends to leave an acidic residue in the soil.

amorphous Describing a solid substance that has no 'long-range' regular arrangement of atoms; i.e. is not crystalline. Amorphous materials can consist of minute particles that possess order over a very short distance. Glasses are also amorphous; the atoms in the solid have a random arrangement. X-ray analysis has shown that many substances that were once described as amorphous are composed of very small crystals. For example, charcoal, coke, and soot (all forms of carbon) are made up of small graphite-like crystals.

amount of substance Symbol: n A measure of the number of entities present in a substance. See mole.

ampere Symbol: A The SI base unit of electric current, defined as the constant current that, maintained in two straight parallel infinite conductors of negligible circular cross section placed one metre apart in vacuum, would produce a force between the conductors of 2×10^{-7} newton per metre.

amphiprotic See solvent.

ampholyte ion See zwitterion.

amphoteric A material that can display both acidic and basic properties. The term is most commonly applied to the oxides and hydroxides of metals that can form both cations and complex anions. For example, zinc oxide dissolves in acids to form zinc salts and also dissolves in alkalis to form zincates, $[Zn(OH)_4]^{2-}$.

amu See atomic mass unit.

amylopectin The water-insoluble fraction of starch. See starch.

amylose The water-soluble fraction of starch. See starch.

anabolism All the metabolic reactions that synthesize complex molecules from more simple molecules. See also metabolism.

analysis The process of determining the constituents or components of a sample. There are two broad major classes of analysis, qualitative analysis – essentially answering the question 'what is it?' and quantitative analysis – answering the question 'how much of such and such a component is present?' There is a vast number of analytical methods which can be applied, depending on the nature of the sample and the purpose of the analysis. These include gravimetric, volumetric, and systematic qualitative analysis (classical wet methods); and instrumental methods, such as spectroscopic, nuclear, fluorescence, and polarographic techniques.

Andrews' experiment An investigation (1861) into the relationship between pressure and volume for a mass of carbon dioxide at constant temperature. The resulting isothermals showed clearly the existence of a critical point and led to greater understanding of the liquefaction of gases.

angstrom Symbol: Å A unit of length defined as 10^{-10} metre. The angstrom is sometimes still used for expressing wavelengths of light or ultraviolet radiation or for the sizes of molecules, although the nanometre is preferred.

15

anhydride A compound formed by removing water from an acid or, less commonly, a base. Many non-metal oxides are anhydrides of acids: for example CO_2 is the anhydride of H_2CO_3; SO_3 is the anhydride of H_2SO_4. Organic anhydrides are formed by removing H_2O from two carboxylic-acid groups, giving compounds with the functional group $-CO.O.CO-$. These form a class of organic compounds called *acid anhydrides* (or *acyl anhydrides*). *See* acid anhydride.

anhydrite *See* calcium sulphate.

aniline (aminobenzene, $C_6H_5NH_2$) A colourless oily substance made by reducing nitrobenzene ($C_6H_5NO_2$). Aniline is used for making dyes, pharmaceuticals, and other organic compounds.

anion A negatively charged ion, formed by addition of electrons to atoms or molecules. In electrolysis anions are attracted to the positive electrode (the anode). *Compare* cation.

anionic detergent *See* detergent.

anionic resin An ion-exchange material that can exchange anions, such as Cl^- and OH^-, for anions in the surrounding medium. Such resins are used for a wide range of analytical and purification purposes.
They are often produced by adding a quaternary ammonium group ($-N(CH_3)_3{}^+Cl^-$) or a phenolic group ($-OH^-$) to a stable polyphenylethene resin. A typical exchange reaction is:

$$resin-N(CH_3)_3{}^+Cl^- + KOH =$$
$$resin-N(CH_3)_3{}^+OH^- + KCl$$

Anionic resins can be used to separate mixtures of halide ions. Such mixtures can be attached to the resin and recovered separately by elution.
See also ion exchange.

anisotropic A term descriptive of certain substances which have one or more physical properties that differ according to direction. Most crystals are anisotropic.

annealing A type of heat treatment applied to metals to change their physical properties. The metal is heated to, and held at, an appropriate temperature, before being cooled at a suitable rate to produce the desired grain structure. Annealing is most commonly used to remove the stresses that have arisen during rolling, to increase the softness of the metal, and to make it easier to machine.

anode In electrolysis, the electrode that is at a positive potential with respect to the cathode. In any electrical system, such as a discharge tube or electronic device, the anode is the terminal at which electrons flow out of the system.

anode sludge *See* electrolytic refining.

anodizing An industrial process for protecting aluminium with an oxide layer formed in an electrolytic cell containing an oxidizing acid (e.g. sulphuric(VI) acid). The layer of Al_2O_3 is porous and can be coloured with certain dyes.

anthracene ($C_{14}H_{10}$) A white crystalline solid used extensively in the manufacture of dyes. Anthracene is found in the heavy- and green-oil fractions of crude oil and is obtained by fractional crystallization. Its structure is benzene-like, having three six-membered rings fused together. The reactions are characteristic of aromatic compounds. *See also* aromatic compound.

antibonding orbital *See* orbital.

***anti*-isomer** *See* isomerism.

antiknock agent A substance added to petrol to inhibit preignition or 'knocking'. A common example is lead tetraethyl.

antimonic Designating an antimony(IV) compound.

antimonous Designating an antimony(III) compound.

antimony A toxic element existing in two allotropic forms; the most stable is a brittle silvery metal. Antimony belongs to group V of the periodic table. It is found in many minerals, principally stibnite (Sb_2S_3). It is used in alloys and semiconductor devices.
Symbol: Sb; m.p. 630.7°C; b.p. 1380°C; r.d. 6.69; p.n. 51; r.a.m. 121.7.

antimony(III) chloride (antimony trichloride, $SbCl_3$) A white deliquescent solid, formerly known as *butter of antimony*. It is prepared by direct combination of antimony and chlorine. It is readily hydrolysed by cold water to form a white precipitate of *antimony(III) chloride oxide (antimonyl chloride, SbOCl)*:

$$SbCl_3 + H_2O = SbOCl + 2HCl$$

antimony(III) chloride oxide *See* antimony(III) chloride.

antimonyl chloride *See* antimony(III) chloride.

antimony(III) oxide (antimony trioxide, Sb_2O_3) A white insoluble solid. It is an amphoteric oxide with a strong tendency to act as a base. It can be prepared by direct oxidation by air, oxygen, or steam and is formed when antimony(III) chloride is hydrolysed by excess boiling water.

antimony(V) oxide (antimony pentoxide, Sb_2O_5) A yellow solid. It is usually formed by the action of concentrated nitric acid on antimony or by the hydrolysis of antimony(V) chloride. Although an acidic oxide, it is only slightly soluble in water.

antimony pentoxide *See* antimony(V) oxide.

antimony trichloride *See* antimony(III) chloride.

antimony trioxide *See* antimony(III) oxide.

apatite A naturally occurring phosphate of calcium, $CaF_2.Ca_3(PO_4)_3$.

aprotic *See* solvent.

aqua regia A mixture of concentrated nitric acid and three to four parts of hydrochloric acid. It dissolves all metals including gold, hence the name. The mixture contains chlorine and NOCl.

aqueous Describing a solution in water.

arene An organic compound containing a benzene ring; i.e. an aromatic hydrocarbon or derivative.

argentic oxide *See* silver(II) oxide.

argentous oxide *See* silver(I) oxide.

arginine *See* amino acids.

argon An inert colourless odourless monatomic element of the rare-gas group. It forms 0.93% by volume of air. Argon is used to provide an inert atmosphere in electric and fluorescent lights, in welding and in extracting titanium and silicon. Symbol: Ar; m.p. −189.2°C; b.p. −185.7°C; d. 1.78 kg m⁻³; p.n. 18; r.a.m. 39.95.

aromatic compound An organic compound containing benzene rings in its structure. Aromatic compounds, such as benzene, have a planar ring of atoms linked by alternate single and double bonds. The characteristic of aromatic compounds is that their chemical properties are not those expected for an unsaturated compound; they tend to undergo nucleophilic substitution of hydrogen (or other groups) on the ring, and addition reactions only occur under special circumstances.
The explanation of this behaviour is that the electrons in the double bonds are delocalized over the ring, so that the six bonds are actually all identical and intermediate between single bonds and double bonds. The π electrons are thus spread in a molecular orbital above and below the ring. The evidence for this delocalization in benzene is that:

(1) The bond lengths between carbon atoms in benzene are all equal and intermediate between single and double bond lengths.

(2) If two hydrogen atoms attached to adjacent carbon atoms are substituted by other groups, the compound has only one structure. If the bonds were different two isomers would exist.

(3) Benzene has a stabilization energy of 150 kJ mol^{-1} over the Kekulé structure.

The delocalization of the electrons in the π orbitals of benzene accounts for the properties of benzene and its derivatives, which differ from the properties of alkenes and other aliphatic compounds. The phenomenon is called *aromaticity*. A definition of aromaticity is that it occurs in compounds that obey the *Hückel rule*: i.e. that there should be a planar ring with a total of $(4n + 2)$ π electrons (where n is any integer). Using this rule as a criterion certain non-benzene rings show aromaticity. Such compounds are called *non-benzenoid aromatics*. Other compounds that have a ring of atoms with alternate double and single bonds, but do not obey the rule (e.g. cyclooctotetraene, which has a nonplanar ring of alternating double and single bonds) are called *pseudoaromatics*. *Compare* aliphatic compound. *See also* resonance.

aromaticity *See* aromatic compound.

Arrhenius equation An equation relating the rate constant of a chemical reaction and the temperature at which the reaction is taking place:

$$k = A\exp(-E_a/RT)$$

where A is a constant, k the rate constant, T the thermodynamic temperature in kelvins, R the gas constant, and E_a the activation energy of the reaction.

Reactions proceed at different rates at different temperatures, i.e. the magnitude of the rate constant is temperature dependent. The Arrhenius equation is often written in a logarithmic form, i.e.

$$\log_e k = \log_e A - E/2.3RT$$

This equation enables the activation energy for a reaction to be determined.

arsenic A toxic metalloid element existing in several allotropic forms; the most stable is a brittle grey metal. It belongs to group V of the periodic table. Arsenic is found native and in several ores including mispickel (FeSAs), realgar (AsS), and orpiment (As_2S_3). It is used in semiconductor devices, alloys, and gun shot. Various compounds are used in medicines and agricultural insecticides and poisons.
Symbol: As; m.p. 817°C (grey) at 3 MPa pressure; sublimes at 613°C (grey); r.d. 5.73 (grey); p.n. 33; r.a.m. 74.92.

arsenic(III) chloride (arsenious chloride, $AsCl_3$) A poisonous oily liquid. It fumes in moist air due to hydrolysis with water vapour:

$$AsCl_3 + 3H_2O = As_2O_3 + 6HCl$$

Arsenic(III) chloride is covalent and exhibits its nonmetallic properties.

arsenic hydride *See* arsine.

arsenic(III) oxide (arsenious oxide, As_2O_3) A colourless crystalline solid that is very poisonous (0.1 g would be a lethal dose). Analysis of the solid and vapour states suggests a dimerized structure of As_4O_6. An amphoteric oxide, arsenic(III) oxide is sparingly soluble in water, producing an acidic solution. It is formed when arsenic is burned in air or oxygen.

arsenic(V) oxide (arsenic oxide, As_2O_5) A white amorphous deliquescent solid. It is an acidic oxide prepared by dissolving arsenic(III) oxide in hot concentrated nitric acid, followed by crystallization then heating to 210°C.

arsenious chloride *See* arsenic(III) chloride.

arsenious oxide *See* arsenic(III) oxide.

arsine (arsenic hydride, AsH_3) A poisonous colourless gas with an unpleasant smell. It decomposes to arsenic and hydrogen at 230°C. It is produced in the analysis for arsenic (Marsh's test).

artificial radioactivity Radioactivity induced by bombarding stable nuclei with high-energy particles. For example:

$$^{27}_{13}Al + ^1_0n \rightarrow ^{24}_{11}Na + ^4_2He$$

represents the bombardment of aluminium with neutrons to produce an isotope of sodium. All the transuranic elements (atomic numbers 93 and above) are artificially radioactive since they do not occur in nature.

aryl group An organic group derived by removing a hydrogen atom from an aromatic hydrocarbon or derivative.

asparagine See amino acids.

aspartic acid See amino acids.

association The combination of molecules of a substance with those of another to form more complex species. An example is a mixture of water and ethanol (which are termed associated liquids), the molecules of which combine via hydrogen bonding.

astatine A radioactive element belonging to the halogen group. It occurs in minute quantities in uranium ores. 20 short-lived radioisotopes are known, all alpha-particle emitters.
Symbol: At; m.p. 302°C; b.p. 337°C; p.n. 85; stablest isotope ^{210}At (half-life 8.3 hours).

asymmetric atom See isomerism, optical activity.

atmolysis The separation of gases by using their different rates of diffusion.

atmosphere A unit of pressure defined as 101 325 pascals (atmospheric pressure). The atmosphere is used in chemistry only for rough values of pressure; in particular, for stating the pressures of high-pressure industrial processes.

atom The smallest part of an element that can exist as a stable entity. Atoms consist of a small dense positively charged nu-

cleus, made up of neutrons and protons, with electrons in a cloud around this nucleus. The chemical reactions of an element are determined by the number of electrons (which is equal to the number of protons in the nucleus). All atoms of a given element have the same number of protons (the proton number). A given element may have two or more isotopes, which differ in the number of neutrons in the nucleus.

The electrons surrounding the nucleus are grouped into *shells* – i.e. main orbits around the nucleus. Within these main orbits there may be sub-shells. These correspond to atomic orbitals. An electron in an atom is specified by four quantum numbers:

(1) The *principal quantum number* (n), which specifies the main energy levels. n can have values 1, 2, etc. The corresponding shells are denoted by letters K, L, M, etc., the K shell ($n = 1$) being the nearest to the nucleus. The maximum number of electrons in a given shell is $2n^2$.

(2) The *orbital quantum number* (1), which specifies the angular momentum. For a given value of n, 1 can have possible values of $n-1$, $n-2$, ... 2, 1, 0. For instance, the M shell ($n = 3$) has three sub-shells with different values of 1 (0, 1, and 2). Sub-shells with angular momentum 0, 1, 2, and 3 are designated by letters s, p, d, and f.

(3) The *magnetic quantum number* (m). This can have values -1, $-(1 - 1)$ ... 0 ... $+ (1 + 1)$, $+ 1$. It determines the orientation of the electron orbital in a magnetic field.

(4) The *spin quantum number* (m_s), which specifies the intrinsic angular momentum of the electron. It can have values $+ \frac{1}{2}$ and $- \frac{1}{2}$.

Each electron in the atom has four quantum numbers and, according to the Pauli exclusion principle, no two electrons can have the same set of quantum numbers. This explains the electronic structure of atoms. *See also* Bohr theory.

atomic heat See Dulong and Petit's law.

19

atomicity The number of atoms per molecule of a compound. Methane, for instance has an atomicity of five (CH_4).

atomic mass unit (amu) Symbol: u A unit of mass used for atoms and molecules, equal to $1/12$ of the mass of an atom of carbon-12. It is equal to $1.660\ 33 \times 10^{-27}$ kg.

atomic number *See* proton number.

atomic orbital *See* orbital.

atomic weight *See* relative atomic mass.

atto- Symbol: a A prefix denoting 10^{-18}. For example, 1 attometre (am) = 10^{-18} metre (m).

Aufbau principle A principle that governs the order in which the atomic orbitals are filled in elements of successive proton number; i.e. a statement of the order of increasing energy. The order is as follows: $1s^2$, $2s^2$, $2p^6$, $3s^2$, $3p^6$, $4s^2$, $3d^{10}$, $4p^6$, $5s^2$, $4d^{10}$, $5p^6$, $6s^2$, $4f^{14}$, $5d^{10}$, $6p^6$, $7s^2$, $5f^{14}$, $6d^{10}$
(the superscript indicates the maximum number of electrons for each level).
Note
(1) Hund's rule of maximum multiplicity applies; i.e. degenerate orbitals are occupied singly before spin pairing occurs.
(2) the unexpected position of the d-levels, which give rise to the first, second, and third transition series.
(3) the unusual position of the f-levels, giving rise to the lanthanoids and actinoids.

auric chloride *See* gold(III) chloride.

aurous Designating a compound of gold(I).

autocatalysis *See* catalyst.

Avogadro constant Symbol: N_A The number of particles in one mole of a substance. Its value is $6.022\ 52 \times 10^{23}\ mol^{-1}$.

Avogadro's law The principle that equal volumes of all gases at the same temperature and pressure contain equal numbers of molecules. It is often called *Avogadro's hypothesis*. It is strictly true only for ideal gases.

azeotrope (azeotropic mixture) A mixture of liquids for which the vapour phase has the same composition as the liquid phase. It therefore boils without change in composition and consequently without progressive change in boiling point.
The composition and boiling points of azeotropes vary with pressure, indicating that they are not chemical compounds. Azeotropes may be broken by distillation in the presence of a third liquid, by chemical reactions, adsorption, or fractional crystallization.
See constant boiling mixture.

azeotropic distillation A method used to separate mixtures of liquids that cannot be separated by simple distillation. Such a mixture is called an azeotrope. A solvent is added to form a new azeotrope with one of the components, and this is then removed and subsequently separated in a second column. An example of the use of azeotropic distillation is the dehydration of 96% ethanol to absolute ethanol using benzene. Azeotropic distillation is not widely used because of the difficulty of finding inexpensive non-toxic non-corrosive solvents that can be easily removed from the new azeotrope.

azide 1. An inorganic compound containing the ion N_3^-.
2. An organic compound of general formula RN_3.

azine An organic heterocyclic compound that has a hexagonal ring containing carbon and nitrogen atoms. Pyridine is the simplest example.

azo compound A type of organic compound of the general formula RN:NR', where R and R' are aromatic groups. Azo compounds can be formed by coupling a diazonium compound with an aromatic phenol or amine. Most are coloured because of the —N:N— (azo group).

azo dye An important type of dye used in acid dyes for wool and cotton. The dyes are azo compounds; usually sodium salts of sulphonic acids.

azoimide *See* hydrazoic acid.

Azo compound

B

back e.m.f. An e.m.f. that opposes the normal flow of electric charge in a circuit or circuit element. In some electrolytic cells a back e.m.f. is caused by the layer of hydrogen bubbles that builds up on the cathode as hydrogen ions pick up electrons and form gas molecules (i.e. as a result of polarization of the electrode).

Bakelite (Trademark) A common thermosetting synthetic polymer formed by the condensation of phenol and formaldehyde.

baking soda *See* sodium hydrogencarbonate.

ball mill A device commonly used in the chemical industry for reducing the size of solid material. Ball mills usually have slowly rotating steel-lined drums, which contain steel balls. The material is crushed by the tumbling action of the contents of the drum. *Compare* hammer mill.

Balmer series A series of lines in the spectrum of radiation emitted by excited hydrogen atoms. The lines correspond to the atomic electrons falling into the second lowest energy level, emitting energy as radiation. The wavelengths (λ) of the radiation in the Balmer series are given by:
$$1/\lambda = R(1/2^2 - 1/n^2)$$
where n is an integer and R is the Rydberg constant. *See* Bohr theory. *See also* spectral series.

banana bond (bent bond) In strained-ring compounds the angles assumed on the basis of hybridization are often not equal to the angles obtained by joining the atomic centres. In these cases it is sometimes assumed that the bonding orbital is bent or banana-like in shape. Cyclopropane is an example, in which geometric considerations imply a bond angle of 60° while sp^3 hybridization implies an interorbital angle of around 100°, giving a banana bond.

band spectrum A spectrum that appears as a number of bands of emitted or absorbed radiation. Band spectra are characteristic of molecules. Often each band can be resolved into a number of closely spaced lines. The bands correspond to changes of electron orbit in the molecules. The close lines seen under higher resolution are the result of different vibrational states of the molecule. *See also* spectrum.

bar A unit of pressure defined as 10^5 pascals. The *millibar* (mb) is more common; it is used for measuring atmospheric pressure in meteorology.

Barft process A process formerly used for protecting iron by heating it in steam, to form a layer of tri-iron tetroxide (Fe_3O_4).

barium A dense, low-melting reactive metal; the fifth member of group II of the periodic table and a typical alkaline-earth element. The electronic configuration is that of xenon with two additional outer 6s electrons. Barium is of low abundance; it is found as witherite ($BaCO_3$) and barytes ($BaSO_4$). The metal is obtained by the electrolysis of the fused chloride using a

Banana bonds in cyclopropane

22

cooled cathode which is slowly withdrawn from the melt. Because of its low melting point barium is readily purified by vacuum distillation. Barium metal is used as a 'getter', i.e., a compound added to a system to seek out the last traces of oxygen; and as an alloy constituent for certain bearing metals.

Barium has a low ionization potential and a large radius. It is therefore strongly electropositive and its properties, and those of its compounds, are very similar to those of the other alkaline-earth elements calcium and strontium. Notable differences in the chemistry of barium from the rest of the group are:

(1) The much higher stability of the carbonate.

(2) The formation of the peroxide below 800°C. Barium peroxide decomposes on strong heating to give oxygen and barium oxide:

$$BaO_2 = BaO + O$$

This equilibrium was the basis of the totally obsolete *Brin process* for the commercial production of oxygen.

Barium is also notable for the very low solubility of the sulphate, which permits its application to gravimetric analysis for either barium or sulphate. Barium compounds give a characteristic green colour to flames which is used in qualitative analysis. Barium salts are all highly toxic with the exception of the most insoluble materials. Metallic barium has the body centred cubic structure.

Symbol: Ba; m.p. 714°C; b.p. 1640°C; r.d. 3.51; p.n. 56; r.a.m. 137.34.

barium bicarbonate *See* barium hydrogencarbonate.

barium carbonate ($BaCO_3$) A white insoluble salt that occurs naturally as the mineral witherite. Barium carbonate can be readily precipitated by adding an alkali carbonate to a barium salt solution. On heating it decomposes reversibly with the formation of the oxide and carbon dioxide:

$$BaCO_3 = BaO + CO_2$$

It is used as a rat poison.

barium chloride ($BaCl_2$) A white solid that can be prepared by dissolving barium carbonate in hydrochloric acid and crystallizing out the dihydrate ($BaCl_2.2H_2O$). Barium chloride is used as the electrolyte in the extraction of barium, as a rat poison, and in the leather industry.

barium hydrogencarbonate (barium bicarbonate, $Ba(HCO_3)_2$) A compound that occurs only in aqueous solution. It is formed by the action of cold water containing carbon dioxide on barium carbonate, to which it reverts on heating:

$$BaCO_3 + CO_2 + H_2O = Ba(HCO_3)_2$$

barium hydroxide (baryta, $Ba(OH)_2$) A white solid usually obtained as the octahydrate, $Ba(OH)_2.8H_2O$. Barium hydroxide is the most soluble of the group IIA hydroxides and can be used in volumetric analysis for the estimation of weak acids using phenolphthalein as an indicator.

barium oxide (BaO) A white powder prepared by heating barium in oxygen or by thermal decomposition of barium carbonate. It has been used in the manufacture of lubricating-oil additives.

barium peroxide (BaO_2) A dense off-white powder that can be prepared by carefully heating barium oxide in oxygen. Barium peroxide is used for bleaching straw and silk and in the laboratory preparation of hydrogen peroxide.

barium sulphate ($BaSO_4$) A white solid that occurs naturally as the mineral barytes. Barium sulphate is very insoluble in water and can be prepared easily as a precipitate by adding sulphuric acid to barium chloride. Barium sulphate is an important industrial chemical. Under the name of 'blanc fixe' it is used as a pigment extender in surface coating compositions. It is also used in the glass and rubber industries and medically (taken orally) to allow radiographs to be taken of the digestive system.

23

barn Symbol: b A unit of area defined as 10^{-28} square metre. The barn is sometimes used to express the effective cross-sections of atoms or nuclei in the scattering or absorption of particles.

barrel A measurement of volume often used in the chemical industry. It is equal to 159 litres (35 UK gallons).

baryta See barium hydroxide.

barytes (heavy spar) A mineral form of barium sulphate ($BaSO_4$).

base A compound that releases hydroxyl ions, OH^-, in aqueous solution. Basic solutions have a pH greater than 7. In the Lowry-Bronsted treatment a base is a substance that tends to accept a proton. Thus OH^- is basic as it accepts H^+ to form water, but H_2O is also a base (although somewhat weaker) because it can accept a further proton to form H_3O^+. In this treatment the ions of classical mineral acids such as $SO_4{}^{2-}$ and $NO_3{}^-$ are weak *conjugate bases* of their respective acids. Electron-pair donors, such as trimethylamine and pyridine, are examples of organic bases.

base-catalysed reaction A reaction catalysed by bases. Typical base-catalysed reactions are the Claisen condensation and the aldol reaction, in which the first step is abstraction of a proton to give a carbanion.

base metal A metal such as iron, lead, or copper, distinguished from a noble metal such as gold or silver.

base unit A unit that is defined in terms of reproducible physical a base unit in the SI, being defined in terms of the wavelength of a particular light emission. See also SI units.

basic Having a tendency to release OH^- ions. Thus any solution in which the concentration of OH^- ions is greater than that in pure water at the same temperature is described as basic; i.e. the pH is greater than 7.

basic aluminium ethanoate See aluminium acetate.

basic salt A compound intermediate between a normal salt and a hydroxide or oxide. The term is often restricted to hydroxyhalides (such as $Pb(OH)Cl$, $Mg_2(OH)_3Cl_2.4H_2O$, and $Zn(OH)F$) and hydroxy-oxy salts (for example, $2PbCO_3$. $Pb(OH)_2$, $Cu(OH)_2.CuCl_2$, $Cu(OH)_2.$ $CuCo_3$). The hydroxyhalides are essentially closely packed assemblies of OH^- with metal or halide ions in the octahedral holes. The hydroxy-oxy salts have more complex structures than those implied by the formulae.

basic slag See slag.

batch process A manufacturing process in which the reactants are fed into the process in fixed quantities (batches), rather than in a continuous flow. At any particular instant all the material, from its preparation to the final product, has reached a definite stage in the process. Baking a cake is an example of a batch process. Such processes present problems of automation and instrumentation and tend to be wasteful of energy. For this reason, batch processing is only economically operated on an industrial scale when small quantities of valuable or strategic materials are required, e.g. drugs, titanium metal. *Compare* continuous process.

battery A number of similar units, such as electric cells, working together. Many dry 'batteries' used in radios, flashlights, etc., are in fact single cells. If a number of identical cells are connected in series, the total e.m.f. of the battery is the sum of the e.m.f.s of the individual cells. If the cells are in parallel, the e.m.f. of the battery is the same as that of one cell, but the current drawn from each is less (the total current is split amongst the cells).

Kekulé structures

Dewar structures

Benzene structures

representation

bauxite A mineral hydrated form of aluminium hydroxide; the principal ore of aluminium.

b.c.c. *See* body-centred cubic crystal.

Beckmann thermometer A type of thermometer designed to measure differences in temperature rather than scale degrees. Beckmann thermometers have a larger bulb than common thermometers and a stem with a small internal diameter, so that a range of 5°C covers about 30 centimetres in the stem. The mercury bulb is connected to the stem in such a way that the bulk of the mercury can be separated from the stem once a particular 5° range

has been attained. The thermometer can thus be set for any particular range.

bent bond *See* banana bond.

benzaldehyde *See* benzenecarbaldehyde.

benzene (C_6H_6) A colourless liquid hydrocarbon with a characteristic odour. Benzene is a highly toxic compound and continued inhalation of the vapour is harmful. It was originally isolated from coal tar and for many years this was the principal source of the compound. Contemporary manufacture is from hexane; petroleum vapour is passed over platinum at 500°C and at 10 atmospheres pressure:

$$C_6H_{14} \rightarrow C_6H_6 + 4H_2$$

Benzene is the simplest aromatic hydrocarbon. It shows characteristic electrophilic substitution reactions, which are difficult to explain assuming a simple unsaturated structure (such as Kekulé's (1865) or Dewar's (1867) formulae). This anomolous behaviour can now be explained by assuming that the six π electrons are delocalized and that benzene is, therefore, a resonance hybrid. It consists of two Kekulé structures and three Dewar structures, the former contributing 80% and the latter 20% to the hybrid. Benzene is usually represented by either a Kekulé structure or a hexagon containing a circle (which denotes the delocalized electrons).

benzenecarbaldehyde (benzaldehyde, C_6H_5CHO) A yellow organic oil with a distinct almond-like odour. Benzenecarbaldehyde undergoes the reactions characteristic of aldehydes and may be synthesized in the laboratory by the usual methods of aldehyde synthesis. It is used as a food flavouring and in the manufacture of dyes and antibiotics, and can be readily manufactured by the chlorination of methylbenzene and the subsequent hydrolysis of (dichloromethyl) benzene:

$$C_6H_5CH_3 + Cl_2 \rightarrow C_6H_5CHCl_2$$
$$C_6H_5CHCl_2 + 2H_2O \rightarrow C_6H_5CH(OH)_2 + 2HCl$$
$$C_6H_5CH(OH)_2 \rightarrow C_6H_5CHO + H_2O$$

benzenecarbonyl chloride (benzoyl chloride, C_6H_5COCl) A liquid acyl chloride used as a benzoylating agent.

benzenecarbonyl group (benzoyl group) The group $C_6H_5.CO-$.

benzenecarboxylic acid (benzoic acid, C_6H_5COOH) A white crystalline carboxylic acid. It is used as a food preservative. The carboxyl group ($-COOH$) directs further substitution onto the benzene ring in the 3 position.

benzenesulphonic acid ($C_6H_5SO_2OH$) A white crystalline acid made by sulphonation of benzene. Any further substitution onto the benzene ring is directed into the 3 position.

benzoic acid *See* benzenecarboxylic acid.

benzoylation *See* acylation.

benzoyl chloride *See* benzenecarbonyl chloride.

benzoyl group *See* benzenecarbonyl group.

benzyl alcohol *See* phenylmethanol.

benzyl group The group $C_6H_5CH_2-$.

Bergius process A process formerly used for making hydrocarbon fuels from coal. A powdered mixture of coal, heavy oil, and a catalyst was heated with hydrogen at high pressure.

berkelium A radioactive transuranic element of the actinoid series of metals, not found naturally on Earth. Several radioisotopes have been synthesized. Symbol: Bk; m.p. 986°C; p.n. 97; stablest isotope ^{247}Bk (half-life 1400 years).

beryl A mineral, $3BeO.Al_2O_3.6SiO_2$, used as a source of beryllium. There are several varieties including the gemstones emerald (coloured green by traces of chromium oxide) and aquamarine (a blue-green colour).

beryllium A light metallic element, similar to aluminium but somewhat harder; the first element in group II of the periodic table. It has the electronic configuration of helium with two additional outer 2s electrons.

Beryllium occurs in a number of minerals such as beryllonite ($NaBePO_4$), chrysoberyl ($Be(AlO_2)_2$), bertrandite ($4BeO.2SiO_2$), and beryl ($3BeO.Al_2O_3.6SiO_2$). The element accounts for only 0.0006% by mass of the Earth's crust. The metal is obtained by conversion of the ore to the sulphate at high temperature and pressure with concentrated sulphuric acid, then to the chloride, followed by electrolysis of the fused chloride. Alternatively, extraction by hydrogen fluoride followed by electrolysis of the fused fluoride may be employed. The metal has a much lower general reactivity than lithium or other elements in group II. It is used as an antioxidant and hardener in some alloys, such as copper and phosphor bronzes.

Beryllium has the highest ionization potential of group II and the smallest size. Consequently it is less electropositive and more polarizing than other members of the group. Thus, Be^{2+} ions do not exist as such in either solids or solutions, and even with the most electronegative elements there is partial covalent character in the bonds. The metal reacts directly with oxygen, nitrogen, sulphur, and the halogens at various elevated temperatures, to form the oxide BeO, nitride Be_3N_2, sulphide BeS, and halides BeX_2, all of which are covalent. Beryllium does not react directly with hydrogen but a polymeric hydride $(BeH_2)_n$ can be prepared by reduction of $(CH_3)_2Be$ using lithium tetrahydridoaluminate.

Beryllium is amphoteric forming *beryllate* species, such as $[Be(OH)_4]^{2-}$ and $[Be(OH)]_3^{3+}$. The hydroxide is only weakly basic. The element does not form a true carbonate; the basic carbonate, $BeCO_3$. $Be(OH)_2$ is formed when sodium carbonate is added to solutions of beryllium compounds.

Beryllium hydride, chloride, and dimethylberyllium form polymeric bridged species but, whereas the bridging in the chloride is via an electron pair on chlorine atoms and can be regarded as an electron-pair donor bond, the bonding in the hydride and in the methyl compound involves two-electron three-centre bonds. Coordination compounds are quite common with beryllium; some examples include $[BeCl_4]^{2-}$, $(R_2O)_2BeCl_2$, and $[Be(NH_3)_4]Cl_2$. Beryllium also forms a number of alkyl compounds, some of which can be stabilized by coordination. Beryllium is extremely toxic.

Symbol: Be; m.p. 1280°C; b.p. 2480°C; r.d. 1.85; p.n. 4; r.a.m. 9.01.

beryllium bicarbonate *See* beryllium hydrogen carbonate.

beryllium carbonate ($BeCO_3$) An unstable solid prepared by prolonged treatment of a suspension of beryllium hydroxide with carbon dioxide. The resulting solution is evaporated and filtered in an atmosphere of carbon dioxide.

beryllium chloride ($BeCl_2$) A white solid obtained by passing chlorine over a heated mixture of beryllium oxide and carbon. It is a poor conductor in the fused state and in the solid form consists of a covalent polymeric structure. It is readily soluble in organic solvents but is hydrolysed in the presence of water to give the hydroxide. The anhydrous salt is used as a catalyst.

beryllium hydrogen carbonate (beryllium bicarbonate, $Be(HCO_3)_2$) A compound formed in solution by the action of carbon dioxide on a suspension of the carbonate, to which it reverts on heating:

$$BeCO_3 + CO_2 + H_2O = Be(HCO_3)_2$$

beryllium hydroxide ($Be(OH)_2$) A solid that can be precipitated from beryllium-containing solutions by an alkali. In an excess of alkali, beryllium hydroxide dissolves to give a beryllate, $Be(OH)_4^{2-}$. In this way, beryllium and aluminium hydroxides are similar and the reaction is an indication that beryllium hydroxide is amphoteric.

beryllium oxide (BeO) A white solid formed by heating beryllium in oxygen or by the decomposition of beryllium hydroxide or carbonate. Beryllium oxide is insoluble in water but it shows basic properties by dissolving in acids to form beryllium salts:

$$BeO + 2H^+ \rightarrow Be^{2+} + H_2O$$

However, beryllium oxide also resembles acidic oxides by reacting with alkalis to form beryllates:

$$BeO + 2OH^- + H_2O \rightarrow Be(OH)_4^{2-}$$

It is thus an amphoteric oxide. Beryllium oxide is used for the production of beryllium and beryllium-copper refractories, for high-output transistors, and for printed circuits. The chemical properties of beryllium oxide are similar to those of aluminium oxide.

beryllium sulphate ($BeSO_4$) An insoluble salt prepared by the reaction of beryllium oxide with concentrated sulphuric acid. On heating, it breaks down to give the oxide. When heated in the presence of water, it readily dissolves to give an acidic solution.

Bessemer process A process for making steel from pig iron. A vertical cylindrical steel vessel (*converter*) is used lined with a refractory material. Air is blown through the molten iron and carbon is oxidized and thus removed from the iron. Other impurities — silicon, sulphur, and phosphorus — are also oxidized. Irons containing large amounts of phosphorus are treated in a converter lined with a basic material, so that a phosphate slag is formed. The required amount of carbon is then added to the iron to produce the desired type of steel.

beta particle An electron emitted by a radioactive substance as it decays.

bi- Prefix used formerly in naming acid salts. For instance, sodium bisulphate ($NaHSO_4$) is sodium hydrogen sulphate, etc.

bicarbonate *See* hydrogen carbonate.

bimolecular Describing a reaction or step that involves two molecules. For example, the decomposition of hydrogen iodide,
$$2HI \rightarrow H_2 + I_2$$
takes place between two molecules and is therefore a bimolecular reaction. Other common examples are:
$$H_2O_2 + I_2 \rightarrow OH^- + HIO$$
$$OH^- + H^+ \rightarrow H_2O$$
All bimolecular reactions are second order, but some second-order reactions are not bimolecular.

binary compound A chemical compound formed from two elements; e.g. Fe_2O_3 or $NaCl$.

biochemistry The study of chemical compounds and reactions occurring in living organisms.

biodegradable *See* pollution.

biosynthesis The reactions by which organisms obtain the various compounds needed for life.

Birkeland-Eyde process An industrial process for fixing nitrogen (as nitrogen monoxide) by passing air through an electric arc:
$$N_2 + O_2 \rightarrow 2NO$$
See also nitrogen fixation.

bismuth A brittle pinkish metallic element belonging to group V of the periodic table. It occurs native and in the ores Bi_2S_3 and Bi_2O_3. Bismuth is widely used in alloys, especially low-melting alloys. Compounds are used in cosmetics and medicines.
Symbol: Bi; m.p. 271.3°C; b.p. 1560°C; r.d. 9.75; p.n. 83; r.a.m. 208.98.

bismuth(III) carbonate dioxide (bismuthyl carbonate, $Bi_2O_2CO_3$) A white solid prepared by mixing solutions of bismuth nitrate and ammonium carbonate. It contains the $(BiO)^+$ ion.

bismuth(III) chloride (bismuth trichloride, $BiCl_3$) A white deliquescent solid. It can be prepared by direct combination of bismuth and chlorine. Bismuth(III) chloride dissolves in excess dilute hydrochloric acid to form a clear liquid, but if diluted it produces a white precipitate of *bismuth(III) chloride oxide (bismuthyl chloride, BiOCl)*:
$$BiCl_3 + H_2O = BiOCl + 2HCl$$
This reaction is often used as an example of a reversible reaction and as a confirmatory test for bismuth during quantitative analysis. Bismuth(V) chloride does not form.

bismuth(III) chloride oxide *See* bismuth(III) chloride.

bismuth(III) nitrate oxide (bismuthyl nitrate, $BiONO_3$) A white insoluble solid, often referred to as *bismuth subnitrate*. It is precipitated when bismuth(III) nitrate is diluted and contains the $(BiO)^+$ ion. Bismuth(III) nitrate oxide is used in pharmaceutical preparations.

bismuth trichloride *See* bismuth(III) chloride.

bismuthyl compound A compound containing the $(BiO)^+$ ion or BiO grouping.

bismuthyl carbonate *See* bismuth(III) carbonate dioxide.

bismuthyl chloride *See* bismuth(III) chloride.

bismuthyl nitrate *See* bismuth(III) nitrate oxide.

bittern The liquid that is left after sodium chloride has been crystallized from sea water.

bitumen A mixture of solid or semisolid hydrocarbons obtained from coal, oil, etc.

bivalent (divalent) Having a valency of two.

blackdamp *See* firedamp.

blanc fixe *See* barium sulphate.

blast furnace A furnace for producing iron from iron(III) oxide. A mixture of the ore with coke and a flux is heated by preheated air blown into the bottom of the furnace. The flux is often calcium oxide (from limestone). Molten pig iron is run off from the bottom of the furnace.

bleaching powder A white solid that can be regarded as a mixture of calcium chlorate(I) (calcium hypochlorite), calcium chloride, and calcium hydroxide. It is prepared on a large scale by passing a current of chlorine through a tilted cylinder down which is passed calcium hydroxide. Bleaching powder has been used for bleaching paper pulps and fabrics and for sterilizing water. Its bleaching power arises from the formation, in the presence of air containing carbon dioxide, of the oxidizing agent chloric(I) acid (hypochlorous acid, HClO):

$$Ca(ClO)_2.Ca(OH)_2.CaCl_2 + 2CO_2 \rightarrow$$
$$2CaCO_3 + CaCl_2 + 2HClO$$

blende A naturally occurring sulphide ore.

blue vitriol Copper(II) sulphate pentahydrate ($CuSO_4.5H_2O$).

boat conformation *See* conformation.

body-centred cubic crystal (b.c.c.) A crystal structure in which the unit cell has an atom, ion, or molecule at each corner of a cube and at the centre of the cube. In this type of structure the coordination number is 8. It is less close-packed than the face-centred cubic structure. The alkali metals form crystals with body-centred cubic structures. *See also* cubic crystal.

Bohr theory A theory introduced by Niels Bohr (1911) to explain the spectrum of atomic hydrogen. The model he used was that of a nucleus with charge $+e$, orbited by an electron with charge $-e$, moving in a circle of radius r. If v is the velocity of the electron, the centripetal force, mv^2/r is equal to the force of electrostatic attraction, $e^2/4\pi\varepsilon_0r^2$. Using this, it can be shown that the total energy of the electron (kinetic and potential) is

$$-e^2/8\pi\varepsilon_0r$$

If the electron is considered to have wave properties, then there must be a whole number of wavelengths around the orbit, otherwise the wave would be a progressive wave. For this to occur

$$n\lambda = 2\pi r$$

where n is an integer, 1, 2, 3, 4, ... The wavelength, λ, is h/mv, where h is the

boiling

Planck constant and mv the momentum. Thus for a given orbit:
$$nh/2\pi = mvr$$
This means that orbits are possible only when the angular momentum (mvr) is an integral number of units of $h/2\pi$. Angular momentum is thus quantized. In fact, Bohr in his theory did not use the wave behaviour of the electron to derive this relationship. He assumed from the beginning that angular momentum was quantized in this way. Using the above expressions it can be shown that the electron energy is given by
$$E = -me^4/8\varepsilon_0^2 n^2 h^2$$
Different values of n (1, 2, 3, etc.) correspond to different orbits with different energies; n is the principal quantum number. In making a transition from an orbit n_1 to another orbit n_2 the energy difference ΔW is given by:
$$\Delta W = W_1 - W^2 = me^4(1/n_2^2 - 1/n_1^2)/8\varepsilon_0^2 h^2$$
This is equal to hv where v is the frequency of radiation emitted or absorbed. Since $v\lambda = c$, then
$$1/\lambda = me^4(1/n_1^2 - 1/n_2^2)/8\varepsilon_0^2 ch^3$$
The theory is in good agreement with experiment in predicting the wavelengths of lines in the hydrogen spectrum, although it is less successful for larger atoms. Different values of n_1 and n_2 correspond to different spectral series, with lines given by the expression:
$$1/\lambda = R(1/n_1^2 - 1/n_2^2)$$
R is the *Rydberg constant*. Its experimental value is $1.096\,78 \times 10^7$ m^{-1}. The value from Bohr theory ($me^4/8\pi\varepsilon_1^2 ch^2$) is $1.097\,00 \times 10^7$ m^{-1}. *See also* atom.

boiling The process by which a liquid is converted into a gas or vapour by heating at its boiling point. At this temperature the vapour pressure of the liquid is equal to the external pressure, and bubbles of vapour can form within the liquid.

boiling point The temperature at which the vapour pressure of a liquid is equal to atmospheric pressure. This temperature is always the same for a particular liquid at a given pressure (for reference purposes usually taken as standard pressure). *See also* elevation of boiling point.

Boiling point – composition diagram

boiling point-composition diagram A diagram for a two-component liquid system representing both the variation of the boiling point and the composition of the vapour phase as the liquid-phase composition is varied. A mixture of A and B at composition L_1 would have a boiling point T_1 and a vapour composition V_1, which when condensed would give a liquid at L_2. L_2 would boil at T_2 to give vapour V_2 equivalent to liquid of composition L_3, and so on. Thus the whole process of either distillation or fractionation of this system will lead to progressive enrichment in component A.

For perfect solutions obeying Raoult's law the curves would coincide but in real cases there will be sufficient intermolecular attraction to cause deviation from this. The separation of the curves as well as the difference in boiling points determines the performance of fractionation columns. *See also* constant-boiling mixture.

Boltzmann constant Symbol: k The constant $1.380\,54$ J K^{-1}, equal to the gas constant (R) divided by the Avogadro constant (N_A). *See also* degrees of freedom.

bomb calorimeter A sealed insulated container, used for measuring energy released during combustion of substances (e.g. foods and fuels). A known amount of the substance is ignited inside the calorim-

30

Borax bead colours

Metal	Oxidizing flame	Reducing flame
chromium	green H & C	green H & C
cobalt	blue H & C	blue H & C
copper	green H, blue C	often opaque
iron	brown-red H, yellow C	green H & C
manganese	violet H & C	colourless H & C
nickel	red-brown C	grey-black C

H hot
C cold

eter in an atmosphere of pure oxygen, and undergoes complete combustion at constant volume. The resultant rise in temperature is related to the energy released by the reaction. Such energy values (*calorific values*) are often quoted in joules per kilogram ($J kg^{-1}$).

bond The interaction between atoms, molecules, or ions holding these entities together. The forces giving rise to bonding may vary from extremely weak intermolecular forces, about $0.1 kJ mol^{-1}$, to very strong chemical bonding, about $10^3 kJ mol^{-1}$. *See* bond energy, coordinate bond, covalent bond, electrovalent bond, hydrogen bond, metallic bond.

bond energy The energy involved in forming a bond. For ammonia, for instance, the energy of the N−H bond is one third of the energy involved for the process

$$NH_3 \rightarrow N + 3H$$

It is thus one third of the heat of atomization.
The *bond dissociation energy* is a different quantity to the bond energy. It is the energy required to break a particular bond in a compound, e.g.:

$$NH_3 \rightarrow NH_2 + H$$

bonding orbital *See* orbital.

boracic acid *See* boric acid.

boranes *See* boron hydrides.

borate *See* boron.

borax *See* disodium tetraborate decahydrate.

borax-bead test A preliminary test in qualitative inorganic analysis that can be a guide to the presence of certain metals. A borax bead is formed by heating a little borax on a loop in a platinum wire. A minute sample is introduced into the bead and the colour observed in both the oxidizing and reducing areas of the bunsen flame. The colour is also noted when the bead is cold.

boric acid (boracic acid, H_3BO_3) A white crystalline solid soluble in water; in solution it is a very weak acid. Boric acid is used as a mild antiseptic eye lotion and was formerly used as a food preservative. It is used in glazes for enamelled objects and is a constituent of Pyrex 'glass'.
Trioxoboric(III) acid is the full systematic name for the solid acid and its dilute solutions; in more concentrated solutions polymerization occurs to give polydioxoboric(III) acid.

boride A compound of boron, especially one with a more electropositive element.

Born-Haber cycle A cycle used in calculating the lattice energies of solids. The steps involved are:
Atomization of sodium:

$$Na(s) \rightarrow Na(g) \; \Delta H_1$$

Ionization of sodium:

$$Na(g) \rightarrow Na^+(g) \; \Delta H_2$$

Atomization of chlorine:

$$Cl_2(g) \rightarrow 2Cl(g) \; \Delta H_3$$

31

boron

Ionization of chlorine:
$$Cl(g) + e^- \rightarrow Cl^-(g) \; \Delta H_4$$
Formation of solid:
$$Na^+(g) + Cl^-(g) \rightarrow NaCl(s) \; \Delta H_5$$
This last step involves the lattice energy ΔH_5. The sum of all these enthalpies is equal to the heat of the reaction:
$$Na(s) + \tfrac{1}{2}Cl_2(g) \rightarrow NaCl(s)$$
See also Hess's law.

boron A hard rather brittle metalloid element of group II of the periodic table. It has the electronic structure $1s^2 2s^2 2p^1$. Boron is of low abundance (0.0003%) but the natural minerals occur in very concentrated form as borax ($Na_2B_4O_7.10H_2O$) and colemanite ($Ca_2B_6O_{11}$). The element is obtained by conversion to boric acid followed by dehydration to B_2O_3 then reduction with magnesium. High-purity boron for semiconductor applications is obtained by conversion to boron trichloride, which can be purified by distillation, then reduction using hydrogen. Only small quantities of elemental boron are needed commercially; the vast majority of boron supplied by the industry is in the form of borax or boric acid.

As boron has a small atom and has a relatively high ionization potential its compounds are predominantly covalent; the ion B^{3+} does not exist.

Boron does not react directly with hydrogen to form boron hydrides or 'boranes' but the hydrolysis of magnesium boride does produce a range of boranes such as B_4H_{10}, B_5H_9, and B_6H_{10}. Thermal decomposition of these higher boranes produces, among other things, the simplest borane, B_2H_6 (diborane). The species 'BH_3' is only a short-lived reaction intermediate.

Finely divided boron burns in oxygen above 600°C to give the oxide, B_2O_3, an acidic oxide which will dissolve slowly in water to give boric acid ($B(OH)_3$) and rapidly in alkalis to give borates such as $Na_2B_4O_7$. A number of polymeric species with B–B and B–O links are known, e.g. a lower oxide $(BO)_x$, and a polymeric acid $(HBO_2)_n$. Although the parent acid is weak, many salts containing *borate* anions are known but their stoichiometry gives little indication of their structure, many of which are cyclic or linear polymers. These contain both 'BO_3' and planar groups and 'BO_4' tetrahedra. Boric acid and the borates give a range of glassy substances on heating; these contain cross-linked B–O–B chains and nets. In the molten state these materials react with metal ions to form borates, which on cooling give characteristic colours to the glass.

Boron reacts with nitrogen on strong heating (1000°C) to give boron nitride, a slippery white solid with a layer structure similar to that of graphite, i.e., hexagonal rings of alternating B and N atoms. The material has an extremely high melting point and is *thermally* very stable but there is sufficient bond polarity in the B–N links to permit slow hydrolysis by water to give ammonia. There is also a 'diamond-like' form of B–N which is claimed to be even harder than diamond.

Elemental boron reacts directly with fluorine and chlorine but for practical purposes the halides are obtained via the BF_3 route from boric acid:
$$B_2O_3 + 3CaF_2 + 3H_2SO_4 \rightarrow$$
$$3CaSO_4 + 3H_2O + 2BF_3$$
$$BF_3 + AlCl_3 \rightarrow AlF_3 + BCl_3$$
These halides are all covalent molecules, which are all planar and trigonal in shape. Boron halides are industrially important as catalysts or promoters in a variety of organic reactions including polymerization and Friedel-Crafts type alkylations. The decomposition of boron halides in atmospheres of hydrogen at elevated temperatures is also used to deposit traces of pure boron in semiconductor devices.

Boron forms a range of compounds with elements that are less electronegative than itself, called *borides*. Borides such as ZrB_2 and TiB_2 are hard refractory substances, which are chemically inert and have remarkably high electrical conductivities. Borides have a wide range of stoichiometries, from M_4B through to MB_6, and can exist in close-packed arrays, chains, and two-dimensional nets. Natural boron consists of two isotopes, ^{10}B (18.83%) and ^{11}B (81.17%). These percentages are high enough for detection by splitting of infrared absorption or by n.m.r. spectroscopy.

Both borax and boric acid are used as mild antiseptics and are not regarded as toxic; boron hydrides are however highly toxic. Symbol: B; m.p. 2300°C; sublimes at 2550°C; r.d. 2.34; p.n. 5; r.a.m. 10.81.

boron hydrides Hydrides of boron ranging from B_2H_6 to $B_{20}H_{16}$, which have complex molecular structures. These compounds are notable for containing B−H−B bonds. Investigations have shown that the boron hydrides are electron deficient and thus the bonding can only be rationalized in terms of multicentre bonds. The compound B_2H_6 can be prepared by reacting boron trichloride with hydrogen. A complex ion, tetrahydridoborate(III) (borohydride, BH_4^-), can be formed; it has very strong reducing properties.

boron nitride (BN) A compound formed by heating boron in nitrogen (1000°C). It has two crystalline forms. *See* boron.

boron oxide (B_2O_3) A glassy hygroscopic solid that eventually forms boric acid. It forms various salts but also exhibits some amphoteric properties.

boron tribromide (BBr_3) A colourless liquid. *See* boron trichloride.

boron trichloride (BCl_3) A fuming liquid made by passing dry chlorine over heated boron. It is rapidly hydrolysed by water:
$$BCl_3 + 3H_2O \rightarrow 3HCl + H_3BO_3$$
As there are only three pairs of shared electrons in the outer shell of the boron atom, boron halides form very stable addition compounds with ammonia by the acceptance of a lone electron pair in a coordinate bond to complete a shared octet.

boron trifluoride (BF_3) A colourless fuming gas made by heating a mixture of boron oxide, calcium fluoride, and concentrated sulphuric acid. *See* boron trichloride.

borosilicates Complex compounds similar to silicates, but containing BO_3 and BO_4 units in addition to the SiO_4 units. Certain crystalline borosilicate minerals are known. In addition, borosilicate glasses can be made by using boron oxide in addition to silicon dioxide. These tend to be tougher and more heat resistant than 'normal' silicate glass.

Bosch process The reaction
$$CO + H_2O \rightarrow CO_2 + H_2$$
Using water gas over a hot catalyst, used to make hydrogen for the Haber process.

bowl classifier A device that separates solid particles in a mixture of solids and liquid into fractions according to particle size. Feed enters the centre of a shallow bowl, which contains revolving blades. The coarse solids collect on the bottom, fine solids at the periphery.

Boyle's law At a constant temperature, the pressure of a fixed mass of a gas is inversely proportional to its volume: i.e.
$$pV = K$$
where K is a constant. The value of K depends on the temperature and on the nature of the gas. The law holds strictly only for ideal gases. Real gases follow Boyle's law at low pressures and high temperatures. *See* gas laws.

Brady's reagent *See* 2,4-dinitrophenylhydrazine.

Bragg equation An equation used to deduce the crystal structure of a material using data obtained from X-rays directed at its surface. The conditions under which a crystal will reflect a beam of X-rays with maximum intensity is:
$$n\lambda = 2d\sin\theta$$
where θ is the angle of incidence and reflection (Bragg angle) that the X-rays make with the crystal planes, n is a small integer, λ is the wavelength of the X-rays, and d is the distance between the crystal planes.

branched chain *See* chain.

brass Any of a group of copper-zinc alloys containing up to 50% of zinc. The colour of brass changes from red-gold to golden to silvery-white with increasing zinc content. Brasses are easy to work and resist corrosion well. Brasses with up to 35% zinc can be worked cold and are specially suited for rolling into sheets, drawing into wire, and making into tubes. Brasses with 35–46% zinc are harder and stronger but less ductile; they require hot working (e.g. forging). The properties of brass can be improved by the addition of other elements; lead improves its ability to be machined, while aluminium and tin increase its corrosion resistance. *See also* bronze, nickel-silver.

bremsstrahlung *See* X-radiation.

brine A concentrated solution of sodium chloride.

Brin process *See* barium.

bromic(I) acid (hypobromous acid, HBrO) A pale yellow liquid made by reacting mercury(II) oxide with bromine water. It has strong bleaching powers. Bromic(I) acid will donate protons to only a small extent and hence is a weak acid in aqueous solution. It is a strong oxidizing agent.

bromic(V) acid (HBrO$_3$) A colourless liquid made by the addition of dilute sulphuric acid to barium bromate. It dissociates extensively in aqueous solution and is a strong acid.

bromide *See* halide.

bromination *See* halogenation.

bromine A deep red, moderately reactive liquid element belonging to the halogens (group VII of the periodic table). It occurs in small amounts in seawater, salt lakes, and salt deposits but is much less abundant than chlorine. Bromine reacts with most metals but generally with less vigour than chlorine. It has less oxidizing power than chlorine and consequently can be released from solutions of bromides by reaction with chlorine gas. The laboratory method is the more convenient oxidation

by manganese dioxide. Industrial methods of production utilize odixation by chlorine or electrolysis with removal from the solution by purging with air. Bromine and its compounds are used in pharmaceuticals, photography, chemical synthesis, fumigants, and in significantly large quantities as 1,2-dibromoethane (which is added to petrol to combine with lead produced from the decomposition of the antiknock agent lead tetraethyl).

The electropositive elements form ionic bromides and the less electropositive elements and metalloids form partly covalent bromides. The non-metals form fully covalent bromides, with stabilities that are usually less than those of the corresponding chlorides. The metal bromides are generally more soluble than the corresponding chlorides (because the larger bromide ion reduces the lattice energies).

Like chlorine, bromine forms oxides, Br$_2$O and BrO$_2$, both of which are unstable. The related oxo-acid anions hypobromite (BrO$^-$) and bromate (BrO$_3{}^-$) are formed by the reaction of bromine with cold aqueous alkali and hot aqueous alkali respectively, but the bromine analogues of chlorite and perchlorate are not known.

Bromine and the interhalogens are highly toxic. Liquid bromine and bromine solutions are also very corrosive and goggles and gloves should always be worn when handling such compounds.

Symbol: Br; m.p. $-7.2°C$; b.p. $58.7°C$; r.d. 3.12; p.n. 35; r.a.m. 79.909.

bromine trifluoride (BrF$_3$) A colourless fuming liquid made by direct combination of fluorine and bromine. It is a very reactive compound, its reactions being similar to those of its component halides.

bromoethane (ethyl bromide, C$_2$H$_5$Br) A colourless volatile compound, used as a refrigerant. It can be made from ethene and hydrogen bromide.

bromoform *See* tribromomethane.

bromomethane (methyl bromide, CH$_3$Br) A colourless volatile compound

34

used as a solvent. It can be made from methane and bromine.

bronze Any of a group of copper-tin alloys usually containing 0.5–10% of tin. They are generally harder, stronger in compression, and more corrosion-resistant than brass. Zinc is often added, as in *gunmetal* (2–4% zinc), to increase strength and corrosion-resistance; British bronze coins now contain more zinc (2.5%) than tin (0.5%). The presence of lead improves its machining qualities.

Some copper-rich alloys containing no tin are also called bronzes. *Aluminium bronzes*, for example, with up to 10% aluminium, are strong, resistant to corrosion and wear, and can be worked cold or hot; *silicon bronzes*, with 1–5% silicon, have high corrosion resistance; *beryllium bronzes*, with about 2% beryllium, are very hard and strong.

Brownian movement The random motion of small particles in a fluid – for example, smoke particles in air. The particles, which may be large enough to be visible with a microscope, move because they are continuously bombarded by the molecules of the fluid.

brown-ring test A qualitative test used for the detection of nitrate. A freshly prepared solution of ferrous sulphate is mixed with the sample and concentrated sulphuric acid is introduced slowly to the bottom of the tube using a dropping pipette so that two layers are formed. A brown ring formed where the liquids meet indicates the presence of nitrate. The brown colour is [Fe(NO)]SO$_4$, which breaks down on shaking.

buffer A solution in which the pH remains reasonably constant when acids or alkalis are added to it; i.e. it acts as a buffer against (small) changes in pH. Buffer solutions generally contain a weak acid and one of its salts derived from a strong base; e.g. a solution of acetic acid and sodium acetate. If an acid is added, the H$^+$ reacts with the acetate ion (from dissociated sodium acetate) to form undissociated acetic

acid; if a base is added the OH$^-$ reacts with the acetic acid to form water and the acetate ion. The effectiveness of the buffering action is determined by the concentrations of the acid-anion pair:

$$K = [H^+][CH_3COO^-]/[CH_3COOH]$$
where K is the dissociation constant. Phosphate, oxalate, tartrate, borate, and carbonate systems can also be used for buffer solutions.

bumping Violent boiling of a liquid caused when bubbles form at pressure above atmospheric pressure.

Bunsen burner A gas burner consisting of a vertical metal tube with an adjustable air-inlet hole at the bottom. Gas is allowed into the bottom of the tube and the gas-air mixture is burnt at the top. With too little air the flame is yellow and sooty. Correctly adjusted, the burner gives a flame with a pale blue inner cone of incompletely burnt gas, and an almost invisible outer flame where the gas is fully oxidized and reaches a temperature of about 1500°C.

Bunsen cell A type of primary cell in which the positive electrode is formed by carbon plates in a nitric acid solution and the negative electrode consists of zinc plates in sulphuric acid solution.

burette A piece of apparatus used for the addition of variable volumes of liquid in a controlled and measurable way. The burette is a long cylindrical graduated tube of uniform bore fitted with a stopcock and a small-bore exit jet, enabling a drop of liquid at a time to be added to a reaction vessel. Burettes are widely used in volumetric analysis. Standard burettes permit volume measurement to 0.005 cm^3 and have a total capacity of 50 cm^3; a variety of smaller *microburettes* is available.

buta-1,3-diene (butadiene, CH$_2$: CHCH:CH$_2$) A colourless gas made by catalytic dehydrogenation of butane. It is polymerized in the manufacture of synthetic rubber. Buta-1,3-diene is conjugated and to some extent the pi electrons are delocalized over the molecule.

butanal (butyraldehyde, C_3H_7CHO) A colourless liquid aldehyde.

butane (C_4H_{10}) A gaseous alkane obtained either from the gaseous fraction of crude oil or by the 'cracking' of heavier fractions. Butane is easily liquefied and its main use is as a portable supply of fuel (bottle gas).
Butane is the fourth member of the homologous series of alkanes.

butanedioic acid (succinic acid) A crystalline carboxylic acid, $HOOC\cdot(CH_2)_2COOH$, that occurs in amber and certain plants. It forms during the fermentation of sugar.

butanoic acid (butyric acid, C_3H_7COOH) A colourless liquid carboxylic acid. Esters of butanoic acid are present in butter.

butanols (C_4H_9OH) Two alcohols that are derived from butane: butan-1-ol ($CH_3(CH_2)_2CH_2OH$) and butan-2-ol ($CH_3CH(OH)CH_2CH_3$). Both are colourless volatile liquids used as solvents.

butanone (methyl ethyl ketone, $CH_3COC_2H_5$) A colourless volatile liquid ketone. It is manufactured by the oxidation of butane and used as a solvent.

butenedioic acid Either of two isomers. *Transbutenedioic acid* (fumaric acid) is a crystalline compound found in certain plants.

Cisbutenedioic acid (maleic acid) is used in the manufacture of synthetic resins. It can be converted into the trans isomer by heating at 120°C.

butyl group The straight chain alkyl group $CH_3(CH_2)_2CH_2-$.

butyraldehyde *See* butanal.

butyric acid *See* butanoic acid.

cis (maleic)

trans (fumaric)

Butenedioic acids

by-product A substance obtained during the manufacture of a main chemical product. For example, propanone is not now manufactured from propan-1-ol, but is obtained as a by-product in the manufacture of phenol.

36

C

cadmium A transition metal obtained as a by-product during the extraction of zinc. It is used to protect other metals from corrosion, as a neutron absorber in nuclear reactors, and in alkali-type batteries.
Symbol: Cd; m.p. 321°C; b.p. 765°C; r.d. 8.64; p.n. 48; r.a.m. 112.41.

cadmium cell *See* Weston cadmium cell.

caesium A soft silvery highly reactive element of the alkali metal group. It is found in several silicate minerals, including pollucite ($CsAlSi_2O_6$). Caesium is used in photocells, as a catalyst, in the caesium atomic clock, and recently in ion-propulsion systems for spacecraft.
Symbol: Cs; m.p. 28.5°C; b.p. 690°C; r.d. 1.87; p.n. 55; r.a.m. 132.91.

caesium-chloride structure A form of crystal structure that consists of alternate layers of caesium ions and chloride ions with the centre of the lattice occupied by a caesium ion in contact with eight chloride ions (i.e. four chloride ions in the plane above and four in the plane below).

cage compounds *See* clathrate.

calcination The formation of a calcium carbonate deposit from hard water.

calcite A mineral form of calcium carbonate occurring in limestone, chalk, and marble.

calcium A moderately soft, low-melting reactive metal; the third element in group II of the periodic table. The electronic configuration is that of argon with an additional pair of 4s electrons.
Calcium is widely distributed in the Earth's crust and is the third most abundant element. Large deposits occur as chalk or marble, $CaCO_3$; gypsum, $CaSO_4.2H_2O$; anhydrite, $CaSO_4$; fluorspar, CaF; and apatite, $CaF_2.Ca_3(PO_4)_3$. However sufficient-

ly large quantities of calcium chloride are available as waste from the Solvay process to satisfy industrial requirements for the metal, which is produced by electrolysis of the fused salt. Large quantities of lime, $Ca(OH)_2$, and quicklime, CaO, are produced by decomposition of the carbonate for use in both building and agriculture. Several calcium minerals are mined for the content of the other mineral. Thus, limestone is a cheap source of carbon dioxide, gypsum and anhydrite are used in the manufacture of sulphuric acid, phosphate rock for phosphoric acid, and fluorspar for a range of fluorochemicals.
Calcium has a low ionization potential and a relatively large atomic radius. It is therefore a very electropositive element. The metal is very reactive and the compounds contain the divalent ion Ca^{2+}. Calcium forms the oxide (CaO), a white ionic solid, on burning in air, but for practical purposes the oxide is best prepared by heating the carbonate, which decomposes at about 800°C. Both the oxide and the metal itself react with water to give the basic hydroxide ($Ca(OH)_2$). On heating with nitrogen, sulphur, or the halogens, calcium reacts to form the nitride (Ca_3N_2), sulphide (CaS), or the halides (CaX_2). Calcium also reacts directly with hydrogen to give the hydride CaH_2 and borides, arsenides, carbides, and silicides can be prepared in a similar way. Both the carbonate and sulphate are insoluble. Calcium salts impart a characteristic brick-red colour to flames which is an aid to qualitative analysis. At ordinary temperatures calcium has the face-centred cubic structure with a transition at 450°C to the close-packed hexagonal structure.
Symbol: Ca; m.p. 838°C; b.p. 1487°C; r.d. 1.54; p.n. 20; r.a.m. 40.08.

calcium acetylide *See* calcium dicarbide.

calcium bicarbonate *See* calcium hydrogencarbonate.

37

calcium carbide See calcium dicarbide.

calcium carbonate ($CaCO_3$) A white solid that occurs naturally in two crystalline forms: calcite and aragonite. These minerals make up the bulk of such rocks as marble, limestone, and chalk. Calcium carbonate also occurs in the mineral dolomite ($CaCO_3.MgCO_3$). It is sparingly soluble in water but dissolves in rainwater containing carbon dioxide to form calcium hydrogencarbonate, which causes temporary hardness of water. Calcium carbonate is a basic raw material in the Solvay process and is used for making glass, mortar, and cement.

calcium chloride ($CaCl_2$) A white solid that occurs in a number of hydrated forms and is readily available as a by-product of the Solvay process. It is readily soluble in water and the solution, known as *brine*, is used in refrigerating plants. Other applications that depend on its water-absorbing property and the low freezing point of the aqueous solution include the suppression of dust on roads and in mines and the melting of snow. Calcium chloride is also used as the electrolyte in the production of calcium.

calcium cyanamide ($CaCN_2$) A solid prepared by heating calcium dicarbide to a temperature in excess of 800°C in an atmosphere of nitrogen. It is used as a fertilizer because ammonia and calcium carbonate are slowly formed when water is added:

$$CaCN_2 + 3H_2O \rightarrow CaCO_3 + 2NH_3$$

Other uses include the defoliation of cotton plants and the production of melamine.

calcium dicarbide (calcium acetylide, calcium carbide, CaC_2) A colourless solid when pure. In countries where electricity is cheap, calcium dicarbide is produced on a large scale by heating calcium oxide with coke at a temperature in excess of 2000°C in an electric-arc furnace. Water is then added to give ethyne (C_2H_2), an important industrial organic chemical:

$$CaC_2 + 2H_2O \rightarrow C_2H_2 + Ca(OH)_2$$

The structure of calcium dicarbide is interesting because the carbon is present as carbide ions (C_2^{2-}). See ethyne.

calcium-fluoride structure (fluorite structure) A form of crystal structure in which each calcium ion is surrounded by eight fluoride ions arranged at the corners of a cube and each fluoride ion is surrounded tetrahedrally by four calcium ions.

calcium hydrogencarbonate (calcium bicarbonate, $Ca(HCO_3)_2$) A solid formed when water containing carbon dioxide dissolves calcium carbonate:

$$CaCO_3 + H_2O + CO_2 \rightarrow Ca(HCO_3)_2$$

Calcium hydrogencarbonate is a cause of temporary hard water. The solid is unknown at room temperature.

calcium hydroxide (slaked lime, $Ca(OH)_2$) A white solid that dissolves sparingly in water to give the alkali known as *limewater*. Calcium hydroxide is manufactured by adding water to the oxide, a process known as *slaking*, which evolves much heat. If just sufficient water is added so that the oxide turns to a fine powder, the product is *slaked lime*. If more water is added, a thick suspension called *milk of lime* is formed. Calcium hydroxide has many uses. As a base, it is used to neutralize acid soil and in industrial processes such as the Solvay process. It is also used in the manufacture of mortar, whitewash, and bleaching powder and for the softening of temporary hard water.

calcium nitrate ($Ca(NO_3)_2$) A deliquescent salt that is very soluble in water. It crystallizes as the hydrate $Ca(NO_3)_2.4H_2O$. When the hydrate is heated, the anhydrous salt is first produced and this subsequently decomposes to give calcium oxide, nitrogen dioxide, and oxygen. Calcium nitrate is used as a nitrogenous fertilizer.

calcium octadecanoate (calcium stearate, $Ca(CH_3(CH_2)_{16}COO)_2$) An insoluble salt of octadecanoic acid. It is formed as 'scum' when soap, containing the soluble salt sodium octadecanoate, is mixed with hard water containing calcium ions. *See* soap.

calcium oxide (quicklime, CaO) A white solid formed by heating calcium in oxygen or, more widely, by the thermal decomposition of calcium carbonate. On a large scale, limestone (calcium carbonate) is heated in a tall tower called a lime kiln to a temperature of 550°C. The reversible reaction:

$$CaCO_3 \rightleftharpoons CaO + CO_2$$

proceeds in a forward direction as the carbon dioxide is carried away by the upward current through the kiln. Calcium oxide is used in extractive metallurgy to produce a slag with the impurities in metal ores; it is also used as a drying agent, and it is an intermediate for the production of calcium hydroxide.

calcium phosphate ($Ca_3(PO_4)_2$) A solid that occurs naturally in the mineral apatite ($CaF_2.Ca_3(PO_4)_3$) and in rock phosphate. It is the chief constituent of animal bones and is used extensively in fertilizers.

calcium stearate *See* calcium octadecanoate.

calcium sulphate (anhydrite, $CaSO_4$) A white solid that occurs abundantly as the mineral anhydrite and as the dihydrate ($CaSO_4.2H_2O$), known as *gypsum* or *alabaster*. When heated, gypsum loses water to form the hemihydrate ($2CaSO_4.H_2O$), which is *plaster of Paris*. If the water is replaced, gypsum reforms and sets as a solid. Plaster of Paris is therefore used for taking casts and for setting broken limbs. Calcium sulphate is sparingly soluble in water and is a cause of permanent hardness. It is used in ceramics, paint, and in paper making.

Calgon (*Tradename*) A substance often added to detergents to remove unwanted chemicals that have dissolved in water and would otherwise react with soap to form a scum. Calgon consists of complicated polyphosphate molecules, which absorb dissolved calcium and magnesium ions. The metal ions become trapped within the Calgon molecules.

caliche An impure commercial form of sodium nitrate.

californium A radioactive transuranic element of the actinoid series of metals, not found naturally on Earth. Several radioisotopes have been synthesized, including californium-252, which is used as an intense source of neutrons.
Symbol: Cf; p.n. 98; stablest isotope ^{251}Cf (half-life 800 years).

calomel *See* mercury(I) chloride.

calomel electrode A half cell having a mercury electrode coated with mercury(I) chloride (called *calomel*), in an electrolyte consisting of potassium chloride and (saturated) mercury(I) chloride solution. Its standard electrode potential against the hydrogen electrode is accurately known (-0.2415 V at 25°C) and it is a convenient secondary standard.

calorie Symbol: cal A unit of energy approximately equal to 4.2 joules. It was formerly defined as the energy needed to raise the temperature of one gram of water by one degree Celsius. Because the specific thermal capacity of water changes with temperature, this definition is not precise. The mean or thermochemical calorie (cal_{TH}) is defined as 4.184 joules. The international table calorie (cal_{IT}) is defined as 4.1868 joules. Formerly the mean calorie was defined as one hundredth of the heat needed to raise one gram of water from 0°C to 100°C, and the 15°C calorie as the heat needed to raise it from 14.5°C to 15.5°C.

calorific value The energy content of a fuel, defined as the energy released in burning unit mass of the fuel.

calorimeter A device or apparatus for measuring thermal properties such as specific heat capacity, calorific value, etc. *See* bomb calorimeter.

calx A metal oxide obtained by heating an ore to high temperatures in air.

candela Symbol: cd The SI base unit of luminous intensity, defined as the intensity (in the perpendicular direction) of the black-body radiation from a surface of $1/600\,000$ square metre at the temperature of freezing platinum and at a pressure of 101 325 pascals.

cane sugar *See* sucrose.

Cannizzaro reaction The reaction of aldehydes to give alcohols and acid anions in the presence of strong bases. The aldehydes taking part in the Cannizzaro reaction lack hydrogen atoms on the carbon attached to the aldehyde group. For instance, in the presence of hot aqueous sodium hydroxide:

$$NaOH + 2C_6H_5CHO \rightarrow$$
$$C_6H_5CH_2OH + C_6H_5COO^-Na^+$$

The reaction is a disproportionation, involving both oxidation (to acid) and reduction (to alcohol). It can also occur with methanal to give methanol and methanoate ions.

$$NaOH + 2HCHO \rightarrow CH_3OH +$$
$$HCOO^-Na^+$$

canonical form *See* resonance.

carbamide *See* urea.

carbanion An intermediate in an organic reaction in which one carbon atom is electron-rich and carries a negative charge. Carbanions are usually formed by abstracting a hydrogen ion from a C–H bond using a base, e.g. from ethanal to form $^-CH_2CHO$, or from organometallic compounds in which the carbon atom is bonded to an electropositive metal.

carbene A transient species of the form RR'C:, with two valence electrons that do not form bonds. The simplest example is *methylene*, H_2C:. Carbenes are intermediates in some organic reactions.

carbide A compound of carbon with a more electropositive element. The carbides of the elements are classified into:
(1) *Ionic carbides*, which contain the carbide ion C^{4-}. An example is aluminium carbide, Al_4C_3. Compounds of this type react with water to give methane (they were formerly also called *methanides*). Another type of ionic carbon compound contains the dicarbide ion $^-C:C^-$. The best-known example is calcium dicarbide, CaC_2, also known as calcium carbide, or simply *carbide*. Compounds of this type give acetylene with water. They were formerly called *acetylides*. Ionic carbides are formed with very electropositive metals. They are crystalline.
(2) *Covalent carbides*, which have giant-molecular structures, as in silicon carbide (SiC) and boron carbide (B_4C_3). These are hard high-melting solids. Other covalent compounds of carbon (CO_2, CS_2, CH_4, etc.) have covalent molecules.
(3) *Interstitial carbides*, which are interstitial compounds of carbon with transition metals. Titanium carbide (TiC) is an example. These are all hard high-melting solids, with metallic properties. Some carbides (e.g. nickel carbide Ni_3C) have properties intermediate between those of interstitial and ionic carbides.

carbocyclic compound A compound, such as benzene or cyclohexane, that contains a ring of carbon atoms in its structure.

carbohydrates A class of compounds occurring widely in nature and having the general formula type $C_x(H_2O)_y$. (Note that although the name suggests a hydrate of carbon these compounds are in no way hydrates and have no similarities to classes of hydrates.) Carbohydrates are generally divided into two main classes: sugars and polysaccharides.

Carbohydrates are both stores of energy and structural elements in living systems; plants having typically 15% carbohydrate and animals about 1% carbohydrate. The body is able to build up polysaccharides from simple units (anabolism) or break the larger units down to more simple units for

Carbon structures

releasing energy (catabolism). *See also* polysaccharide, sugar.

carbolic acid A former name for phenol (hydroxybenzene, C_6H_5OH).

carbon The first element of group IV of the periodic table. Carbon is a universal constituent of living matter and the principal deposits of carbon compounds are derived from living sources; i.e., carbonates (chalk and limestone) and fossil fuels (coal, oil, and gas). It also occurs in the mineral dolomite. The element forms only 0.032% by mass of the Earth's crust. Minute quantities of elemental carbon also occur as the allotropes graphite and diamond.

The industrial demand for graphite is such that it is manufactured in large quantities using the Acheson process in which coke and small amounts of asphalt or clay are raised to high temperatures. Large quantities of impure carbon are also consumed in the reductive extraction of metals. Apart from the demand for diamond as a gemstone there is a large industrial demand for low grade small diamonds for drilling and grinding machinery. High-temperature studies show that graphite and diamond can be inter-converted at 3000°C and extremely high pressures but the commercial exploitation would not be viable. Carbon burns in oxygen to form carbon

dioxide and carbon monoxide, CO. Carbon dioxide is soluble in water forming the weakly acidic carbonic acid, the parent acid of the metal carbonates. In contrast CO is barely soluble in water but it will react with alkali to give the methanoate (formate) ion;

$$CO + OH^- \rightarrow HCO_2^-$$

Carbon will react readily with sulphur at red heat to form CS_2 but it does not react directly with nitrogen. Cyanogen, $(CN)_2$, must be prepared by heating covalent metal cyanides such as CuCN. Carbon will also react directly with many metals at elevated temperatures to give carbides. Carbides can also be obtained by heating the metal oxide with carbon or heating the metal with a hydrocarbon. There is a bewilderingly wide range of metal carbides, both salt-like with electropositive elements (CaC_2) and covalent with the metalloids (SiC), and there are also many interstitial carbides formed with metals such as Cr, Mn, Fe, Co, and Ni.

The compounds with C–N bonds form a significant branch of inorganic chemistry of carbon, these are hydrogen cyanide (HCN) and the cyanides, cyanic acid (HNCO) and the cyanates, and thiocyanic acid (HNCS) and the thiocyanates.

Naturally occurring carbon has the isotopic composition ^{12}C (98.89%), ^{13}C (1.11%) and ^{14}C (minute traces in the upper atmosphere produced by slow neutron capture by ^{14}N atoms). ^{14}C is used for radiocarbon

dating because of its long half-life of 5730 years.
Symbol: C; m.p. 3570°C; b.p. 4827°C; r.d. 3.51; p.n. 6; r.a.m. 12.01.

carbonate A salt of carbonic acid (containing the ion CO_3^{2-}).

carbonation 1. The solution of carbon dioxide in a liquid under pressure, as in carbonated soft drinks.
2. The addition of carbon dioxide to compounds, e.g. the insertion of carbon dioxide into Grignard reagents.

carbon dating (radiocarbon dating) A method of dating – measuring the age of (usually archaeological) materials that contain matter of living origin. It is based on the fact that ^{14}C, a beta emitter of half-life approximately 5730 years, is being formed continuously in the atmosphere as a result of cosmic-ray action.
The ^{14}C becomes incorporated into living organisms. After death of the organism the amount of radioactive carbon decreases exponentially by radioactive decay. The ratio of ^{12}C to ^{14}C is thus a measure of the time elapsed since the death of the organic material.
The method is most valuable for specimens of up to 20 000 years old, though it has been modified to measure ages up to 70 000 years. For ages of up to about 8000 years the carbon time scale has been calibrated by dendrochronology; i.e. by measuring the $^{12}C:^{14}C$ ratio in tree rings of known antiquity.

carbon dioxide (CO_2) A colourless odourless nonflammable gas formed when carbon burns in excess oxygen. It is also produced by respiration. Carbon dioxide is present in the atmosphere (0.03% by volume) and is converted in plants to carbohydrates by photosynthesis. In the laboratory it is made by the action of dilute acid on metal carbonates. Industrially, it is obtained as a by-product in certain processes, such as fermentation or the manufacture of lime. The main uses are as a refrigerant (solid carbon dioxide, called

dry ice) and in fire extinguishers and carbonated drinks.
Carbon dioxide is the anhydride of the weak acid carbonic acid, which is formed in water:
$$CO_2 + H_2O \rightarrow H_2CO_3$$

carbon disulphide (CS_2) A colourless poisonous flammable liquid made from methane (natural gas) and sulphur. It is a solvent. The pure compound is virtually odourless, but CS_2 usually has a revolting smell because of the presence of other sulphur compounds.

carbon fibres Fibres of graphite which are used, for instance, to strengthen polymers. They are made by heating stretched textile fibres and have an orientated crystal structure.

carbonium ion An intermediate in an organic reaction in which one carbon atom is electron-deficient and carries a positive charge. Carbonium ions are formed by breaking a covalent bond between a carbon atom and an electronegative atom:
$$C_2H_5Br \rightarrow C_2H_5^+ + Br^-$$
They are high-energy intermediates and can only be stabilized and isolated under special conditions.

carbonize (carburize) To convert an organic compound into carbon by incomplete oxidation at high temperature.

carbon monoxide (CO) A colourless flammable toxic gas formed by the incomplete combustion of carbon. In the laboratory it can be made by dehydrating methanoic acid with concentrated sulphuric acid
$$HCOOH - H_2O \rightarrow CO$$
Industrially, it is produced by the oxidation of carbon or of natural gas, or by the water-gas reaction. It is a powerful reducing agent and is used in metallurgy.
Carbon monoxide is neutral and only sparingly soluble in water. It is not the anhydride of methanoic acid, although under extreme conditions it can react with sodium hydroxide to form sodium methanoate. It forms metal carbonyls with

transition metals, and its toxicity is due to its ability to form a complex with haemoglobin.

carbon tetrachloride (tetrachloromethane, CCl_4) A colourless nonflammable liquid made by the chlorination of methane. Its main use is as a solvent.

carbonyl A complex in which carbon monoxide ligands are coordinated to a metal atom. A common example is tetracarbonyl nickel(0), $Ni(CO)_4$.

carbonyl group The group $C=O$. It occurs in aldehydes (RCO.H), ketones (RR'CO), carboxylic acids (RCO.OH), and in carbonyl complexes of transition metals.

carborundum See silicon carbide.

carboxyl group The organic group $-CO.OH$, present in carboxylic acids.

Carboxylic acid

carboxylic acid A type of organic compound with the general formula RCOOH. Many carboxylic acids occur naturally in plants and (in the form of esters) in fats and oils, hence the alternative name *fatty acids*. Carboxylic acids with one COOH group are called *monobasic*, those with two, *dibasic*, and those with three, *tribasic*. The methods of preparation are:
(1) Oxidation of a primary alcohol or an aldehyde:
$$RCH_2OH + 2[O] \rightarrow RCOOH + H_2O$$
(2) Hydrolysis of a nitrile using dilute hydrochloric acid:
$$RCN + HCl + 2H_2O \rightarrow RCOOH + NH_4Cl$$
The acidic properties of carboxylic acids are due to the carbonyl group, which attracts electrons from the C–O and O–H bonds. The carboxylate ion formed, R–

COO^-, is also stabilized by delocalization of electrons over the O–C–O grouping. Other reactions of carboxylic acids include the formation of esters and the reaction with phosphorus(V) chloride to form acyl halides.

carburizing See case hardening.

carbylamine reaction See isocyanide test.

carcinogen An agent that causes cancer in animals, such as tobacco smoke or ionizing radiation.

carnallite A mineral chloride of potassium and magnesium, $KCl.MgCl_2.6H_2O$.

Carius method A method in quantitative analysis for identifying halogens, phosphorus, and sulphur in organic compounds. The compound is heated with concentrated nitric acid, which oxidizes the organic matter and leaves the element in a form in which it can be detected by normal qualitative analysis.

Carnot cycle The idealized reversible cycle of four operations occurring in a perfect heat engine. These are the successive adiabatic compression, isothermal expansion, adiabatic expansion, and isothermal compression of the working substance. The cycle returns to its initial pressure, volume, and temperature, and transfers energy to or from mechanical work. The efficiency of the Carnot cycle is the maximum attainable in a heat engine. See Carnot's principle.

Carnot's principle (Carnot theorem) The efficiency of any heat engine cannot be greater than that of a reversible heat engine operating over the same temperature range. It follows directly from the second law of thermodynamics, and means that all reversible heat engines have the same efficiency, independent of the working substance. If heat is absorbed at temperature T_1 and given out at T_2, then the Carnot efficiency is $(T_1 - T_2)/T_1$.

Caro's acid *See* peroxosulphuric(VI) acid.

carrier gas The gas used to carry the sample in gas chromatography.

case hardening Processes for increasing the hardness of the surface or 'case' of the steel used to make such components as gears and crankshafts. The oldest method is *carburizing*, in which the carbon content of the surface layer is increased by heating in a carbon-rich environment. *Nitriding* involves the diffusion of nitrogen into the surface layer of steel forming intensely hard nitride particles in the structure. A combination of both carburizing and nitriding is sometimes employed.

cast iron Alloys of iron and carbon made by remelting the crude iron produced in a blast furnace. The carbon content is usually between 2.4 and 4.0% and may be present as iron carbide (white cast iron) or as graphite (grey cast iron). Metals can be added to improve the properties of the alloy. Additional elements such as phosphorus, sulphur, and manganese are also present as impurities. Cast iron is a cheap metal with an extensive range of possible properties and it has been used on a very large scale.

catabolism All the metabolic reactions that break down complex molecules to simpler compounds. The function of catabolic reactions is to provide energy. *See also* metabolism.

catalyst A substance that alters the rate of a chemical reaction without itself being changed chemically in the reaction. The catalyst can, however, undergo physical change; for example, large lumps of catalyst can, without loss in mass, be converted into a powder. Small amounts of catalyst are often sufficient to increase the rate of reaction considerably. A *positive catalyst* increases the rate of a reaction and a *negative catalyst* reduces it. *Homogeneous catalysts* are those that act in the same phase as the reactants (i.e. in gaseous and liquid systems). For example, nitrogen(II) oxide gas will catalyse the reaction between sulphur(IV) oxide and oxygen in the gaseous phase. *Heterogeneous catalysts* act in a different phase from the reactants. For example, finely divided nickel (a solid) will catalyse the hydrogenation of oil (liquid).

The function of a catalyst is to provide a new pathway for which the rate-determining step has a lower activation energy than in the uncatalysed reaction. A catalyst does not change the products in an equilibrium reaction and their concentration is identical to that in the uncatalysed reaction; i.e. the position of the equilibrium remains unchanged. The catalyst simply increases the rate at which equilibrium is attained. In *autocatalysis*, one of the products of the reaction itself acts as a catalyst. In this type of reaction the reaction rate increases with time to a maximum and finally slows down. For example, in the hydrolysis of ethyl acetate, the acetic acid produced catalyses the reaction.

catenation The formation of chains of atoms in molecules.

cathode In electrolysis, the electrode that is at a negative potential with respect to the anode. In any electrical system, such as a discharge tube or electronic device, the cathode is the terminal at which electrons enter the system.

cation A positively charged ion, formed by removal of electrons from atoms or molecules. In electrolysis, cations are attracted to the negatively charged electrode (the cathode). *Compare* anion.

cationic detergent *See* detergent.

cationic resin An ion exchange material that can exchange cations, such as H^+ and Na^+, for ions in the surrounding medium. Such resins are used for a wide range of purification and analytical purposes.

They are often produced by adding a sulphonic acid group ($-SO_3^-H^+$) or a carboxylate group ($-COO^-H^+$) to a stable polyphenylethene resin. A typical exchange reaction is:

resin$-SO_3^-H^+$ + NaCl =
resin$-SO_3^-Na^+$ + HCl

They have been used to great effect to separate mixtures of cations of similar size having the same charge. Such mixtures can be attached to cationic resins and progressive elution will recover them in order of decreasing ionic radius. Promethium was first isolated using this technique. *See also* ion exchange.

caustic potash *See* potassium hydroxide.

caustic soda *See* sodium hydroxide.

celestine A mineral sulphate of strontium, $SrSO_4$.

cell A system having two plates (electrodes) in a conducting liquid (electrolyte). An *electrolytic cell* is used for producing a chemical reaction by passing a current through the electrolyte (i.e. by electrolysis). A *voltaic* (or *galvanic*) *cell* produces an e.m.f. by chemical reactions at each electrode. Electrons are transferred to or from the electrodes, giving each a net charge.

There is a convention for writing cell reactions in voltaic cells. The Daniell cell, for instance, consists of a zinc electrode in a solution of Zn^{2+} ions connected (through a porous pot) to a solution of Cu^{2+} ions in which is placed a copper electrode. The reactions at the electrodes are

$$Zn \rightarrow Zn^{2+} + 2e$$

i.e. oxidation of the zinc to zinc(II), and

$$Cu^{2+} + 2e \rightarrow Cu$$

i.e. reduction of copper(II) to copper. A cell reaction of this type is written:

$$Zn|Zn^{2+}(aq)|Cu^{2+}(aq)|Cu$$

The e.m.f. is the potential of a lead on the right minus the potential of a lead on the left. Copper is positive in this case and the e.m.f. of the cell is stated as $+1.10$ volts. *See also* accumulator, Daniell cell, electrolysis, Leclanché cell.

cellulose A polysaccharide $(C_6H_{10}O_5)_n$ of glucose, which is the main constituent of the cell walls of plants. It is obtained from wood pulp.

cellulose acetate A polymeric substance made by acetylating cellulose. It is used in plastics, in acetate film, and in acetate rayon.

cellulose nitrate (gun cotton, nitrocellulose) A highly flammable substance made by treating cellulose with a nitric-sulphuric acid mixture. Cellulose nitrate is used in explosives and in lacquers. It is an ester of nitric acid (not a nitro compound).

Celsius scale A temperature scale in which the temperature of melting pure ice is taken as $0°$ and the temperature of boiling water $100°$ (both at standard pressure). The *degree Celsius* ($°C$) is equal to the kelvin. This was known as the *centigrade scale* until 1948, when the present name became official. Celsius' original scale was inverted (i.e. had $0°$ as the steam temperature and $100°$ as the ice temperature).
See also temperature scale.

cement A powdered mixture of calcium silicates and aluminates, which is made by heating limestone $(CaCO_3)$ with clay, and grinding the result. When mixed with water reactions occur with the water (hence the name *hydraulic cement*), and a hard solid aluminosilicate is formed.

cementite (Fe_3C) A constituent of certain cast irons and steels. The presence of cementite increases the hardness of the metal.

centi- Symbol: c A prefix denoting 10^{-2}. For example, 1 centimetre (cm) = 10^{-2} metre (m).

centigrade scale *See* Celsius scale.

centrifuge An apparatus for rotating a container at high speeds, used to increase the rate of sedimentation of suspensions or the separation of two immiscible liquids. *See also* ultracentrifuge.

centrifugal pump A device commonly used for transporting fluids around a chemical plant. Centrifugal pumps usually have 6–12 blades rotating inside a fixed circular casing. As the blades rotate, the fluid is impelled out of the pump along a pipe. Centrifugal pumps do not produce high pressures but they have the advantage of being relatively cheap because they are simple in design, have no valves, and work at high speeds. In addition they are not damaged if a blockage develops. *Compare* displacement pump.

ceramics Useful high-melting inorganic materials. Ceramics include silicates and aluminosilicates, refractory metal oxides, and metal nitrides, borides, etc. Pottery and porcelain are examples of ceramics.

cerium A ductile malleable silvery element of the lanthanoid series of metals. It occurs in association with other lanthanoids in many minerals, including monazite and bastnasite. Cerium is used in several alloys (especially for lighter flints), as a catalyst, and in compound form in carbon-arc searchlights, etc., and in the glass industry.
Symbol: Ce; m.p. 799°C; b.p. 3430°C; r.d. 6.7; p.n. 58; r.a.m. 140.12.

cerussite A naturally occurring form of lead(II) carbonate that is an important lead ore. It forms orthorhombic crystals and is often found together with galena (PbS).

CFC *See* chlorofluorocarbon.

c.g.s. system A system of units that uses the centimetre, the gram, and the second as the base mechanical units. Much early scientific work used this system, but it has now almost been abandoned.

chain When two or more atoms form bonds with each other in a molecule, a chain of atoms results. This chain may be a *straight chain*, in which each atom is added to the end of the chain, or it may be a *branched chain*, in which the main chain of atoms has one or more smaller *side chains* branching off it.

chain reaction A self-sustaining chemical reaction consisting of a series of steps, each of which is initiated by the one before it. An example is the reaction between hydrogen and chlorine:

$$Cl_2 \rightarrow 2Cl\cdot$$
$$H_2 + \cdot Cl\cdot \rightarrow HCl + H\cdot$$
$$H\cdot + Cl_2 \rightarrow HCl + Cl\cdot$$
$$2H\cdot \rightarrow H_2$$
$$2Cl\cdot \rightarrow Cl_2$$

The first stage, chain initiation, is the dissociation of chlorine molecules into atoms; this is followed by two chain propagation reactions. Two molecules of hydrogen chloride are produced and the ejected chlorine atom is ready to react with more hydrogen. The final steps, chain termination, stop the reaction.

Induced nuclear fission reactions depend on chain reactions; the fission reaction is maintained by the two or three neutrons set free in each fission.

chair conformation *See* conformation.

chalcogens *See* group VI elements.

chalk A natural form of calcium carbonate ($CaCO_3$) formed originally by marine organisms. (Blackboard chalk is calcium sulphate, $CaSO_4$.)

charcoal An amorphous form of carbon made by heating wood or other organic material in the absence of air. *Activated charcoal* is charcoal heated to drive off absorbed gas. It is used for absorbing gases and for removing impurities from liquids.

Charles' law For a given mass of gas at constant pressure, the volume increases by a constant fraction of the volume at 0°C for each Celsius degree rise in temperature. The constant fraction (α) has almost the same value for all gases – about 1/273 – and Charles' law can be written in the form

$$V = V_0(1 + \alpha_v\theta)$$

where V is the volume at temperature $\theta\,^{\circ}$C and V_0 the volume at $0\,^{\circ}$C. The constant α_v is the thermal expansivity of the gas. For an ideal gas its value is 1273.15.

A similar relationship exists for the pressure of a gas heated at constant volume:

$$p = p_0(1 + \alpha_p\theta)$$

Here, α_p is the pressure coefficient. For an ideal gas

$$\alpha_p = \alpha_v$$

although they differ slightly for real gases. It follows from Charles' law that for a gas heated at constant pressure,

$$V/T = K$$

where T is the thermodynamic temperature and K is a constant. Similarly, at constant volume, p/T is a constant.

Charles' volume law is sometimes called *Gay-Lussac's law* after its independent discoverer. *See also* absolute temperature, gas laws.

chelate A metal coordination complex in which one ligand coordinates at two or more points to the same metal ion. The resulting complex contains rings of atoms that include the metal atom. An example of a chelating agent is 1,2-diaminoethane ($H_2NCH_2CH_2NH_2$), which can coordinate both its amine groups to the same atom. It is an example of a *bidentate ligand* (having two 'teeth'). Edta, which can form up to six bonds, is another example of a chelating agent. The word chelate comes from the Greek word meaning 'claw'.

chemical bond A link between atoms that leads to an aggregate of sufficient stability to be regarded as an independent molecular species. Chemical bonds include covalent bonds, electrovalent bonds, coordinate bonds, and metallic bonds. Some authors also include hydrogen bonds and van der Waals forces.

chemical combination, laws of A group of chemical laws developed during the late 18th and early 19th centuries, which arose from the recognition of the importance of quantitative (as opposed to qualitative) study of chemical reactions. The laws are:

(1) the law of conservation of mass (matter) (Lavoisier, 1774)
(2) the law of constant (definite) proportions (Proust, 1799)
(3) the law of multiple proportions (Dalton, 1804).

These laws played a significant part in Dalton's development of his atomic theory (1808). *See* conservation of mass (law of), constant proportions (law of), multiple proportions (law of).

chemical engineering The branch of engineering concerned with the design and maintenance of a chemical plant and its ability to withstand extremes of temperature and pressure, corrosion, and wear. It enables laboratory processes producing grams of material to be converted into a large-scale plant producing tonnes of material. Chemical engineers plan large-scale chemical processes by linking together the appropriate unit processes and by studying such parameters as heat and mass transfer, separations, and distillations.

chemiluminescence The emission of light during a chemical reaction.

chemisorption *See* adsorption.

Chile saltpetre *See* sodium nitrate.

china clay (kaolin) A white powder obtained from the decomposition of granites. It is used in the paper and pottery industries.

chiral centre An atom in a molecule that is attached to four different groups. *See* isomerism, optical activity.

chirality The property of existing in left- and right-handed forms. In chemistry the term is applied to the existence of optical isomers. *See* isomerism, optical activity.

chloral *See* trichloroethanal.

chloral hydrate *See* trichloroethanal.

chloramine (NH_2Cl) A colourless liquid made by reacting ammonia with sodium chlorate(I) (NaOCl). It is formed as an intermediate in the production of hydrazine. Chloramine is unstable and changes explosively into ammonium chloride and nitrogen trichloride.

chlorate A salt of chloric acid.

chloric(I) acid (hypochlorous acid, HClO) A colourless liquid produced when chlorine is dissolved in water. It is a bleach and gives chlorine water its disinfectant properties. To increase the yield of acid, the chlorine water can be shaken with a small amount of mercury(II) chloride. The Cl–O bond is broken more easily than the O–H bond in aqueous solution; the acid is consequently a poor proton donor and hence, a weak acid.

chloric(III) acid (chlorous acid, $HClO_2$) A pale yellow liquid produced by reacting chlorine dioxide with water. It is a weak acid and oxidizing agent.

chloric(V) acid (chloric acid, $HClO_3$) A colourless liquid with a pungent odour, formed by the action of dilute sulphuric acid on barium chlorate. It is a strong acid and has bleaching properties. Chloric(V) acid is a strong oxidizing agent and in concentrated solution it will ignite organic substances, such as paper and sugar.

chloric(VII) acid (perchloric acid, $HClO_4$) A colourless liquid that fumes strongly in moist air. It is made by vacuum distillation of a mixture of potassium perchlorate and concentrated sulphuric acid. In contact with organic material it is dangerously explosive.
The hydrate ($HClO_4.H_2O$) of chloric(VII) acid is a white crystalline solid at room temperature and has an ionic lattice structure of the form $(H_3O)^+(ClO_4)^-$.

chloric acid See chloric(V) acid.

chloride See halide.

chlorination 1. Treatment with chlorine; for instance, the use of chlorine to disinfect water.

2. See halogenation.

chlorine A green reactive gaseous element belonging to the halogens (group VII of the periodic table). It occurs in seawater, salt lakes, and underground deposits of halite, NaCl. It accounts for about 0. 055% of the Earth's crust. Chlorine is strongly oxidizing and can only be liberated from its salts by strong oxidizing agents, such as manganese dioxide, potassium permanganate, or potassium dichromate; note that sulphuric acid is not sufficiently oxidizing to release chlorine from chlorides. Industrially, chlorine is prepared by the electrolysis of brine and in some processes chlorine is recovered by the high-temperature oxidation of waste hydrochloric acid. Chlorine is used in large quantities, both as the element, to produce chlorinated organic solvents, and for the production of polyvinyl chloride (PVC), the major thermoplastic in use today, and in the form of hypochlorites for bleaching. Chlorine reacts directly and often vigorously with many elements; it reacts explosively with hydrogen in sunlight to form hydrogen chloride, HCl, and combines with the electropositive elements to form metal chlorides. The metals of main groups I and II form ionic chlorides but an increase in the metal charge/size ratio leads to the chlorides becoming increasingly covalent. For example, CsCl is totally ionic, $AlCl_3$ has a layer lattice, and $TiCl_4$ is essentially covalent. The electronegative elements form volatile molecular chlorides characterized by the single covalent bond to chlorine. With the exception of Pb^{2+}, Ag^+, and Hg_2^{2+}, the ionic chlorides are soluble in water, dissolving to give the hydrated metal ion and the chloride ion Cl^-. Chlorides of metals other than the most electropositive are hydrolysed if aqueous solutions are evaporated; for example,

$$ZnCl_2 + H_2O \rightarrow Zn(OH)Cl + HCl$$
$$FeCl_3 + 3H_2O \rightarrow Fe(OH)_3 + 3HCl$$

Chlorine forms four oxides, chlorine monoxide, Cl_2O; chlorine dioxide, ClO_2; chlorine hexoxide, Cl_2O_6; and chlorine heptoxide, Cl_2O_7; all of which are highly reactive and explosive. Chlorine dioxide finds commercial application as an active oxidizing agent but because of its explosive nature it is usually diluted by air or other gases. The chloride ion is able to function as a ligand with a large variety of metal ions forming such species as $[FeCl_4]^-$, $[CuCl_4]^{3-}$, and $[Co(NH_3)_4Cl_2]^+$. The formation of anionic chloro-complexes is applied to the separation of metals by anion-exchange methods.

Because of the hydrolysis of many metal chlorides when solutions are evaporated, special techniques must be used to prepare anhydrous chlorides. These are:
(1) reaction of dry chlorine with the hot metal
(2) reaction of dry hydrogen chloride with the hot metal (lower valencies)
(3) reaction of dry hydrogen chloride on the hydrated chloride.

The solubility of inorganic metal chlorides is such that they are not an environmental problem unless the metal ion itself is toxic but many organochlorine compounds are sufficiently stable for the accumulated residues of chlorine containing pesticides to present a severe problem in some areas. This arises because they can accumulate in food chains and concentrate in the tissues of higher animals. Chlorine and hydrogen chloride are both highly toxic. Thus chlorides that hydrolyse to release HCl should also be regarded as toxic. As organochlorine compounds are frequently even more toxic, they should not be handled without gloves and precautions should be taken against inhalation.
Symbol: Cl; m.p. $-101\,°C$; b.p. $-34.6\,°C$; r.d. 1.2; p.n. 17; r.a.m. 35.453.

chlorine dioxide (ClO_2) An orange gas formed by the action of concentrated sulphuric acid on potassium chlorate. It is a powerful oxidizing agent and its explosive properties in the presence of a reducing agent were used to make one of the first matches. It is widely used in the purification of water and as a bleach in the flour and wood pulp industry. On an industrial scale an aqueous solution of chlorine dioxide is made by passing nitrogen dioxide up a tower packed with a fused mixture of aluminium oxide and clay, down which a solution of sodium chlorate flows.

chlorine monoxide *See* dichlorine oxide.

chlorite A chlorate(III) salt; i.e. a salt of chloric(III) acid (chlorous acid).

chloroacetic acid *See* chloroethanoic acid.

chlorobenzene (monochlorobenzene, C_6H_5Cl) A colourless liquid made by the catalytic reaction of chlorine with benzene. It can be converted to phenol by reaction with sodium hydroxide under extreme conditions ($300\,°C$ and 200 atmospheres pressure). It is also used in the manufacture of other organic compounds.

2-chlorobuta-1,3-diene (chloroprene, $H_2C{:}CH.CCl{:}CH_2$) A colourless liquid derivative of butadiene used in the manufacture of neoprene rubber.

chloroethane (ethyl chloride, C_2H_5Cl) A gaseous compound made by the addition of hydrogen chloride to ethene. It is used as a refrigerant and a local anaesthetic.

chloroethanoic acid (chloroacetic acid, $CH_2ClCOOH$) A colourless crystalline solid made by substituting one of the hydrogen atoms of the methyl group of ethanoic acid with chlorine, using red phosphorus. It is a stronger acid than ethanoic acid because of the electron-withdrawing effect of the chlorine atom. *Dichloroethanoic acid* (dichloroacetic acid, $CHCl_2COOH$) and *trichloroethanoic acid* (trichloroacetic acid, CCl_2COOH) are made in the same way. The acid strength increases with the number of chlorine atoms present.

chloroethene (vinyl chloride, $H_2C{:}CHCl$) A gaseous organic compound used in the manufacture of PVC (polyvinyl chloride). Vinyl chloride is manufactured by the reaction between ethyne and hydrogen chloride using a mercury chloride catalyst:

$$C_2H_2 + HCl \rightarrow H_2C{:}CHCl$$

An alternative source, making use of the ready supply of ethene, is via dichloroethane:

$$H_2C{:}CH_2 + Cl_2 \rightarrow CH_2Cl.CH_2Cl \rightarrow$$
$$H_2C{:}CHCl$$

chloroform *See* trichloroethane.

chlorofluorocarbon (CFC) Any of various compounds formed from hydrocarbons by replacing some hydrogen atoms by chlorine and some by fluorine. They are mostly unreactive compounds used as refrigerants and aerosol propellants and in making plastic foams. It is thought that CFCs in the atmosphere may break down to chlorine and attack the ozone layer. For this reason many countries have banned or restricted their use.

chloromethane (methyl chloride, CH_3Cl) A colourless gaseous compound made by the chlorination of methane. It is used as a refrigerant and local anaesthetic.

chlorophylls The two green pigments present in plants which act as catalysts in the photosynthesis of carbohydrates from carbon dioxide and water.

chloroplatinic acid (platinic chloride, H_2PtCl_6) A reddish compound prepared by dissolving platinum in aqua regia. When crystallized from the resulting solution, crystals of the hexahydrate (H_2PtCl_6. $6H_2O$) are obtained. The crystals are needle-shaped and deliquesce on exposure to moist air. Chloroplatinic acid is a relatively strong acid, giving rise to the family of chloroplatinates.

chloroprene *See* 2-chlorobuta-1,3-diene.

chlorous acid *See* chloric(III) acid.

chromate Any oxygen-containing derivative of chromium. Usually the term is reserved for the chromate(VI) species.

chromatography A technique used to separate or analyse complex mixtures. A number of related techniques exist; the most important are *column chromatography, paper chromatography, thin-layer chromatography, gas chromatography,* and *ion exchange.* These chromatographic techniques require a sample to be placed at one end of a column of a solid or a liquid, which is held on an inert packing material. These columns are known as the stationary phase. A liquid or a gas, known as the mobile phase, is passed through the column. The components of the sample will dissolve in, and be carried along by, the mobile phase. Some of the components, however, will tend to cling to the stationary phase more than others and their progress through the column will be slower. The mixture will separate as the solvent moves through the stationary phase. In chromatography used for purification, separated components are collected as they emerge from the end of the column. For analysis purposes, however, the different components can often be identified simply by comparing their rate of progress through the column with that of known standards.

Components of the mixture are held back by the stationary phase either by adsorption (e.g. on the surface of alumina) or because they dissolve in it (e.g. in the moisture within chromatography paper). *See* partition law.

chrome alum *See* alum.

chromic anhydride *See* chromium(VI) oxide.

chromic oxide *See* chromium(III) oxide.

chromium A transition metal that occurs naturally as chromite ($FeO.Cr_2O_3$), large deposits of which are found in Zimbabwe. The ore is converted into sodium dichromate(VI) and then reduced with carbon to chromium(III) oxide and finally to metallic chromium with aluminium. Chromium is used in strong alloy steels and stainless steel and for plating articles. It is a hard silvery metal that resists corrosion at normal temperatures. It reacts slowly with dilute hydrochloric and sulphuric acids to give hydrogen and blue chromium(II) compounds, which quickly oxidize in air to green chromium(III) ions. The oxidation states are $+6$ in chromates (CrO_4^{2-}) and dichromates ($Cr_2O_7^{2-}$), $+3$ (the most stable), and $+2$. In acidic solutions the yellow chromate(VI) ion changes to the orange dichromate(VI) ion. Dichromates are used as oxidizing agents in the laboratory; for example as a test for sulphur dioxide and to oxidize alcohols. Symbol: Cr; m.p. 1900°C; b.p. 2642°C; r.d. 7.19; p.n. 24; r.a.m. 52.00.

chromium(II) oxide (chromous oxide, CrO) A black powder prepared by the oxidation of chromium amalgam with dilute nitric acid. At high temperatures (around 1000°C) chromium(II) oxide is reduced by hydrogen.

chromium(III) oxide (chromic oxide, chromium sesquioxide, Cr_2O_3) A green powder that is almost insoluble in water. It is isomorphous with iron(III) oxide and aluminium(III) oxide. Chromium(III) oxide is prepared by gently heating chromium(III) hydroxide or by heating ammonium dichromate. Alternative preparations include the heating of a mixture of ammonium chloride and potassium dichromate or the decomposition of chromyl chloride by passing it through a red hot tube. Chromium(III) oxide is used as a pigment in the paint and glass industries.

chromium(IV) oxide (chromium dioxide, CrO_2) A black solid prepared by heating chromium(III) hydroxide in oxygen at a temperature of 300–350°C. Chromium(IV) oxide is very unstable.

chromium(VI) oxide (chromium trioxide, chromic anhydride, CrO_3) A red crystalline solid formed when concentrated sulphuric acid is added to a cold saturated solution of potassium dichromate. The long prismatic needle-shaped crystals that are produced are extremely deliquescent. Chromium(VI) oxide is readily soluble in water, forming a solution that contains several of the polychromic acids. On heating it decomposes to give chromium(III)

oxide. Chromium(VI) oxide is used as an oxidizing reagent.

chromophore A group of atoms in a molecule that is responsible for the colour of the compound.

chromous oxide *See* chromium(II) oxide.

chromyl chloride (CrO_2Cl_2) A dark red covalent liquid prepared either by distilling a dry mixture of potassium dichromate and sodium chloride with concentrated sulphuric acid or by the action of concentrated sulphuric acid on chromium(VI) oxide dissolved in concentrated hydrochloric acid. Chromyl chloride is hydrolysed by water and with solutions of alkalis it undergoes immediate hydrolysis to produce chromate ions. It is used as an oxidizing agent in organic chemistry. Chromyl chloride oxidizes methyl groups at the ends of aromatic side chains to aldehyde groupings (Étard's reaction).

cinnabar A red mineral form of mercury(II) sulphide. It is the principal ore of mercury.

cis- Designating an isomer with groups that are adjacent. *See* isomerism.

cis-trans isomerism *See* isomerism.

citric acid A white crystalline carboxylic acid important in plant and animal cells. It is present in many fruits. The systematic name is 2-hydroxypropane-1,2,3-tricarboxylic acid. The formula is:

$$HO_2CCH_2C(OH)(CO_2H)CH_2CO_2H$$

Claisen condensation A reaction in which two molecules of ester combine to give a keto-ester – a compound containing a ketone group and an ester group. The reaction is base catalysed by sodium ethoxide; the reaction of ethyl ethanoate refluxed with sodium ethoxide gives:

$$2CH_3.CO.OC_2H_5 \rightarrow$$
$$CH_3.CO.CH_2.CO.OC_2H_5 + C_2H_5OH$$

The mechanism is similar to that of the aldol reaction, the first step being formation of a carbanion from the ester:

$$CH_3COC_2H_5 + {}^-OC_2H_5 \rightarrow$$
$${}^-CH_2COC_2H_5 + C_2H_5OH$$

This attacks the carbon atom of the carbonyl group on the other ester molecule, forming an intermediate anion that decomposes to the keto-ester and the ethanoate ion.

Clark cell A type of cell formerly used as a standard source of e.m.f. It consists of a mercury cathode coated with mercury sulphate, and a zinc anode. The electrolyte is zinc sulphate solution. The e.m.f. produced is 1.4345 volts at 15°C. The Clark cell has been superseded as a standard by the Weston cadmium cell.

clathrate (enclosure compound) A substance in which small (guest) molecules are trapped within the lattice of a crystalline (host) compound. Clathrates are formed when suitable host compounds are crystallized in the presence of molecules of the appropriate size. Although the term 'clathrate compound' is often used, they are not true compounds; no chemical bonds are formed, and the guest molecules interact by weak van der Waals forces. The clathrate is maintained by the cagelike lattice of the host. The host lattice must be broken down, for example, by heating or dissolution in order to release the guest. This should be compared with zeolites, in which the holes in the host lattice are large enough to permit entrance or emergence of the guest without breaking bonds in the host lattice. Quinol forms many clathrates, e.g. with SO_2; water (ice) forms a clathrate with xenon.

clays Naturally occurring aluminosilicates which form pastes and gels with water.

cleavage The splitting of a crystal along planes of atoms, to form smooth surfaces.

close packing The arrangement of particles (usually atoms) in crystalline solids in which each particle has 12 nearest neighbours: six in the same layer (or plane) as itself and three each in the layer above and below. This arrangement provides the most economical use of space. The two principal types are cubic close packing and hexagonal close packing.

coagulation The association of particles (e.g. in colloids) into clusters.

coal tar Tar produced by heating coal in the absence of oxygen. It is a mixture of many organic compounds (e.g. benzene, toluene, and naphthalene) and also contains free carbon.

cobalt A transition metal occurring in Canada in association with nickel. It is used in alloys for magnets, cutting tools, and electrical heating elements. Symbol: Co; m.p. 1495°C; b.p. 2870°C; r.d. 8.92; p.n. 27; r.a.m. 58.93.

cobaltic oxide See cobalt(III) oxide.

cobaltous oxide See cobalt(II) oxide.

cobalt(II) oxide (cobaltous oxide, CoO) A green powder prepared by the action of heat on cobalt(II) hydroxide in the absence of air. Alternatively it may be prepared by the thermal decomposition of cobalt(II) sulphate, nitrate, or carbonate, also in the absence of air. Cobalt(II) oxide is a basic oxide, reacting with acids to give solutions of cobalt(II) salts. It is stable in air up to temperatures of around 600°C, after which it absorbs oxygen to form tricobalt tetroxide. Cobalt(II) oxide can be reduced to cobalt by heating in a stream of carbon monoxide or hydrogen. It is used in the pottery industry and in the production of vitreous enamels.

cobalt(III) oxide (cobaltic oxide, Co_2O_3) A dark grey powder formed by the thermal decomposition of either cobalt(II) nitrate or carbonate in air. If heated in air, cobalt(III) oxide undergoes further oxidation to give tricobalt tetroxide, Co_3O_4.

coenzymes Small organic molecules (compared to the size of an enzyme) that enable enzymes to carry out their catalytic activity. Examples include nicotinamide adenine dinucleotide (NAD) and ubiquinone (coenzyme Q). Some coenzymes are capable of catalysing reactions in the absence of an enzyme but the rate of reaction is never as high as when a catalyst is present. A coenzyme is not a true catalyst as it undergoes chemical change during the reaction.

coherent units A system or sub-set of units (e.g. SI units) in which the derived units are obtained by multiplying or dividing together base units, with no numerical factor involved.

coinage metals A group of malleable metals forming a subgroup of group I of the periodic table. The metals all have an outer s^1 electronic configuration but they differ from the alkali metals (also outer s^1) in having inner d-electrons. They are copper (Cu), silver (Ag), and gold (Au). The coinage metals all have much higher ionization potentials than the alkali metals and high positive standard electrode potentials and are therefore much more difficult to oxidize. A further significant difference from the alkali metals is the variety of oxidation states observed for the coinage metals. Thus copper in aqueous chemistry is the familiar (hydrated) blue divalent Cu^{2+} ion but colourless copper(I) compounds can be prepared with groups that stabilize low valencies such as CN^-, e.g. CuCN. A few compounds of copper(III) have also been prepared. The common form of silver is Ag(I), e.g. $AgNO_3$, with a few Ag(II) compounds stable in the solid only. The most common oxidation state of gold is Au(III), although cyanide ion again stabilizes Au(I) compounds, e.g., $[Au(CN)_4]^{3-}$. The metals all form a large number of coordination compounds (unlike group IA) and are generally treated with the other elements of the appropriate transition series.

coke A dense substance obtained from the carbonization of coal. It is used as a fuel and as a reducing agent.

colligative properties A group of properties of solutions that depends on the number of particles present, rather than the nature of the particles. Colligative properties include:
(1) The lowering of vapour pressure.
(2) The elevation of boiling point.
(3) The lowering of freezing point.
(4) Osmotic pressure.
Colligative properties are all based upon empirical observation. The explanation of these closely related phenomena depends on intermolecular forces and the kinetic behaviour of the particles, which is qualitatively similar to those used in deriving the kinetic theory of gases.

collimator An arrangement for producing a parallel beam of radiation for use in a spectrometer or other instrument. A system of lenses and slits is utilized.

colloid A heterogeneous system in which the interfaces between phases, though not visibly apparent, are important factors in determining the system properties. The three important attributes of colloids are:
(1) They contain particles, commonly made up of large numbers of molecules, forming the distinctive unit or *disperse phase*.
(2) The particles are distributed in a continuous medium (the *continuous phase*).
(3) There is a stabilizing agent, which has an affinity for both the particle and the medium; in many cases the stabilizer is a polar group.
Particles in the disperse phase typically have diameters in the range $10^{-6} - 10^{-4}$ mm. Milk, rubber, and emulsion paints are typical examples of colloids. *See also* sol.

colorimetric analysis Quantitative analysis in which the concentration of a coloured solute is measured by the intensity of the colour. The test solution can be compared against standard solutions.

columbium The former name (in the US) for niobium.

column chromatography A technique widely used for the separation of mixtures of neutral substances with similar chemical and physical properties. Column chromatography employs a solid stationary phase, such as alumina or silica gel, packed in a vertical tube. The sample mixture is poured in at the top and one or more of its components become attached to the solid phase. These adsorbed fractions can be washed off the column and collected by using an appropriate solvent. Alumina, which has been activated by heating to remove adsorbed gases, is a strongly polar substance and the ease with which fractions can be removed from the column will depend on the polarity of their molecules. Careful choice of the solvent (eluent) used for the washing will allow the fractions to be selectively removed from the column. A non-polar solvent is used initially and is gradually replaced by a more polar solvent.
Compare thin-layer chromatography.

combustion A reaction with oxygen with the production of heat and light. The combustion of solids and liquids occurs when they release flammable vapour, which reacts with oxygen in the gas phase. Combustion reactions usually involve a complex sequence of free-radical chain reactions. The light is produced by excited atoms, molecules, or ions. In highly luminous flames it comes from small incandescent particles of carbon.
Sometimes the term is also applied to slow reactions with oxygen, and also to reactions with other gases (for example, certain metals 'burn' in chlorine).

common salt *See* sodium chloride.

complex A type of compound in which molecules or ions form coordinate bonds with a metal atom or ion. The coordinating species (called *ligands*) have lone pairs of electrons, which they can donate to the metal atom or ion. They are molecules such as ammonia or water, or negative ions such as Cl^- or CN^-. The resulting complex may be neutral or it may be a *complex ion*. For example:

$$Cu^{2+} + 4NH_3 \rightarrow [Cu(NH_3)_4]^{2+}$$
$$Fe^{3+} + 6CN^- \rightarrow [Fe(CN)_6]^{3-}$$
$$Fe^{2+} + 6CN^- \rightarrow [Fe(CN)_6]^{4-}$$

The formation of such coordination complexes is typical of transition metals. Often the complexes contain unpaired electrons and are paramagnetic and coloured. *See also* chelate.

component One of the separate chemical substances in a mixture in which no chemical reactions are taking place. For example, a mixture of ice and water has one component; a mixture of nitrogen and oxygen has two components. When chemical reactions occur between the substances in a mixture, the number of components is defined as the number of chemical substances present minus the number of equilibrium reactions taking place. Thus, the system: $N_2 + 3H_2 = 2NH_3$ is a two-component system. *See also* phase rule.

compound A chemical combination of atoms of different elements to form a substance in which the ratio of combining atoms remains fixed and is specific to that substance. The constituent atoms cannot be separated by physical means; a chemical reaction is required for the compounds to be formed or to change.
The existence of a compound does not necessarily imply that it is stable. Many compounds have lifetimes of less than a second.
See also mixture.

concentrated Denoting a solution in which the amount of solute in the solvent is relatively high. The term is always relative; for example, whereas concentrated sulphuric acid may contain 96% H_2SO_4, concentrated potassium chlorate may contain as little as 10% $KClO_3$. *Compare* dilute.

concentration The amount of substance per unit volume or mass in a solution. *Molar concentration* is amount of substance (in moles) per cubic decimetre (litre). *Mass concentration* is mass of solute per unit volume. *Molal concentration*

octahedral

square-planar

tetrahedral

trigonal-bipyramid

Shapes of complexes

is amount of substance (in moles) per kilogram of solute.

condensation The conversion of a gas or vapour into a liquid or solid by cooling.

condensation polymerization *See* polymerization.

condensation reaction A reaction in which addition of two molecules occurs followed by elimination of a smaller molecule, usually water. Condensation reactions (addition-elimination reactions) are characteristic of aldehydes and ketones reacting with a range of nucleophiles. There is typically nucleophilic addition at the C

atom of the carbonyl group followed by elimination of water.

conducting polymer A type of organic polymer that conducts electricity like a metal. Conducting polymers are crystalline substances containing conjugated unsaturated carbon-carbon bonds. In principle, they provide lighter and cheaper alternatives to copper and other metallic conductors. They tend, however, to be unstable and, at present, are not in commercial use.

conductiometric titration A titration in which measurement of the electrical conductance is made continuously throughout the addition of the titrant and

Condensation reactions with aldehydes and ketones

nucleophile	product	name

NH_3
ammonia

$$R' - \underset{\underset{OH}{|}}{\overset{\overset{R}{|}}{C}} - NH_2$$

hydroxy amine

H_2NNH_2
hydrazine

$$\overset{R}{\underset{R'}{\diagup}} C = N - NH_2$$

hydrazone

$C_6H_5NH\,NH_2$
phenyl hydrazine

$$\overset{R}{\underset{R'}{\diagup}} C = N - N \overset{H}{\underset{C_6H_5}{\diagdown}}$$

phenyl hydrazone

$HONH_2$
hydroxlamine

$$\overset{R}{\underset{R'}{\diagup}} C = N \diagup^{OH}$$

oxime

In the above the reactant is RCOR'

well beyond the equivalence point. This is in place of traditional end-point determination by indicators. The operation is carried out in a conductance cell, which is part of a resistance bridge circuit. The method depends on the fact that ions have different ionic mobilities, H^+ and OH^- having particularly high values. The method is especially useful for weak acid-strong base and strong acid-weak base titrations for which colour-change titrations are unreliable.

configuration 1. The arrangement of electrons about the nucleus of an atom. Configurations are represented by symbols, which contain:
(1) An integer, which is the value of the principal quantum number (shell number).
(2) A lower-case letter representing the value of the azimuthal quantum number (1), i.e.

s means $1 = 0$, p means $1 = 1$,

d means $1 = 2$, f means $1 = 3$.

chair conformation   boat conformation

Conformations of cyclohexane

eclipsed   staggered   gauche

Conformations for rotation about a single bond

Conformations

(3) A numerical superscript giving the number of electrons in that particular set; for example, $1s^2$, $2p^3$, $3d^5$.
The ground state electronic configuration (i.e. the most stable or lowest energy state) may then be represented as follows, for example, He, $1s^2$; N, $1s^2 2s^2 2p^5$. These are identical to the 'electron box' models, which are often met. However, elements are commonly abbreviated by using an inert gas to represent the 'core', e.g. Zr, $[Kr]4d^2 5s^2$.
2. The arrangement of atoms or groups in a molecule.

conformation A particular shape of molecule that arises through the normal rotation of its atoms or groups about single bonds. Any of the possible conformations that may be produced is called a *conformer*, and there will be an infinite number of these possibilities, differing in the angle between certain atoms or groups on adjacent carbon atoms. Under normal conditions conformers are non-isolable.

conjugate acid *See* acid.

conjugate base *See* base.

conjugated Describing compounds that have alternating double and single bonds in their structure. But-1,3-ene (H_2C:-$CHCH$:CH_2) is a conjugated compound. In such compounds there is delocalization of the electrons in the double bonds.

conservation of energy, law of Enunciated by Helmholtz in 1847, this law states that in all processes occurring in an isolated system the energy of the system

57

Constant – boiling mixtures

remains constant. The law does of course permit energy to be converted from one form to another (including mass, since energy and mass are equivalent).

conservation of mass, law of Formulated by Lavoisier in 1774, this law states that matter cannot be created or destroyed. Thus in a chemical reaction the total mass of the products equals the total mass of the reactants (the term 'mass' must include any solids, liquids, and gases – including air – that participate).

Constantan (*Trademark*) A copper-nickel (cupronickel) alloy containing 45% nickel. It has a high electrical resistivity and very low temperature coefficient of resistance and is therefore used in thermocouples and resistors.

constant-boiling mixture A general observation for most liquids is that the vapour phase above a liquid is richer in the more volatile component (deviation from Raoult's law). Consequently most liquid mixtures show a regular increase in the boiling point as the liquid is progressively distilled. The boiling point-composition curve shows that distillation of the liquid of composition L_1 gives a vapour richer in A and represented by composition V_1. Further distillation leads to the liquid composition moving towards B. For certain mixtures in which there are strong inter-molecular attractions the boiling point-composition curves show minima or maxima. Fractional distillation of the former leads to initial changes in the vapour until a distillate of composition L_2 is reached at

which point a constant boiling mixture or azeotrope distils over. Further attempts to fractionate the distillate do not lead to a change in composition. An example of an azeotropic mixture of minimum boiling point is water (b.p. 100°C) and ethanol (b.p. 78.3°C), the azeotrope being 4.4% water and boiling at 78.1°C.

Mixtures that display a maximum in the boiling point-composition curve can lead to initial separation of pure A on fractionation but as the composition of the liquid moves towards B and reaches the maximum, a constant boiling mixture L is reached that will distil over unchanged. An example of an azeotropic mixture of maximum boiling point is water (b.p. 100°C) and hydrogen chloride (b.p. −80°C), the azeotrope being 80% water and boiling at 108.6°C. *See also* azeotrope.

constant composition, law of *See* constant proportions, law of.

constant proportions, law of Formulated by Proust in 1779 after the analysis of a large number of compounds, the principle that the proportion of each element in a compound is fixed or constant. It follows that the composition of a pure chemical compound is independent of the method of preparation. It is also called the *law of definite proportions* and the *law of constant composition*.

contact process An industrial process for the manufacture of sulphuric acid. Sulphur(IV) oxide and air are passed over a heated catalyst (vanadium(V) oxide or platinum) and sulphur(VI) oxide is produced:

$$2SO_2 + O_2 \rightarrow 2SO_3$$

The sulphur(VI) oxide is dissolved in sulphuric acid

$$SO_3 + H_2SO_4 \rightarrow H_2S_2O_7$$

The resulting oleum is diluted to give sulphuric acid.

continuous phase See colloid.

continuous process A manufacturing process in which the raw materials are constantly fed into the plant. These react as they flow through the equipment to give a continuing flow of product. At any point, only a small amount of material is at a particular stage in the process but material at all stages of the reaction is present. The fractional distillation of crude oil is an example of a continuous process. Such processes are relatively easy to automate and can therefore manufacture a product cheaply. The disadvantages of continuous processing are that it usually caters for a large demand and the plant is expensive to install and cannot normally be used to make other things. Compare batch process.

continuous spectrum A spectrum composed of a continuous range of emitted or absorbed radiation. Continuous spectra are produced in the infrared and visible regions by hot solids. See also spectrum.

coordinate bond (dative bond) A covalent bond in which the bonding pair is visualized as arising from the donation of a lone pair from one species to another species, which behaves as an electron acceptor. The definition includes such examples as the 'donation' of the lone pair of the ammonia molecule to H^+ (an acceptor) to form NH_4^+ or to Cu^{2+} to form $[Cu(NH_3)_4]^{2+}$.

The donor groups are known as Lewis bases and the acceptors are either hydrogen ions or Lewis acids. Simple combinations, such as $H_3N \rightarrow BF_3$, are known as adducts.

coordination number The number of coordinate bonds formed to a metal atom or ion in a complex.

copolymerization See polymerization.

copper A transition metal occurring in nature principally as the sulphide. It is extracted by roasting the ore in a controlled air supply and purified by electrolysis of copper(II) sulphate solution using impure copper as the anode and pure copper as the cathode. Copper is used in electrical wires and in such alloys as brass and bronze.

The metal itself is golden-red in colour; copper(I) compounds are white (except the oxide, which is red), and copper(II) compounds are blue in solution. Copper is unreactive to dilute acids (except nitric acid). Copper(I) compounds are unstable in solution and decompose to copper and copper(II) ions. Both copper(I) and copper(II) ions form complexes, copper(II) ions being identified by the dark blue complex $[Cu(NH_3)_4]^{2+}$ formed with excess ammonia solution.

Symbol: Cu; m.p. 1083°C; b.p. 2582°C; r.d. 8.9; p.n. 29; r.a.m. 63.55.

copper(I) chloride (cuprous chloride, CuCl) A white solid, insoluble in water, prepared by heating copper(II) chloride in concentrated hydrochloric acid with excess copper turnings. When the solution is colourless, it is poured into air-free water (or water containing sulphur(IV) oxide) and a white precipitate of copper(I) chloride is obtained. On exposure to air this precipitate turns green due to the formation of basic copper(II) chloride. Copper(I) chloride absorbs carbon monoxide gas. It is used as a catalyst in the rubber industry. The chloride is essentially covalent in structure. In the vapour phase both dimeric and trimeric forms exist. It is used in organic chemistry in the Sandmeyer reactions.

copper(II) chloride (cupric chloride, CuCl$_2$) A compound prepared by dissolving excess copper(II) oxide or copper(II) carbonate in dilute hydrochloric acid. On crystallization, emerald green crystals of the dihydrate ($CuCl_2.2H_2O$) are obtained. The anhydrous chloride may be prepared as a brown solid by burning copper in excess chlorine. Alternatively the dihydrate may be dehydrated using concentrated sulphuric acid. Dilute solutions of copper(II) chloride are blue, concentrated solutions are green, and solutions in the presence of excess hydrochloric acid are yellow.

copper(II) nitrate (cupric nitrate, $Cu(NO_3)_2$) A compound prepared by dissolving either excess copper(II) oxide or copper(II) carbonate in dilute nitric acid. On crystallization, deep blue crystals of the trihydrate ($Cu(NO_3)_2.3H_2O$) are obtained. The crystals are prismatic in shape and are extremely deliquescent. On heating, copper(II) nitrate decomposes to give copper(II) oxide, nitrogen(IV) oxide, and oxygen.
The white anhydrous salt is prepared by adding a solution of dinitrogen pentoxide in nitric acid to crystals of the trihydrate.

copper(I) oxide (cuprous oxide, CuO) An insoluble red powder prepared by the heating of copper with copper(II) oxide or the reduction of an alkaline solution of copper(II) sulphate. Copper(I) oxide is easily reduced by hydrogen when heated; it is oxidized to copper(II) oxide when heated in air. It is used in the glass industry. Copper(I) oxide undergoes disproportionation in acid solutions, producing copper(II) ions and copper. The oxide will dissolve in concentrated hydrochloric acid due to the formation of the complex ion $[CuCl_2]^-$. Copper(I) oxide is a covalent solid.

copper(II) oxide (cupric oxide, CuO) A black solid prepared by the action of heat on copper(II) nitrate, hydroxide, or carbonate. It is a basic oxide and reacts with dilute acids to form solutions of copper(II) salts. Copper(II) oxide can be reduced to copper by heating in a stream of hydrogen or carbon monoxide. It can also be reduced by mixing with carbon and heating the mixture. Copper(II) oxide is stable up to its melting point, after which it decomposes to give oxygen, copper(I) oxide, and eventually copper.

copper(II) sulphate (cupric sulphate, $CuSO_4$) A compound prepared as the hydrate by the action of dilute sulphuric acid on copper(II) oxide or copper(II) carbonate. On crystallization, blue triclinic crystals of the pentahydrate ($CuSO_4.5H_2O$) are formed. Industrially copper(II) sulphate is prepared by passing air through a hot mixture of dilute sulphuric acid and scrap copper. The solution formed is recycled until the concentration of the copper(II) sulphate is sufficient. Copper(II) sulphate is readily soluble in water. The monohydrate ($CuSO_4.H_2O$) is formed at $100°C$ and the anhydrous salt at $250°C$. Anhydrous copper(II) sulphate is white; it is extremely hygroscopic and turns blue on absorption of water. It decomposes on heating to give copper(II) oxide and sulphur(VI) oxide.
Copper(II) sulphate is used as a wood preservative, a fungicide (in Bordeaux mixture), and in the dyeing and electroplating industries.

corn rule *See* optical activity.

corrosion Reaction of a metal with an acid, oxygen, or other compound with destruction of the surface of the metal. Rusting is a common form of corrosion.

corundum (emery, Al_2O_3) A naturally occurring form of aluminium oxide that sometimes contains small amounts of iron and silicon(IV) oxide. Ruby and sapphire are impure crystalline forms. It is used in various polishes, abrasives, and grinding wheels.

coulomb Symbol: C The SI unit of electric charge, equal to the charge transported by an electric current of one ampere flowing for one second. $1\,C = 1\,A\,s$.

coulombmeter (coulometer, voltameter) A device for determining electric charge or electric current using electrolysis. The mass m of material released is measured and this can be used to calculate the charge (Q) and the current (I) from the electrochemical equivalent of the element, using $Q = m/z$ or $I = m/zt$.

coupling A chemical reaction in which two groups or molecules join together. An example is the formation of azo compounds.

covalent bond A bond formed by the sharing of an electron pair between two atoms. The covalent bond is conventionally represented as a line, thus H−Cl indicates that between the hydrogen atom and the chlorine atom there is an electron pair formed by electrons of opposite spin and implying that the binding forces are strongly localized between the two atoms. Molecules are combinations of atoms bound together by covalent bonds; covalent bonding energies are of the order 10^3 kJ mol^{-1}.
Modern bonding theory treats the electron pairing in terms of the interaction of electron (atomic) orbitals and describes the covalent bond in terms of both 'bonding' and 'anti-bonding' molecular orbitals.

covalent crystal A crystal in which the atoms present are covalently bonded. They are sometimes referred to as giant lattices or macromolecules. The commonest completely covalent crystal is diamond.

covalent radius The radius an atom is assumed to have when involved in a covalent bond. For homonuclear diatomic molecules (e.g. Cl_2) this is simply half the measured internuclear distance. For heteroatomic molecules substitutional methods are used. For example, the internuclear distance of bromine fluoride (BrF) is about 180 pm, therefore using 71 pm for the covalent radius of fluorine (from F_2) we get 109 pm for bromine. The accepted value is 114 pm.

cracking A process whereby petroleum distillates are thermally decomposed to lighter fractions. The process can be pure-ly thermal, although *catalytic cracking* is by far the most widely used method.

cresols *See* methylphenols.

critical point The conditions of temperature and pressure under which a liquid being heated in a closed vessel becomes indistinguishable from the gas or vapour phase. At temperatures below the critical temperature (T_c) the substance can be liquefied by applying pressure; at temperatures above T_c this is not possible. For each substance there is one critical point; for example, for carbon dioxide it is at 31.1°C and 73.0 atmospheres.

critical pressure The lowest pressure needed to bring about liquefaction of a gas at its critical temperature.

critical temperature The temperature below which a gas can be liquefied by applying pressure and above which no amount of pressure is sufficient to bring about liquefaction. Some gases have critical temperatures above room temperature (e.g. carbon dioxide 31.1°C and chlorine 144°C) and have been known in the liquid state for many years. Liquefaction proved much more difficult for those gases (e.g. oxygen −118°C and nitrogen −146°C) that have very low critical temperatures.

critical volume The volume of one mole of a substance at its critical point.

cross linkage An atom or short chain joining two longer chains in a polymer.

crude oil *See* petroleum.

cryohydrate A eutectic mixture of ice and certain salts. *See also* eutectic.

cryolite *See* sodium hexafluoro-aluminate.

cryoscopic constant *See* depression of freezing point.

crystal A solid substance that has a definite geometric shape. A crystal has fixed angles between its faces, which have distinct edges. The crystal will sparkle if the faces are able to reflect light. The constant angles are caused by the regular arrangements of particles (atoms, ions, or molecules) in the crystal. If broken, a large crystal will form smaller crystals.

In crystals, the atoms, ions, or molecules of the substance form a distinct regular array in the solid state. The faces and their angles bear a definite relationship to the arrangement of these particles.

crystal habit The shape of a crystal. The habit depends on the way in which the crystal has grown; i.e. the relative rates of development of different faces.

crystalline Denoting a substance that forms crystals. Crystalline substances have a regular internal arrangement of atoms, even though they may not exist as geometrically regular crystals. For instance, lead (and other metals) are crystalline. Such substances are composed of accumulations of tiny crystals. *Compare* amorphous.

crystallite A small crystal that has the potential to grow larger. It is often used in mineralogy to describe specimens that contain accumulations of many minute crystals of unknown chemical composition and crystal structure.

crystallization The process of forming crystals. When a substance cools from the gaseous or liquid state to the solid state, crystallization occurs. Crystals will also form from a solution saturated with a solute.

crystallography The study of the formation, structure, and properties of crystals. *See also* X-ray crystallography.

crystal structure The particular repeating arrangement of atoms (molecules or ions) in a crystal. 'Structure' refers to the internal arrangement of particles, not the external appearance.

crystal system A classification of crystals based on the shapes of their unit cell. If the unit cell is a parallelopiped with lengths a, b, and c and the angles between these edges are α (between b and c), β (between a and c), and γ (between a and b), then the classification is:
cubic: $a = b = c$; $\alpha = \beta = \gamma = 90°$
tetragonal: $a = b \neq c$; $\alpha = \beta = \gamma = 90°$
orthorhombic: $a \neq b \neq c$; $\alpha = \beta = \gamma = 90°$
hexagonal: $a = b \neq c$; $\alpha = \beta = 90°$; $\gamma = 120°$
trigonal: $a = b \neq c$; $\alpha = \beta = \gamma \neq 90°$
monoclinic: $a \neq b \neq c$; $\alpha = \gamma = 90° \neq \beta$
triclinic: $a \neq b \neq c$; $\alpha \neq \beta \neq \gamma$
The orthorhombic system is also called the *rhombic* system.

cubic Denoting a crystal in which the unit cell is a cube. In a *simple cubic* crystal the particles are arranged at the corners of a cube. *See also* body-centred cubic crystal, face-centred cubic crystal, crystal system.

cubic close packed *See* face-centred cubic crystal.

cumene process An industrial process for the manufacture of phenol from isopropylbenzene (cumene), which is itself made by passing benzene vapour and propene over a phosphoric acid catalyst (250°C and 30 atmospheres).

$$C_6H_6 + CH_2{:}CH(CH_3) \rightarrow$$
$$C_6H_5CH(CH_3)_2$$

The isopropylbenzene is oxidized by air to a 'hydroperoxide':

$$C_6H_5C(CH_3)_2-O-O-H$$

This is hydrolysed by dilute acid to phenol (C_6H_5OH) and propanone (CH_3COCH_3), which is a valuable by-product.

cupellation A technique used for the separation of silver and/or gold from impurities by the use of an easily oxidizable metal, such as lead. The impure metal is heated in a furnace and a blast of hot air is directed upon it. The lead or other metal forms an oxide, leaving the silver or gold behind.

Cubic crystal structures

face-centred

body-centred

simple cubic

cupric chloride *See* copper(II) chloride.

cupric nitrate *See* copper(II) nitrate.

cupric oxide *See* copper(II) oxide.

cupric sulphate *See* copper(II) sulphate.

cupronickel A series of strong copper-nickel alloys containing up to 45% nickel. Those containing 20% and 30% nickel are very malleable, can be worked cold or hot, and are very corrosion-resistant. They are used, for example, in condenser tubes in power stations. Cupronickel with 25% nickel is widely used in coinage. *See also* Constantan.

cuprous chloride *See* copper(I) chloride.

cuprous oxide *See* copper(I) oxide.

curie Symbol: Ci A unit of radioactivity, equivalent to the amount of a given radioactive substance that produces 3.7×10^{10} disintegrations per second, the number of disintegrations produced by one gram of radium.

curium A highly toxic radioactive silvery element of the actinoid series of metals. A transuranic element, it is not found naturally on Earth but is synthesized from plutonium. Curium-244 and curium-242 have been used in thermoelectric power generators.

63

Symbol: Cm; m.p. 1340°C; b.p. $\simeq$ 3550°C; r.d. 13.5 (calc.); p.n. 96; stablest isotope ^{247}Cm (half-life 1.6×10^7 years).

cyanamide process An industrial process for fixing nitrogen by heating calcium dicarbide in air.
$$CaC_2 + N_2 \rightarrow CaCN_2$$
The product, calcium cyanamide, hydrolyses to ammonia and can be used as a fertilizer. The process is expensive due to the high cost of calcium dicarbide. *See also* nitrogen fixation.

cyanide 1. A salt of hydrogen cyanide, containing the cyanide ion (CN^-).
2. *See* nitrile.

cyanide process A technique used for the extraction of gold from its ores. After crushing the gold ore to a fine powder it is agitated with a very dilute solution of potassium cyanide. The gold is dissolved by the cyanide to form potassium aurocyanide $KAu(CN)_2$. This complex is reduced with zinc, filtered, and melted to obtain the pure gold metal.

cyanogen (C_2N_2) A toxic flammable gas prepared by heating mercury cyanide. *See also* pseudohalogen.

cyanohydrin An addition compound formed between an aldehyde or ketone and hydrogen cyanide. The general formula is RCH(OH)(CN) (from an aldehyde) or RR'C(OH)(CN) (from a ketone). Cyanohydrins are easily hydrolysed to hydroxycarboxylic acids. For instance, the compound 2-hydroxypropanonitrile ($CH_3CH(OH)(CN)$) is hydrolysed to 2-hydroxypropanoic acid ($CH_3CH(OH)$-(COOH)).

cyclic compound A compound containing a ring of atoms. If the atoms forming the ring are all the same the compound is *homocyclic*; if different atoms are involved it is *heterocyclic*.

cyclization A reaction in which a straight-chain compound is converted into a cyclic compound.

cyclohexane (C_6H_{12}) A colourless liquid alkane that is commonly used as a solvent and in the production of hexanedioic acid (adipic acid) for the manufacture of nylon. Cyclohexane, itself, is manufactured by the reformation of longer chain hydrocarbons present in crude-oil fractions. It is also interesting from a structural point of view, existing as a 'puckered' six-membered ring, having all bonds between carbon atoms at 109.9° (the tetrahedral angle). The molecule undergoes rapid interconversion between two 'chair-like' conformations, which are energetically equivalent, passing through a 'boat-like' structure of higher energy. It is commonly represented by a hexagon. *See also* conformation.

cyclopentadiene A cyclic hydrocarbon made by cracking petroleum. The molecules have a five-membered ring containing two carbon-carbon double bonds and one CH_2 group. It forms the negative *cyclopentadienyl ion* $C_5H_5^-$, present in sandwich compounds, such as ferrocene.

cyclopentadienyl ion *See* cyclopentadiene.

cysteine *See* amino acids.

D

Dalton's atomic theory A theory explaining the formation of compounds by elements first published by John Dalton in 1803. It was the first modern attempt to describe chemical behaviour in terms of atoms. The theory was based on certain postulates:

(1) All elements are composed of small particles, which Dalton called atoms.

(2) All atoms of the same element are identical.

(3) Atoms can neither be created nor destroyed.

(4) Atoms combine to form 'compound atoms' (i.e. molecules) in simple ratios.

Dalton also suggested symbols for the elements. The theory was used to explain the law of conservation of mass and the laws of chemical combination.

Dalton's law (of partial pressures) The pressure of a mixture of gases is the sum of the partial pressures of each individual constituent. (The *partial pressure* of a gas in a mixture is the pressure that it would exert if it alone were present in the container.) Dalton's law is strictly true only for ideal gases.

from aldehyde

from ketone

Cyanohydrins

Daniell cell A type of primary cell consisting of two electrodes in different electrolytes separated by a porous partition. The positive eletrode is copper immersed in copper(II) sulphate solution. The negative electrode is zinc-mercury amalgam in either dilute sulphuric acid or zinc sulphate solution. The porous pot prevents mixing of the electrolytes, but allows ions to pass. With sulphuric acid the e.m.f. is about 1.08 volts; with zinc sulphate it is about 1.10 volts.

At the copper electrode copper ions in solution gain electrons from the metal and are deposited as copper atoms:

$$Cu^{2+} + 2e^- \rightarrow Cu$$

The copper electrode thus gains a positive charge. At the zinc electrode, zinc atoms from the electrode lose electrons and dissolve into solution as zinc ions, leaving a net negative charge on the electrode

$$Zn \rightarrow 2e^- + Zn^{2+}$$

dative bond See coordinate bond.

d-block elements The transition elements of the first, second, and third long periods, i.e. Sc to Zn, Y to Cd, and La to Hg. They are so called because in general they have inner d-levels with configurations of the type $(n-1)d^x ns^2$ where $x = 1-10$. See transition elements.

deactivation A process in which the reactivity of a substance is lessened, or even totally removed. Usually this deactivation is unwanted, as in the poisoning of catalysts.

de Broglie wave A wave associated with a particle, such as an electron or proton. In 1924, Louis de Broglie suggested that, since electromagnetic waves can be described as particles (photons), particles of matter could also have wave properties. The wavelength (λ) has the same relationship to momentum (p) as in electromagnetic radiation:

$$\lambda = h/p$$

where h is the Planck constant. *See also* quantum theory.

debye Symbol: D A unit of electric dipole moment equal to $3.335\,64 \times 10^{-30}$ coulomb metre. It is used in expressing the dipole moments of molecules.

deca- Symbol: da A prefix denoting 10. For example, 1 decametre (dam) = 10 metres (m).

decahydrate A crystalline solid containing ten molecules of water of crystallization per molecule of compound.

deci- Symbol: d A prefix denoting 10^{-1}. For example, 1 decimetre (dm) = 10^{-1} metre (m).

decomposition A chemical reaction in which a compound is broken down into compounds with simpler molecules.

decrepitation Heating a crystalline solid so that it emits a crackling noise, usually as a result of loss of water of crystallization.

defect An irregularity in the ordered arrangement of particles in a crystal lattice. There are two main types of defect in crystals. *Point defects* occur at single lattice points and there are two types. A *vacancy* is a missing atom; i.e. a vacant lattice point. Vacancies are sometimes called *Schottky defects*. An *interstitial* is an atom that is in a position that is not a normal lattice point. If an atom moves off its lattice point to an interstitial position the result (vacancy plus interstitial) is a *Frenkel defect*. All solids above absolute zero have a number of point defects, the concentration of defects depending on the temperature. Point defects can also be produced by strain or by irradiation.
Dislocations (or *line defects*) are also produced by strain in solids. They are irregularities extending over a number of lattice points along a line of atoms.

definite proportions, law of *See* constant proportions, law of.

degassing Removal of dissolved or absorbed gases from liquids or solids.

degenerate Describing different quantum states that have the same energy. For instance, the five d orbitals in a transition-metal atom all have the same energy but different values of the magnetic quantum number m. Differences in energy occur if a magnetic field is applied or if the arrangement of ligands around the atom is not symmetrical. The degeneracy is then said to be 'lifted'.

degradation A chemical reaction involving the decomposition of a molecule into simpler molecules, usually in stages. The Hofmann degradation of amides is an example.

degrees of freedom 1. The independent ways in which particles can take up energy. In a monatomic gas, such as helium or argon, the atoms have three translational degrees of freedom (corresponding to motion in three mutually perpendicular directions). The mean energy per atom for each degree of freedom is $kT/2$, where k is the Boltzmann constant and T the thermodynamic temperature; the mean energy per atom is thus $3kT/2$.
A diatomic gas also has two rotational degrees of freedom (about two axes perpendicular to the bond) and one vibrational degree (along the bond). The rotations also each contribute $kT/2$ to the average energy. The vibration contributes kT ($kt/2$ for kinetic energy and $kT/2$ for potential energy). Thus, the average energy per molecule for a diatomic molecule is $3kT/2$ (translation) + kT (rotation) + kT (vibration) = $7kT/2$.
Linear triatomic molecules also have two significant rotational degrees of freedom; non-linear molecules have three. For non-linear polyatomic molecules, the number of vibrational degrees of freedom is $3N - 6$, where N is the number of atoms in the molecule.

The molar energy of a gas is the average energy per molecule multiplied by the Avogadro constant. For a monatomic gas it is $3RT/2$, etc. **2.** The independent physical quantities (e.g. pressure, temperature, etc.) that define the state of a given system. *See* phase rule.

dehydration 1. Removal of water from a substance. **2.** Removal of the elements of water (i.e. hydrogen and oxygen in a 2:1 ratio) to form a new compound. An example is the dehydration of propanol to propene over hot pumice:

$$C_3H_7OH \rightarrow CH_3CH{:}CH_2 + H_2O$$

deliquescent Describing a solid compound that absorbs water from the atmosphere, eventually forming a solution. *See also* hygroscopic.

delocalization A spreading out of bonding electrons in a molecule over the molecule.

delocalized bond (non-localized bond) A type of bonding in molecules that occurs in addition to sigma bonding. The electrons forming the delocalized bond are no longer regarded as remaining between two atoms; i.e. the electron density of the delocalized electrons is spread over several atoms and may spread over the whole molecule.
The electron density of the delocalized bond is spread by means of a delocalized molecular orbital and may be regarded as a series of pi bonds extending over several atoms, for example the pi bonds in butadiene and the $C-O$ pi bonds in the carbonate ion.

denaturation The changes in structure which occur when a protein is heated. These changes are irreversible and affect the properties of the protein.

denaturing The process of adding small amounts of contaminants to ethanol to make it unfit for drinking. Denatured alcohol (such as methylated spirits) is not subject to the excise duty on alcoholic drinks.

dendritic growth Growth of crystals in a branching habit.

density Symbol: ρ The mass per unit volume of a given substance. The units are $g\ dm^{-3}$, etc. *See also* relative density.

depolarizer A substance used in a voltaic cell to prevent polarization. Hydrogen bubbles forming on the electrode can be removed by an oxidizing agent, such as manganese(IV) oxide.

Apparatus for measuring depression of freezing point

depression of freezing point A colligative property of solutions in which the freezing point of a given solvent is lowered by the presence of a solute. The amount of the reduction is proportional to the molal concentration of the solute. The depression depends only on the concentration and is independent of solute composition. The proportionality constant, K_f, is called the *freezing point constant* or sometimes the *cryoscopic constant*. $\Delta t = K_f C_M$, where Δt is the lowering of the temperature and C_M is the molal concentration; the unit of K_f is kelvin kilogram mole^{-1} (K kg mol^{-1}). Although closely related to the property of boiling-point elevation, the cryogenic method can be applied to measurement of relative molecular mass with considerable precision. A known weight of pure solvent is slowly frozen, with stirring, in a suitable cold bath and the freezing temperature measured using a Beckmann thermometer. A known weight of solute of known molecular mass is introduced, the solvent thawed out, and the cooling process and measurement repeated. The addition is repeated several times and an average value of K_f for the solvent obtained by plotting Δt against C_M. The whole process is then repeated using the unknown solute and its relative molecular mass determined using the value of K_f previously obtained.

The effect is applied to more precise measurement of relative molecular mass by using a pair of Dewar flasks (pure solvent and solution) and measuring Δt by means of thermocouples. The theoretical explanation is similar to that for lowering of vapour pressure. The freezing point of the solvent is that point at which the curve representing the vapour pressure above the liquid phase intersects the curve representing the vapour pressure above the frozen solvent. The addition of solute depresses the former curve but as the solid phase that separates is always pure solvent (above the eutectic point), there is no attendant depression of the latter curve. Consequently the point of intersection is depressed, resulting in a lowering of the freezing point. *See also* lowering of vapour pressure.

derived unit A unit defined in terms of base units, and not directly from a standard value of the quantity it measures. For example, the newton is a unit of force defined as a kilogram metre second^{-2} (kg m s^{-2}). *See also* SI units.

desiccation Removal of moisture from a substance.

desiccator A laboratory apparatus for drying solids or for keeping solids free of moisture. It is a container in which is kept a hygroscopic material (e.g. calcium oxide, or silica gel).

destructive distillation The process of heating an organic substance such that it wholly or partially decomposes in the absence of air to produce volatile products,

which are subsequently condensed. The destructive distillation of coal was the process for manufacturing coal gas and coal tar. Formerly, methanol was made by the destructive distillation of wood.

detergents A group of substances that improve the cleansing action of solvents, particularly water. The majority of detergents, including soap, have the same basic structure. Their molecules have a hydrocarbon chain (tail) that does not attract water molecules. The tail is said to be hydrophobic (water hating). Attached to this tail is a small group (head) that readily ionizes and attracts water molecules. It is said to be hydrophilic (water loving). Detergents reduce the surface tension of water and thus improve its wetting power. Because the detergent ions have their hydrophilic heads anchored in the water and their hydrophobic tails protruding above it, the water surface is broken up, enabling the water to spread over the material to be cleaned and penetrate between the material and the dirt. With the assistance of agitation, the dirt can be floated off. The hydrophobic tails of the detergent molecules 'dissolve' in grease and oils. The protruding hydrophilic heads repel each other causing the oil to roll up and form a drop, which floats off into the water as an emulsion. Soap is the oldest detergent. More recently synthetic detergents, often derived from petrochemicals, have been developed. Unlike soaps these detergents do not form soluble scums with hard water.

Synthetic detergents are of three types. *Anionic detergents* form ions consisting of a hydrocarbon chain to which is attached either a sulphonate group, $-SO_2-O^-$, or a sulphate group, $-O-SO_2-O^-$. The corresponding metal salts are soluble in water. *Cationic detergents* have organic positive ions of the type RNH_3^+, in which R has a long hydrocarbon chain. Non-ionic detergents are complex chemical compounds called ethoxylates. They owe their detergent properties to the presence of a number of oxygen atoms in one part of the molecule, which are capable of forming hydrogen bonds with the surface water

molecules, thus reducing the surface tension of the water.
See also soap.

deuterated compound A compound in which one or more ${}^1H$ atoms have been replaced by deuterium (${}^2H$) atoms.

deuterium Symbol: D A naturally occurring stable isotope of hydrogen in which the nucleus contains one proton and one neutron. The atomic mass is thus approximately twice that of ${}^1H$; deuterium is known as 'heavy hydrogen'. Chemically it behaves almost identically to hydrogen, forming analogous compounds, although reactions of deuterium compounds are often slower than those of the corresponding ${}^1H$ compounds. This is made use of in kinetic studies where the rate of a reaction may depend on transfer of a hydrogen atom (i.e. a kinetic isotope effect).

deuteron The nucleus of the deuterium atom, ${}^2D^+$.

Dewar flask (vacuum flask) A double-walled container of thin glass with the space between the walls evacuated and sealed to stop conduction and convection of energy through it. The glass is often silvered to reduce radiation.

Dewar structure A representation of the structure of benzene in which there is a single bond between two opposite corners of the hexagonal ring and two double bonds at the sides of the ring. The Dewar structures contribute to the resonance hybrid of benzene. *See* benzene.

dextro-form *See* optical activity.

dextrorotatory *See* optical activity.

dextrose (grape-sugar) Naturally occurring glucose belongs to the stereochemical series D and is dextrorotatory indicated by the symbol (+). Thus the term *dextrose* is used to indicate D-(+)-glucose. As other stereochemical forms of glucose have no significance in biological systems the term 'glucose' is often used interchangeably with dextrose in biology. *See also* glucose.

D-form *See* optical activity.

d-form *See* optical activity.

diagonal relationship There is a general trend in the periodic table for electronegativity to increase from left to right along a period and decrease down a group. Thus a move of 'one across and one down', i.e. a diagonal move in the table, gives rise to effects that tend to cancel each other. There is a similar combined effect for size, which decreases along a period but increases down a group. This diagonal relationship gives rise to similarities in chemical properties, which are particularly noticeable for the following pairs: Li—Mg; Be—Al; B—Si.
Li—Mg:
(1) both have carbonates that give CO_2 on heating,
(2) both burn in air to give the normal oxide only,
(3) both form a nitride,
(4) both form hydrated chlorides that hydrolyse slowly.
Be—Al:
(1) both give hydrogen with alkalis,
(2) both give water-insoluble hydroxides that dissolve in alkali,
(3) both form complex ions of the type MCl_3^-,
(4) both have covalently bridged chlorides.
B—Si:
(1) both form acidic oxides of a giant covalent-molecule type with high melting points,
(2) both form low-stability hydrides that ignite in air,
(3) both form readily hydrolysable chlorides that fume in air,
(4) both have amorphous and crystalline forms and both form glasses with basic oxides, such as Na_2O.

diamagnetism *See* magnetism.

1,6-diaminohexane ($H_2N(CH_2)_6NH_2$) An organic compound used as a starting material in the production of nylon. It is manufactured from cyclohexane. *See* nylon.

diamond An allotrope of carbon. It is the hardest naturally occurring substance and is used for jewellery and, industrially, for drilling equipment. Each carbon atom is surrounded by four equally spaced carbon atoms arranged tetrahedrally. The carbon atoms form a three-dimensional network with each carbon-carbon bond equal to 0.154 nm and at an angle of 109.5° with its neighbours. In diamonds millions of atoms are covalently bonded to form a giant molecular structure, the great strength of which results from the strong covalent bonds. Diamonds can be formed synthetically from graphite in the presence of a catalyst and under extreme temperature and pressure; although small, such diamonds are of adequate size for many industrial uses. *See also* carbon.

diatomic Describing a molecule that consists of two atoms. Hydrogen (H_2), oxygen (O_2), nitrogen (N_2), and the halogens are examples of diatomic elements.

diazotization The reaction of an aromatic amine (e.g. aniline) with nitrous acid at low temperatures (below 5°C).
$$[C_6H_5NH_2 + HNO_2 \rightarrow C_6H_5N^+N + OH^- + H_2O]$$
The acid is prepared *in situ* by reaction between nitric acid and sodium nitrite. The resulting diazonium ion is susceptible to attack by nucleophiles and provides a method of nucleophilic substitution onto the benzene ring.

diazonium salt A compound of the type $RN_2^+X^-$, where R is an aromatic group and X^- a negative ion. Diazonium salts are made by diazotization. They can be isolated but are very unstable, and are usually prepared in solution. The $-N_2^+$ group renders the benzene ring susceptible to nucleophilic substitution (rather than electrophilic substitution). Typical reactions are:
(1) Reaction with water on warming the solution:
$$RN_2^+ + H_2O \rightarrow ROH + N_2 + H^+$$

(2) Reaction with halogen ions (CuCl catalyst for chloride ions):
$$RN_2^+ + I^- \rightarrow RI + N_2$$
Diazonium ions can also act as electrophiles and substitute other benzene rings (*diazo coupling*). *See* azo compound.

dibasic acid An acid which has two active protons, such as sulphuric acid.

dichlorine oxide (chlorine monoxide, Cl_2O) An orange gas made by passing chlorine over mercury(II) oxide. It is a strong oxidizing agent and dissolves in water to give chloric(I) acid.

dichloroacetic acid *See* chloroethanoic acid.

dichloroethanoic acid *See* chloroethanoic acid.

dichromate(VI) A salt containing the ion $Cr_2O_7^-$. Dichromates are strong oxidizing agents. *See* potassium dichromate(VI).

diene An organic compound containing two carbon–carbon double bonds.

diethyl ether *See* ethoxyethane.

diffusion Movement of a gas, liquid, or solid as a result of the random thermal motion of its particles (atoms or molecules). A drop of ink in water, for example, will slowly spread throughout the liquid. Diffusion in solids occurs very slowly at normal temperatures. *See also* Graham's law.

dihydrate A crystalline compound with two molecules of water of crystallization per molecule of compound.

dihydric alcohol *See* diol.

diiodine hexachloride (iodine trichloride, I_2Cl_6) A yellow crystalline solid made by reacting excess chlorine with iodine. It is a strong oxidizing agent and at 70°C it dissociates into iodine monochloride and chlorine.

dilead(II) lead(IV) oxide (red lead, Pb_3O_4) A powder made by heating lead(II) oxide or lead(II) carbonate hydroxide at 400°C. It is black when hot and red or orange when cold. On heating it decomposes to lead(II) oxide and oxygen. Dilead(II) lead (IV) oxide is used as a pigment and in glass making. In practice it tends to have less oxygen than denoted in the formula.

dilute Denoting a solution in which the amount of solute is low relative to that of the solvent. The term is always relative and includes dilution at trace level as well as the common term 'bench dilute acid', which usually means a 2M solution. *Compare* concentrated.

dimer A compound (or molecule) formed by combination or association of two molecules of a monomer. For instance, aluminium chloride ($AlCl_3$) is a dimer (Al_2Cl_6) in the vapour.

dimethylbenzene (xylene, C_6H_4-$(CH_3)_2$) An organic hydrocarbon present in the light-oil fraction of crude oil. It is used extensively as a solvent. There are three isomeric compounds with this name and formula, distinguished as 1,2-, 1,3- and 1,4-dimethylbenzene according to the positions of the methyl groups on the benzene ring.

dimorphism *See* polymorphism.

dinitrogen oxide (nitrous oxide, N_2O) A colourless gas with a faintly sweet odour and taste. It is appreciably soluble in water (1.3 volumes in 1 volume of water at 0°C) but more soluble in ethanol. It is prepared commercially and in the laboratory by the *careful* heating of ammonium nitrate:

$$NH_4NO_3(s) = N_2O(g) + 2H_2O(g)$$

Dinitrogen oxide is fairly easily decomposed on heating to temperatures above 520°C, giving nitrogen and oxygen. The gas is used as a mild anaesthetic in medicine and dentistry, being marketed in small steel cylinders. It is sometimes called *laughing gas* because it induces a feeling of elation.

dinitrogen tetroxide (N_2O_4) A colourless gas that becomes a pale yellow liquid below 21°C and solidifies below −11°C. On heating, the gas dissociates to nitrogen dioxide molecules:

$$N_2O_4(g) = 2NO_2(g)$$

This dissociation is complete at 140°C. Liquid dinitrogen tetroxide has good solvent properties and is used as a nitrating agent.

2,4-dinitrophenylhydrazine (Brady's reagent, $C_6H_6N_4O_4$) An orange solid commonly used in solution with methanol and sulphuric acid to produce crystalline derivatives by condensation with aldehydes and ketones. The derivatives, known as 2,4-dinitrophenylhydrazones, can easily be purified by recrystallization and have characteristic melting points, used to identify the original aldehyde or ketone.

diol (dihydric alcohol, glycol) An alcohol that has two hydroxyl groups (−OH) per molecule of compound.

diphosphane (diphosphine, P_2H_4) A yellow liquid that can be condensed out from phosphine in a freezing mixture. It ignites spontaneously in air.

diphosphine *See* diphosphane.

dipole moment Symbol: μ A quantitative measure of polarity in either a bond (bond moment) or a molecule as a whole (molecular dipole moment). The unit is the debye (equivalent to 3.34×10^{-30} coulomb metre). Molecules such as HF, H_2O, NH_3, and $C_6H_5NH_2$ possess dipole moments; CCl_4, N_2, C_6H_6, and PF_5 do not. The molecular dipole moment can be estimated by vector addition of individual bond moments if the bond angles are known. The possession of a dipole moment permits direct interaction with electric fields or interaction with the electric component of radiation.

direct dyes A group of dyes that are mostly azo-compounds derived from benzidene or benzidene derivatives. They

71

are used to dye cotton, viscose rayon, and other cellulose fibres directly, using a neutral bath containing sodium chloride or sodium sulphate as a mordant.

disaccharide A sugar with molecules composed of two monosaccharide units. Sucrose and maltose are examples. These are linked by a $-O-$ linkage (*glycosidic link*). *See also* sugar.

dislocation *See* defect.

disodium hydrogen phosphate(V) (disodium hydrogen orthophosphate, Na_2HPO_4) A white solid prepared by titrating phosphoric acid with sodium hydroxide solution using phenolphthalein as the indicator. On evaporation the solution yields efflorescent monoclinic crystals of the dodecahydrate, $Na_2HPO_4.12H_2O$. The effloresced salt contains $7H_2O$. Disodium hydrogen phosphate is used in the textile industry.

disodium decahydrate tetraborate (borax, $Na_2B_4O_7.10H_2O$) A white crystalline solid, sparingly soluble in cold water but readily soluble in hot water. It occurs naturally as salt deposits in dry lake beds, especially in California. It is an important industrial material, being used in the manufacture of enamels and heat-resistant glass, as a paper glaze, etc.
Its full systematic name is sodium(I) heptaoxotetraborate(III)-10-water. In solution hydrolysis occurs:
$$B_4O_7^{2-} + 7H_2O = 2OH^- + 4H_3BO_3$$
The borax-bead test gives characteristic colours with certain cations.

disperse dyes Water-insoluble dyes, which, when held in fine suspension, can be applied to acetate rayon fabrics. The dye, together with a dispersing agent, is warmed to a temperature of $45-50°C$ and the fabric added. By modifying the method of application it is possible to dye Tricel, as well as polyacrylic and polyester fibres. The yellow/orange shades are nitroarylamine derivatives and the green to bluish shades are derivatives of 1-amino anthraquinone. Certain azo compounds

are disperse dyes and these give a range of colours.

disperse phase *See* colloid.

dispersing agent A compound used to assist emulsification or dispersion.

dispersion force A weak type of intermolecular force. *See* van der Waals forces.

displacement pump A commonly used device for transporting liquids and gases around chemical plants. It works on the principle of the bicycle pump: a piston raises the pressure of the fluid and, when it is high enough, a valve opens and the fluid is discharged through an outlet pipe. As the piston moves back the pressure falls and the cycle continues. Displacement pumps can be used to generate very high pressures (e.g. in the synthesis of ammonia) but because of the system of valves, they are more expensive than other types of pump. *Compare* centrifugal pump.

displacement reaction A chemical reaction in which an atom or group displaces another atom or group from a molecule. A common example is the displacement of hydrogen from acids by metals:
$$Zn + 2HCl \rightarrow ZnCl_2 + H_2$$

disproportionation A chemical reaction in which there is simultaneous oxidation and reduction of the same compound. An example is the reaction of copper(I) chloride to copper and copper(II) chloride:
$$2CuCl \rightarrow Cu + CuCl_2$$
in which there is simultaneous oxidation (to Cu(II)) and reduction (to Cu(0)).
Another example is the reaction of chlorine with water:
$$Cl_2 + H_2O \rightarrow 2H^+ + Cl^- + ClO^-$$
in which there is reduction to Cl^- and oxidation to ClO^-.

dissociation Breakdown of a molecule into two molecules, atoms, radicals, or ions. Often the reaction is reversible, as in

the ionic dissociation of weak acids in water:
$$CH_3COOH + H_2O = CH_3COO^- + H_3O^+$$

dissociation constant The equilibrium constant of a dissociation reaction. For example, the dissociation constant of a reaction:
$$AB = A + B$$
is given by:
$$K = [A][B]/[AB]$$
where the brackets denote concentration (activity).

Often the degree of dissociation is used — the fraction (α) of the original compound that has dissociated at equilibrium. For an original amount of AB of n moles in a volume V, the dissociation constant is given by:
$$K = \alpha^2 n/(1 - \alpha)V$$
Note that this expression is for dissociation into two molecules.

Acid dissociation constants (or *acidity constants*, symbol: K_a) are dissociation constants for the dissociation into ions in solution.
$$HA + H_2O \rightarrow H_3O^+ + A^-$$
The concentration of water can be taken as unity, and the acidity constant is given by:
$$K_a = [H_3O^+][A^-]/[HA]$$
The acidity constant is a measure of the strength of the acid. Base dissociation constants (K_b) are similarly defined. The expression:
$$K = \alpha^2 n/(1 - \alpha)V$$
applied to an acid is known as *Ostwald's dilution law*. In particular if α is small (a weak acid) then $K = \alpha^2 n/V$, or $\alpha = C\sqrt{V}$, where C is a constant. The degree of dissociation is then proportional to the square root of the dilution.

distillation The process of boiling a liquid and condensing the vapour. Distillation is used to purify liquids or to separate components of a liquid mixture. *See also* destructive distillation, fractional distillation, steam distillation, vacuum distillation.

disulphur dichloride (sulphur monochloride, S_2Cl_2) A red fuming liquid with a strong smell. It is prepared by passing chlorine over molten sulphur and is used to harden rubber.

divalent (bivalent) Having a valence of two.

dl-form *See* optical activity.

Döbereiner's triads Triads of chemically similar elements in which the central member, when placed in order of increasing atomic weight, has an atomic weight approximately equal to the average of the outer two. Other chemical and physical properties of the central member also lie between those of the first and last members of the triad. Döbereiner noted this relationship in 1817; the triads are now recognized as consecutive members of a group of the periodic table; e.g. Ca, Sr, and Ba and Cl, Br, and I.

dolomite A mineral, $CaCO_3.MgCO_3$, used as an ore of magnesium and for the lining of open-hearth steel furnaces and Bessemer converters.

donor The atom, ion, or molecule that provides the pair of electrons in forming a covalent bond.

doping The incorporation of impurities within the crystal lattice of a solid so as to alter its physical properties. For instance, silicon, when doped with boron, becomes semiconducting.

double bond A covalent bond between two atoms that includes two pairs of electrons, one pair being the single bond equivalent (the sigma pair) and the other forming an additional bond, the pi bond (π bond). It is conventionally represented by two lines, for example $H_2C=O$. *See* multiple bond, orbital.

double salt When equivalent quantities of certain salts are mixed in aqueous solution and the solution evaporated, a salt may form, e.g. $FeSO_4.(NH_4)_2SO_4.6H_2O$. In aqueous solution the salt behaves as a mixture of the two individuals. These salts are called double salts to distinguish them from complex salts, which yield complex ions in solution.

Dow process An industrial method whereby magnesium is extracted from seawater via the precipitation of $Mg(OH)_2$ by $Ca(OH)_2$, followed by solvation of the precipitated hydroxide by hydrochloric acid.

Downs process An electrolytic process for making chlorine from fused sodium chloride. The anode is a central piece of graphite, surrounded by a cylindrical steel cathode. A grid and dome between the two keeps the products separate.

dry cell A voltaic cell in which the electrolyte is in the form of a jelly or paste. The Leclanché dry cell is extensively used for flashlights and other portable applications. *See also* Leclanché cell.

dryers Devices used in chemical processes to remove a liquid from a solid by evaporation. Drying equipment is classified by the method of transferring heat to a wet solid. This can be by direct contact between hot gases and the solid (direct dryers), heat transfer by conduction through a retaining metallic wall (indirect dryers), or infrared rays (infrared dryers).

dry ice Solid carbon dioxide, used as a refrigerant.

Dulong and Petit's law The molar thermal capacity of a solid element is approximately equal to $3R$, where R is the gas constant (25 J K^{-1} mol^{-1}). The law applies only to elements with simple crystal structures at normal temperatures. At lower temperatures the molar heat capacity falls with decreasing temperature (it is proportional to T^3). Molar thermal capacity was formerly called *atomic heat* — the product of the atomic weight and specific thermal capacity.

Dumas' method 1. A method for determining the vapour densities of volatile liquids. The method utilizes a Dumas bulb; i.e. a glass bulb with a narrow entrance tube. The bulb is weighed 'empty' (i.e. full of air) then the sample is introduced and the bulb immersed in a heating bath so that the sample boils and expels all the air. When the surplus vapour has been expelled, the bulb is sealed off, cooled, dried, and weighed. The tip of the tube is then broken under water so that the water completely fills the tube and the whole weighed again. This enables the volume of the bulb to be calculated from the known density of water, and knowing the density of air one can compute the vapour density of the sample. *See also* Hofmann's method, Victor Meyer's method. **2.** A method of finding the amount of nitrogen in an organic compound by heating the compound with copper oxide to convert the nitrogen into nitrogen oxides. These are reduced by passing them over hot copper and the volume of nitrogen collected is measured.

Duralumin (*Trademark*) A strong lightweight aluminium alloy containing 3–4% copper, with small amounts of magnesium, manganese, and sometimes silicon. It is widely used in aircraft bodies.

dye A colouring material for fabric, leather, etc. Most dyes are now synthetic organic compounds (the first such was the dye *mauve* synthesized from aniline in 1856 by W. H. Perkin). Dyes are often unsaturated organic compounds containing conjugated double bonds — the bond system responsible for the colour is called the *chromophore*. *See also* azo compound.

dyne Symbol: dyn The former unit of force used in the c.g.s. system. It is equal to 10^{-5} N.

dysprosium A soft malleable silvery element of the lanthanoid series of metals. It occurs in association with other lanthanoids. One of its few uses is as a neutron absorber in nuclear reactors.
Symbol: Dy; m.p. 1409°C; b.p. 2335°C; r.d. 8.56; p.n. 66; r.a.m. 162.50.

E

ebullioscopic constant *See* elevation of boiling point.

ebullition The boiling or bubbling of a liquid.

eclipsed conformation *See* conformation.

Edison cell *See* nickel-iron accumulator.

edta (ethylenediamine tetraacetic acid) A compound with the formula

$(HOOC\ CH_2)_2N(CH_2)_2N(CH_2COOH)_2$

It is used in forming chelates of transition metals. *See* chelate.

efficiency Symbol: η A measure used for processes of energy transfer; the ratio of the useful energy produced by a system or device to the energy input. For a reversible heat engine, the efficiency is given by

$$\eta = (T_1 - T_2)/T_1$$

T_1 being the source temperature and T_2 the 'sink' temperature.

efflorescence The process in which a crystalline hydrated solid loses water of crystallization to the air. A powdery deposit is gradually formed.

einsteinium A radioactive transuranic element of the actinoid series, not found naturally on Earth. Several short-lived isotopes have been synthesized. Symbol: Es; p.n. 99; stablest isotope ^{254}Es (half-life 276 days).

elastomer An elastic substance, e.g. a natural or synthetic rubber.

electrochemical equivalent Symbol: z The mass of an element released from a solution of its ion when a current of one ampere flows for one second during electrolysis.

electrochemical series (electromotive series) A series giving the activities of metals for reactions that involve ions in solution. In decreasing order of activity, the series is

K, Na, Ca, Mg, Al, Zn, Fe
Pb, H, Cu, Hg, Ag, Pt, Au

Any member of the series will displace ions of a lower member from solution. For example, zinc metal will displace Cu^{2+} ions:

$Zn(s) + Cu^{2+}(aq) \rightarrow Zn^{2+}(aq) + Cu(s)$

Zinc has a greater tendency than copper to form positive ions in solution. Similarly, metals above hydrogen displace hydrogen from acids:

$Zn + 2HCl \rightarrow ZnCl_2 + H_2$

The series is based on electrode potentials, which measure the tendency to form positive ions. The series is one of increasing electrode potential for half cells of the type $M^{n+}|M$. Thus, copper ($E^{\ominus}$ for $Cu^{2+}|Cu = +0.34$ V) is lower than zinc ($E^{\ominus}$ for $Zn^{2+}|Zn = -0.76$V). The hydrogen half cell has a value $E^{\ominus} = 0$.

electrochemistry The study of the formation and behaviour of ions in solutions. It includes electrolysis and the generation of electricity by chemical reactions in cells.

electrode Any part of an electrical device or system that emits or collects electrons or other charge carriers. An electrode may also be used to deflect charged particles by the action of the electrostatic field that it produces. *See also* half cell.

electrodeposition The process of depositing a layer of solid (metal) on an electrode by electrolysis. Positive ions in solution gain electrons at the cathode and are deposited as atoms. Copper, for instance, can be deposited on a metal cathode from copper sulphate solution.

electrode potential Symbol: E A measure of the tendency of an element to form ions in solution. For example, a metal in a solution containing M^+ ions may dissolve in the solution as M^+ ions; the metal then has an excess of electrons and the solution an excess of positive ions — thus, the metal becomes negative with respect to the solution. Alternatively, the positive ions may gain electrons from the metal and be deposited as metal atoms. In this case, the metal becomes positively charged with respect to the solution. In either case, a potential difference is developed between solid and solution, and an equilibrium state will be reached at which further reaction is prevented. The equilibrium value of this potential difference would give an indication of the tendency to form aqueous ions.

It is not, however, possible to measure this for an isolated half cell — any measurement requires a circuit, which sets up another half cell in the solution. Therefore, electrode potentials (or *reduction potentials*) are defined by comparison with a hydrogen half cell, which is connected to the half cell under investigation by a salt bridge. The e.m.f. of the full cell can then be measured.

In referring to a given half cell the more reduced form is written on the right for a half-cell reaction. For the half cell $Cu^{2+}|Cu$, the half-cell reaction is a reduction:

$$Cu^{2+}(aq) + 2e \rightarrow Cu$$

The cell formed in comparison with a hydrogen electrode is:

$$Pt(s)H_2(g)|H^+(aq)|Cu^{2+}(aq)|Cu$$

The e.m.f. of this cell is $+0.34$ volt measured under standard conditions. Thus, the standard electrode potential (symbol: $E^{\ominus}$) is $+0.34$ V for the half cell $Cu^{2+}|Cu$. The standard conditions are 1.0 molar solutions of all ionic species, standard pressure, and a temperature of 298 K.

Half cells can also be formed by a solution of two different ions (e.g. Fe^{2+} and Fe^{3+}). In such cases, a platinum electrode is used under standard conditions.

electrolysis The production of chemical change by passing electric charge through certain conducting liquids (electrolytes). The current is conducted by mi-

gration of ions — positive ones (cations) to the cathode (negative electrode), and negative ones (anions) to the anode (positive electrode). Reactions take place at the electrodes by transfer of electrons to or from them.

In the electrolysis of water (containing a small amount of acid to make it conduct adequately) hydrogen gas is given off at the cathode and oxygen is evolved at the anode. At the cathode the reaction is:

$$H^+ + e^- \rightarrow H$$
$$2H \rightarrow H_2$$

At the anode:

$$OH^- \rightarrow e^- + OH$$
$$2OH \rightarrow H_2O + O$$
$$2O \rightarrow O_2$$

In certain cases the electrode material may dissolve. For instance, in the electrolysis of copper(II) sulphate solution with copper electrodes, copper atoms of the anode dissolve as copper ions

$$Cu \rightarrow 2e^- + Cu^{2+}$$

electrolyte A liquid containing positive and negative ions, that conducts electricity by the flow of those charges. Electrolytes can be solutions of acids or metal salts ('ionic compounds'), usually in water. Alternatively they may be molten ionic compounds — again the ions can move freely through the substance. Liquid metals (in which conduction is by free electrons rather than ions) are not classified as electrolytes. *See also* electrolysis.

electrolytic Relating to the behaviour or reactions of ions in solution.

electrolytic cell *See* cell, electrolysis.

electrolytic corrosion Corrosion by electrochemical reaction. *See* rusting.

electrolytic refining A method of purifying metals by electrolysis. Copper is purified by making the impure metal the anode in an electrolytic cell with copper sulphate electrolyte. The cathode is a thin strip of pure copper. The copper at the anode dissolves (as Cu^{2+} ions) and pure copper is deposited on the cathode. In this particular process, gold and silver are ob-

Electromagnetic spectrum
(note: the figures are only approximate)

Radiation	Wavelength (m)	Frequency (Hz)
gamma radiation	$- 10^{-10}$	$10^{19} -$
X-rays	$10^{-12} - 10^{-9}$	$10^{17} - 10^{20}$
ultraviolet radiation	$10^{-9} - 10^{-7}$	$10^{15} - 10^{18}$
visible radiation	$10^{-7} - 10^{-6}$	$10^{14} - 10^{15}$
infrared radiation	$10^{-6} - 10^{-4}$	$10^{12} - 10^{14}$
microwaves	$10^{-4} - 1$	$10^{9} - 10^{13}$
radio waves	$1 -$	$- 10^{9}$

tained as by-products, deposited as an *anode sludge* on the bottom of the cell.

electromagnetic radiation Energy propagated by vibrating electric and magnetic fields. Electromagnetic radiation forms a whole electromagnetic spectrum, depending on frequency and ranging from high-frequency radio waves to low-frequency gamma rays.
Electromagnetic radiation can be thought of as waves (*electromagnetic waves*) or as streams of photons. The frequency and wavelength are related by
$$\lambda v = c$$
where c is the speed of light. The energy carried depends on the frequency.

electromotive series See electrochemical series.

electron An elementary particle of negative charge ($-1.602\,192$ C) and rest mass $9.109\,558$ kg. Electrons are present in all atoms in shells around the nucleus.

electron affinity Symbol: A The energy released when an atom (or molecule or group) gains an electron in the gas phase to form a negative ion. It is thus the energy of:
$$A + e^- \rightarrow A^-$$
A positive value of A (often in electronvolts) indicates that heat is given out. Often the molar enthalpy is given for this process of electron attachment (ΔH). Here the units are joules per mole (J mol^{-1}), and, by the usual convention, a negative value indicates that energy is released.

electron-deficient compounds Compounds in which the number of electrons available for bonding is insufficient for the bonds to consist of conventional two-electron covalent bonds. Diborane, B_2H_6, is an example in which each boron atom has two terminal hydrogen atoms bound by conventional electron-pair bonds and in addition the molecule has two hydrogen atoms bridging the boron atoms (B–H–B). In each bridge there are only two electrons for the bonding orbital. *See also* multicentre bond.

electron diffraction A technique used to determine the structure of substances, principally the shapes of molecules in the gaseous phase. A beam of electrons directed through a gas at low pressure produces a series of concentric rings on a photographic plate. The dimensions of these rings are related to the interatomic distances in the molecules. *See also* X-ray diffraction.

electronegativity A measure of the tendency of an atom in a molecule to attract electrons to itself. Elements to the right-hand side of the periodic table are strongly electronegative (values from 2.5 to 4); those on the left-hand side have low electronegativities (0.8–1.5) and are sometimes called electropositive elements. Different electronegativities of atoms in the same molecule give rise to polar bonds and sometimes to polar molecules.
As the concept of electronegativity is not precisely defined it cannot be precisely measured and several electronegativity scales exist. Although the actual values dif-

fer the scales are in good relative agreement.
See electron affinity, ionization potential.

electronic energy levels See energy levels.

electronic transition The demotion or promotion of an electron between electronic energy levels in an atom or molecule.

electron pair Two electrons in one orbital with opposing spins (*spin paired*), such as the electrons in a covalent bond or lone pair.

electron spin See atom.

electron spin resonance (ESR) A similar technique to nuclear magnetic resonance, but applied to unpaired electrons in a molecule (rather than to the nuclei). It is a powerful technique for studying free radicals and transition-metal complexes.

electronvolt Symbol: eV A unit of energy equal to $1.602\,191\,7 \times 10^{-19}$ joule. It is defined as the energy required to move an electron charge across a potential difference of one volt. It is often used to measure the kinetic energies of elementary particles or ions, or the ionization potentials of molecules.

electrophile An electron-deficient ion or molecule that takes part in an organic reaction. The electrophile can be either a positive ion (H^+, NO_2^+) or a molecule that can accept an electron pair (SO_3, O_3). The electrophile attacks negatively charged areas of molecules, which usually arise from the presence in the molecule of an electronegative atom or group or of π-bonds. *Compare* nucleophile.

electrophilic addition A reaction involving the addition of a small molecule to an unsaturated organic compound, across the atoms held by a double or triple bond. The reaction is initiated by the attack of an electrophile on the electron-rich area of

the molecule. The mechanism of electrophilic addition is thought to be ionic, as in the addition of HBr to ethene:
$$H_2C:CH_2 + H^+Br^- \rightarrow$$
$$H_3CCH_2^+ + Br^- \rightarrow H_3CCH_2Br$$
In the case of higher alkenes (more than two carbon atoms) several isomeric products are possible. The particular isomer produced depends on the stability of the alternative intermediates and this is summarized empirically by Markovnikoff's rule. *See also* addition reaction.

electrophilic substitution A reaction involving substitution of an atom or group of atoms in an organic compound with an electrophile as the attacking substituent. Electrophilic substitution is very common in aromatic compounds, in which electrophiles are substituted onto the ring. An example is the nitration of benzene:
$$C_6H_6 + NO_2^+ \rightarrow C_6H_5NO_2 + H^+$$
The nitronium ion (NO_2^+) is formed by mixing concentrated nitric and sulphuric acids:
$$HNO_3 + H_2SO_4 \rightarrow H_2NO_3^+ + HSO_4^-$$
$$H_2NO_3^+ \rightarrow NO_2^+ + H_2O$$
The accepted mechanism for a simple electrophilic substitution on benzene involves an intermediate of the form $C_6H_5HNO_2^+$. *See also* substitution reaction.

electroplating The process of coating a solid surface with a layer of metal by means of electrolysis (i.e. by electrode-position).

electrovalent bond (ionic bond) A binding force between the ions in compounds in which the ions are formed by complete transfer of electrons from one element to another element or radical. For example, Na + Cl becomes $Na^+ + Cl^-$. The electrovalent bond arises from the excess of the net attractive force between the ions of opposite charge over the net repulsive force between ions of like charge. The magnitude of electrovalent interactions is of the order 10^2-10^3 kJ mol^{-1} and electrovalent compounds are generally solids with rigid lattices of closely packed ions. The strengths of electrovalent bonds vary with the reciprocal of the inter-ionic distances and are discussed in terms of lattice energies.

element A substance that cannot be chemically decomposed into more simple substances. The atoms of an element all have the same proton number (and thus the same number of electrons, which determines the chemical activity). There are about 92 naturally occurring elements on Earth. Those with high relative atomic masses (greater than 93) are made by bombardment of other nuclei with high-energy particles. *See also* element 104.

element 104 The first transactinide element, atomic number 104, i.e. following after the actinoid series. Its synthesis is claimed by Soviet and American groups with several very short-lived isotopes being reported; proposed names (and symbols) are *kurchatovium* (Ku) and *rutherfordium* (Rf), however neither of these names has been adopted and the name *unnilquadium* (Unq) has now been accepted (un- = 1, -nil- = 0, -quad = 4). There are also strong claims for the discoveries of *element 105* and *element 106*, which are named *unnilpentium* and *unnilhexium*, respectively.

elevation of boiling point A colligative property of solutions in which the boiling point of a solution is raised relative to that of the pure solvent. The elevation is directly proportional to the number of solute molecules introduced rather than to any specific aspect of the solute composition. The proportionality constant, k_B, is called the boiling-point elevation constant or sometimes the *ebulioscopic constant*. The relationship is

$$\Delta t = k_B C_M$$

where Δt is the rise in boiling point and C_M is the molal concentration; the units of k_B are kelvins kilograms moles^{-1} (K kg mol^{-1}). The property permits the measurement of relative molecular mass of involatile solutes. An accurately weighed amount of pure solvent is boiled until the temperature is steady, a known weight of solute of known molecular mass is quickly

Beckmann thermometer

condenser

tube for introducing solute

heater

Apparatus for measuring elevation of boiling point

introduced, the boiling continued, and the elevation measured using a Beckmann thermometer. The process is repeated several times and the average value of k_B obtained by plotting Δt against C_M. The whole process is then repeated with the unknown material and its relative molecular mass obtained using the value of k_B previously obtained.

There are several disadvantages with this method and it is therefore used largely for demonstration purposes. The main problem is that the exact amount of solvent remaining in the liquid phase is unknown and varies with the rate of boiling. The theoretical explanation of the effect is identical to that for the lowering of vapour pressure; the boiling points are those temperatures at which the vapour pressure equals the atmospheric pressure.

elimination reaction A reaction involving the removal of a small molecule, often water or hydrogen chloride, from an organic molecule to given an unsaturated compound. An example is the elimination of a water molecule from an alcohol to produce an alkene.

An elimination reaction is often in competition with a substitution reaction and the predominant product will depend on the reaction conditions. The reaction of bromoethane with sodium hydroxide could yield either ethene (by elimination of HBr) or ethanol (by substitution of the Br with OH). The former product predominates if the reaction is carried out in an alcoholic solution and the latter if the solution is aqueous.

elution The removal of an adsorbed substance in a chromatography column or ion-exchange column using a solvent (*eluent*). The chromatography column can selectively adsorb one or more components from the mixture. To ensure efficient recovery of these components graded elution is used. The eluent is changed in a regular manner starting with a non-polar solvent and gradually replacing it by a more polar one. This will wash the strongly polar components from the column.
See column chromatography.

emission spectrum *See* spectrum.

empirical formula The formula of a compound showing the simplest ratio of the atoms present. The empirical formula is the formula obtained by experimental analysis of a compound and it can only be related to a molecular formula if the molecular weight is known. For example P_2O_5 is the empirical formula of phosphorus(V) oxide although its molecular formula is P_4O_{10}. *Compare* molecular formula, structural formula.

emulsion A colloid in which a liquid phase (small droplets with a diameter range $10^{-5}-10^{-7}$ cm) is dispersed or suspended in a liquid medium. Emulsions are classed as lyophobic (solvent-repelling and generally unstable) and lyophilic (solvent-attracting and generally stable).

enantiomer (enantiomorph) A compound whose structure is not superimposable on its mirror image: one of any pair of optical isomers. *See also* isomerism, optical activity.

enantiotropy The existence of different stable allotropes of an element at different temperatures; sulphur, for example, exhibits enantiotropy. The phase diagram for an enantiotropic element has a point at which all the allotropes can coexist in a stable equilibrium. At temperatures above or below this point, one of the allotropes will be more stable than the other(s). *See also* allotropy, monotropy.

enclosure compound *See* clathrate.

endothermic Describing a process in which heat is absorbed (i.e. heat flows from outside the system, or the temperature falls). The dissolving of a salt in water, for instance, is often an endothermic process. *Compare* exothermic.

end point *See* equivalence point, volumetric analysis.

energy Symbol: W A property of a system; a measure of its capacity to do work. Energy and work have the same unit: the joule (J). It is convenient to divide energy into *kinetic energy* (energy of motion) and *potential energy* ('stored' energy). Names are given to many different forms of energy (chemical, electrical, nuclear, etc.); the only real difference lies in the system under discussion. For example, chemical energy is the kinetic and potential energies of electrons in a chemical compound.

energy level One of the discrete energies that an atom or molecule, for instance, can have according to quantum theory. Thus in an atom there are certain definite orbits that the electrons can be in, corresponding to definite *electronic energy levels* of the atom. Similarly, a vibrating or rotating molecule can have discrete vibrational and rotational energy levels.

energy profile A diagram that traces the changes in the energy of a system during the course of a reaction. Energy profiles are obtained by plotting the potential energy of the reacting particles against the reaction coordinate. To obtain the reaction coordinate the energy of the total interacting system is plotted against position for the molecules. The *reaction coordinate* is the pathway for which the energy is a minimum.

The energy profile for the hydrogenation of an alkene with and without a catalyst has certain interesting features: the activation energy without the catalyst is very much larger than that with the catalyst and any point along the horizontal axis represents the changes that the colliding particles experience as atoms are separated in bond breaking and as atoms are brought together in bond formation.

enol An organic compound containing the C:CH(OH) group; i.e. one in which a hydroxyl group is attached to the carbon of a double bond. *See* keto-enol tautomerism.

enthalpy Symbol: H The sum of the internal energy (U) and the product of pressure (p) and volume (V) of a system:
$$H = U + pV$$
In a chemical reaction carried out at constant pressure, the change in enthalpy measured is the internal energy change plus the work done by the volume change:
$$\Delta H = \Delta U + p\Delta V$$

entropy Symbol: S In any system that undergoes a reversible change, the change of entropy is defined as the heat absorbed divided by the thermodynamic temperature:
$$\delta S = \delta Q / T$$
A given system is said to have a certain entropy, although absolute entropies are seldom used: it is change in entropy that is important. The entropy of a system measures the availability of energy to do work. In any real (irreversible) change in a closed system the entropy increases. Although the total energy of the system has not changed (first law of thermodynamics) the available energy is less — a consequence of the second law of thermodynamics.

The concept of entropy has been widened to take in the general idea of disorder —

the higher the entropy, the more disordered the system. For instance, a chemical reaction involving polymerization may well have a decrease in entropy because there is a change to a more ordered system. The 'thermal' definition of entropy is a special case of this idea of disorder — here the entropy measures how the energy transferred is distributed amongst the particles of matter.

enzyme A protein, with a molecular weight between 10^5 and 10^7, that catalyses a specific biochemical reaction. An enzyme mediates the conversion of one substance (the *substrate*) to another (the *product*) by combining with the substrate to form an intermediate complex. The enzyme molecule is very much larger than the substrate molecule. Enzymes differ from inorganic catalysts in the following ways:

(1) they have a high degree of specificity — enzymes can even distinguish between enantiomorphs;

(2) an increase in temperature can cause a decrease in the rate of reaction as a result of the enzyme being denatured;

(3) they are destroyed by too great a change in pH;

(4) they are inactivated by low concentrations of heavy-metal ions.

epimerism A form of isomerism exhibited by carbohydrates in which the isomers (*epimers*) differ in the positions of −OH groups. The α- and β- forms of glucose are epimers. *See* sugar.

epoxide A type of organic compound containing a three-membered ring in which two carbon atoms are each bonded to the same oxygen atom.

epoxyethane (ethylene oxide, C_2H_4O) A colourless gaseous ether. Epoxyethane has a cyclic structure with a three-membered ring. It is made by oxidation of ethene over a silver catalyst. The ring is strained and the compound is consequently highly reactive. It polymerizes to epoxy polymers. Hydrolysis produces 1,2-ethanediol.

Epsom salt

$$CH_2\!\!-\!\!CH_2$$
$$\diagdown_{\;O\;}\diagup$$

Epoxyethane

Epsom salt *See* magnesium sulphate.

equilibrium In a reversible chemical reaction:

$$A + B = C + D$$

The reactants are forming the products:

$$A + B \rightarrow C + D$$

which also react to give the original reactants:

$$C + D \rightarrow A + B$$

The concentrations of A, B, C, and D change with time until a state is reached at which both reactions are taking place at the same rate. The concentrations (or pressures) of the components are then constant — the system is said to be in a state of chemical equilibrium. Note that the equilibrium is a dynamic one; the reactions still take place but at equal rates. The relative proportions of the components determine the 'position' of the equilibrium, which may be *displaced* by changing the conditions (e.g. temperature or pressure).

equilibrium constant In a chemical equilibrium of the type

$$xA + yB = zC + wD$$

The expression:

$$[A]^x[B]^y / [C]^z[D]^w$$

where the square brackets indicate concentrations, is a constant (K_c) when the system is at equilibrium. K_c is the equilibrium constant of the given reaction; its units depend on the stoichiometry of the reaction. For gas reactions, pressures are often used instead of concentration. The equilibrium constant is then K_p, where $K_p = K_c^n$. Here n is the number of moles of product minus the number of moles of reactant; for instance, in

$$3H_2 + N_2 = 2NH_3$$

n is $2 - (1 + 3) = -2$.

equipartition of energy The principle that the total energy of a molecule is, on average, equally distributed among the available degrees of freedom. It is only approximately true in most cases. *See* degrees of freedom.

equivalence point The point in a titration at which the reactants have been added in equivalent proportions, so that there is no excess of either. It differs slightly from the *end point*, which is the observed point of complete reaction, because of the effect of the indicator, errors, etc.

equivalent weight A measure of 'combining power' formerly used in calculations for chemical reactions. The equivalent weight of an element is the number of grams that could combine with or displace one gram of hydrogen (or 8 grams of oxygen or 35.5 grams of chlorine). It is the atomic weight divided by the valence. For a compound the equivalent weight depends on the reaction considered. An acid, for instance, in acid-base reactions has an equivalent weight equal to its molecular weight divided by the number of acidic hydrogen atoms.

erbium A soft malleable silvery element of the lanthanoid series of metals. It occurs in association with other lanthanoids. Erbium has uses in the metallurgical, nuclear, and glass industries.
Symbol: Er; m.p. 1522°C; b.p. 2510°C; r.d. 9.01; p.n. 68; r.a.m. 167.26.

erg A former unit of energy used in the c.g.s. system. It is equal to 10^{-7} joule.

Erlenmeyer flask A conical glass laboratory flask with a narrow neck.

ester A type of organic compound formed by reaction between an acid and an alcohol. If the acid is a carboxylic acid, the formula is RCOOR', where R and R' are organic groups. Esters of low-molecular weight acids and alcohols are volatile fragrant compounds. Esters of some long-chain carboxylic acids occur naturally in

Ethanal trimer

Ethanal tetramer

fats and oils. Esters can be prepared by direct reaction of acids and alcohols, with a dehydrating agent to take out the water and displace the equilibrium. Reaction of alcohols with acyl halides also gives esters. *See also* carboxylic acid, saponification, glyceride.

esterification The reaction of an acid with an alcohol to form an ester and water. The reaction is an equilibrium of the form:

$$CH_3COOH + C_2H_5OH =$$
$$CH_3COOC_2H_5 + H_2O$$

In the preparation of esters the ester is distilled off or the water is removed to increase the yield. The reverse reaction is hydrolysis.

ethanal (acetaldehyde, CH_3CHO) A water-soluble liquid aldehyde used as a starting material in the manufacture of several other compounds. Ethanal can be prepared by the oxidation of ethanol. It is manufactured by the catalytic oxidation of ethyne with oxygen using copper(II) chloride and palladium(II) chloride as catalysts. The mixture of gases is bubbled through an aqueous solution of the catalysts; the reaction involves formation of an intermediate organometallic complex with Pd^{2+} ions. With dilute acids ethanal polymerizes to *ethanal trimer* ($C_3O_3H_3(CH_3)_3$, formerly called *paraldehyde*), which is a sleep-inducing drug. Below 0°C *ethanal tetramer* is formed ($C_4O_4H_4(CH_3)_4$, formerly called *metaldehyde*), which is used as a slug poison and a fuel in small portable stoves.

ethane (C_2H_6) A gaseous alkane obtained either from the gaseous fraction of crude oil or by the 'cracking' of heavier fractions. Ethane is the second member of the homologous series of alkanes.

ethane-1,2-diol (ethylene glycol, glycol, $CH_2(OH)CH_2(OH)$) A syrupy organic liquid commonly used as antifreeze and as a starting material in the manufacture of Terylene. The compound is manufactured from ethene by oxidation over suitable catalysts to form epoxyethane, with subsequent hydrolysis to the diol.

ethanol (ethyl alcohol, alcohol, C_2H_5OH) A colourless volatile liquid alcohol. Ethanol occurs in intoxicating drinks, in which it is produced by fermentation of a sugar:

$$C_6H_{12}O_6 \rightarrow 2C_2H_5OH + 2CO_2$$

Yeast is used to cause the reaction. At about 15% alcohol concentration (by volume) the reaction stops because the yeast is killed. Higher concentrations of alcohol are produced by distillation.

Apart from its use in drinks, alcohol is used as a solvent and to form ethanal. Formerly, the main source was by fermentation of molasses, but now catalytic hydration of ethene is used to manufacture industrial ethanol. *See also* methylated spirits.

ethanoyl chloride (acetyl chloride, CH_3COCl) A liquid acyl chloride used as an acetylating agent.

ethanoyl group (acetyl group) The group RCO−.

ethene (ethylene, C_2H_4) A gaseous alkene. Ethene is not normally present in the gaseous fraction of crude oil but can be obtained from heavier fractions by catalytic cracking. This is the principal industrial source. The compound is important as a starting material in the organic-chemicals industry (e.g. in the manufacture of ethanol) and as the starting material for the production of polythene. Ethene is the first member of the homologous series of alkenes.

ether A type of organic compound containing the group −O−. Simple ethers have the formula R−O−R', where R and R' are alkyl or aryl groups, which may or may not be the same. They are either gases or very volatile liquids and are very flammable. The commonest example is ethoxyethane (diethylether, $C_2H_5OC_2H_5$) used formerly as an anaesthetic. Ethers now find application as solvents. Ethers are prepared in the laboratory by the dehydration of alcohols with concentrated sulphuric acid. An excess of alcohol is used to ensure that only one molecule of water is removed from each pair of alcohol molecules. They are generally unreactive, but the C−O bond can be cleaved by reaction with HI or PCl_5.

ethoxyethane (ether, diethylether, $C_2H_5OC_2H_5$) A colourless volatile liquid. Ether is well known for its characteristic smell and anaesthetic properties, also for its extreme flammability. It still finds some application as an anaesthetic when more modern materials are unsuitable; it is also an excellent solvent. Its manufacture is an extension of the laboratory synthesis: ethanol vapour is passed into a mixture of excess ethanol and concentrated sulphuric acid at 140°C.

$$C_2H_5OH + H_2SO_4 \rightarrow$$
$$C_2H_5.O.SO_2.OH + H_2O$$
$$C_2H_5O.SO_2.OH + C_2H_5OH \rightarrow$$
$$C_2H_5OC_2H_5 + H_2SO_4$$

ethyl acetate *See* ethyl ethanoate.

ethyl alcohol *See* ethanol.

ethylamine (aminoethane, $C_2H_5NH_2$) A colourless liquid amine. It can be prepared from chloroethane heated with concentrated aqueous ammonia:

$$C_2H_5Cl + NH_3 \rightarrow C_2H_5NH_2 + HCl$$

It is used in manufacturing certain dyes.

ethyl bromide *See* bromoethane.

ethyl chloride *See* chloroethane.

ethylene *See* ethene.

Ether

ethylenediamine tetraacetic acid
See edta.

ethylene oxide *See* epoxyethane.

ethyl ethanoate (ethyl acetate, $C_2H_5OOCCH_3$) An ester formed from ethanol and ethanoic acid. It is a fragrant liquid used as a solvent for plastics and in flavouring and perfumery.

ethyl iodide *See* iodoethane.

ethyne (acetylene, C_2H_2) A gaseous alkyne. Traditionally ethyne has found use in oxy-acetylene welding torches, since its combustion with oxygen produces a flame of very high temperature. It is also important in the organic chemicals industry for the production of chloroethene (vinyl chloride), which is the starting material for the production of polyvinyl chloride (PVC), and for the production of other vinyl compounds. Until recently, ethyne was manufactured by the synthesis and subsequent hydrolysis of calcium dicarbide, a very expensive procedure. Modern methods increasingly employ the cracking of alkanes.

eudiometer An apparatus for the volumetric analysis of gases.

europium A silvery element of the lanthanoid series of metals. It occurs in association with other lanthanoids. Its main use is in a mixture of europium and yttrium oxides widely employed as the red phosphor in television screens.
Symbol: Eu; m.p. 822°C; b.p. 1597°C; r.d. 5.24; p.n. 63; r.a.m. 151.96.

eutectic A mixture of two substances in such proportions that no other mixture of the same substances has a lower freezing point. If a solution containing the proportions of the eutectic is cooled no solid is deposited until the freezing point of the eutectic is reached, when two solid phases are simultaneously deposited in the same proportions as the mixture; these two solid phases form the eutectic. If one of the substances is water the eutectic can also be referred to as a *cryohydrate*.

evaporation 1. A change of state from liquid to gas (or vapour). Evaporation can take place at any temperature, the rate increasing with temperature. Some molecules in the liquid have enough energy to escape into the gas phase (if they are near the surface and moving in the right direction). Because these are the molecules with higher kinetic energies, evaporation results in a cooling of the liquid.
2. A change from solid to vapour, especially occurring at high temperatures close to the melting point of the solid. Thin films of metal can be evaporated onto a surface in this way.

exa- Symbol: E A prefix denoting 10^{18}

excitation The process of producing an excited state of an atom, molecule, etc.

excitation energy The energy required to change an atom, molecule, etc. from one quantum state to a state with a higher energy. The excitation energy (sometimes called *excitation potential*) is the difference between two energy levels of the system.

excited state A state of an atom, molecule, or other system, with an energy greater than that of the ground state. *Compare* ground state.

exclusion principle The principle, enunciated by Pauli in 1925, that no two electrons in an atom could have an identical set of quantum numbers.

exothermic Denoting a chemical reaction in which heat is evolved (i.e. heat flows from the system or the temperature rises). Combustion is an example of an exothermic process. *Compare* endothermic.

explosive A substance or mixture which can rapidly decompose upon detonation producing large amounts of heat and gases. The three most important classes of

explosive

explosives are: (1) *Propellants* which burn steadily, and are used as fuels. (2) *Initiators* which are very sensitive and are used in small amounts to detonate less sensitive explosives. (3) *High explosives* which need an initiator, but are very powerful.

face-centred cubic crystal (f.c.c.) A crystal structure in which the unit cell has atoms, ions, or molecules at each corner and each face centre of a cube (also called *cubic close-packed*). It has a coordination number of 12. The structure is close-packed and made up of layers of atoms in which each atom is surrounded by six others arranged hexagonally. Copper and aluminium have face-centred cubic structures. The layers are packed one on top of the other, with the second layer fitting into the holes of the first layer. The third layer is in a different set of holes formed by the second layer. If the layers are designated A, B, C, the packing is ABCABC in contrast to hexagonal close packing, in which the layers are arranged ABABAB.

Fahrenheit scale A temperature scale in which the ice temperature is taken as 32° and the steam temperature is taken as 212° (both at standard pressure). The scale is not used for scientific purposes. To convert between degrees Fahrenheit (F) and degrees Celsius (C) the formula
$$C/5 = (F-32)/9$$
is used. *See also* temperature scale.

Fajan's rules Rules that deal with variations in the degree of covalent character in ionic compounds in terms of polarization effects. Decrease in size and increase in charge for cations is regarded as making them more polarizing and for anions an increase in both charge and size makes them more polarizable. Thus covalent character is said to increase with increasing polarizing power of the cation and/or increasing polarizability of the anion (originally rules 1 and 2). Fajan's third rule states that covalent character is greater for cations with a non-inert gas configuration than for those with a complete octet, charge and size being approximately equal. For example, Cu^+ is more polarizing than Na^+ (approximately the same size) because the d-electrons around Cu^+ do not shield the nucleus as effectively as the complete octet in Na^+.

farad Symbol: F The SI unit of capacitance. When the plates of a capacitor are charged by one coulomb and there is a potential difference of one volt between them, then the capacitor has a capacitance of one farad. One farad = 1 coulomb volt^{-1} (CV^{-1}).

faraday Symbol: F A unit of electric charge equal to the charge required to discharge one mole of a singly charged ion. One faraday is $9.648\,670 \times 10^4$ coulombs. *See* Faraday's laws.

Faraday constant *See* Faraday's laws.

Faraday's laws (of electrolysis) Two laws resulting from the work of Michael Faraday on electrolysis: (1) The amount of chemical change produced is proportional to the electric charge passed. (2) The amount of chemical change produced by a given charge depends on the ion concerned. More strictly it is proportional to the relative ionic mass of the ion, and the charge on it. Thus the charge Q needed to deposit m grams of an ion of relative ionic mass, M, carrying a charge Z is given by:
$$Q = FmZ/M$$
F, the *Faraday constant*, has a value of one faraday, i.e. $9.648\,670 \times 10^4$ coulombs.

fat *See* glyceride.

fatty acid *See* carboxylic acid.

f.c.c. *See* face-centred cubic crystal.

Fehling's solution A solution used to test for the aldehyde group (−CHO). It is a freshly made mixture of copper(II) sulphate solution with alkaline potassium sodium 2,3-dihydroxybutanedioate (tartrate). The aldehyde, when heated with the mixture, is oxidized to a carboxylic acid, and a red precipitate of copper(I) oxide and copper metal is produced. The tartrate is present to complex with the original copper(II) ions to prevent precipitation of copper(II) hydroxide.

femto- Symbol: f A prefix denoting 10^{-15}. For example, 1 femtometre (fm) = 10^{-15} metre (m).

fermentation A chemical reaction produced by microorganisms (moulds, bacteria, or yeasts). A common example is the formation of ethanol from sugars:

$$C_6H_{12}O_6 \rightarrow 2C_2H_5OH + 2CO_2$$

fermi A unit of length equal to 10^{-15} metre. It was formerly used in atomic and nuclear physics.

fermium A radioactive transuranic element of the actinoid series, not found naturally on Earth. Several short-lived isotopes have been synthesized.
Symbol: Fm; p.n. 100; stablest isotope ^{257}Fm (half-life 80 days).

ferric chloride See iron(III) chloride.

ferric oxide See iron(III) oxide.

ferric sulphate See iron(III) sulphate.

ferrocene ($Fe(C_5H_5)_2$) An orange crystalline solid. It is an example of a sandwich compound, in which an iron(II) ion is coordinated to two cyclopentadienyl ions. The bonding involves overlap of d orbitals on the iron with the pi electrons in the cyclopentadienyl ring. The compound melts at 173°C, is soluble in organic solvents, and can undergo substitution reactions on the rings. It can be oxidized to the blue cation $(C_5H_5)_2Fe^+$. The systematic name is di-π-cyclopentadienyl iron(II).

ferromagnetism See magnetism.

ferrosoferric oxide See triiron tetroxide.

ferrous chloride See iron(II) chloride.

ferrous oxide See iron(II) oxide.

ferrous sulphate See iron(II) sulphate.

filler A solid material used to modify the physical properties or reduce the cost of synthetic compounds, such as rubbers, plastics, paints, and resins. Slate powder, glass fibre, mica, and cotton wool are all used as fillers.

filter pump A type of vacuum pump in which a jet of water forced through a nozzle carries air molecules out of the system. Filter pumps cannot produce pressures below the vapour pressure of water. They are used in the laboratory for vacuum filtration, distillation, and similar techniques requiring a low-grade vacuum.

filtration The process of removing suspended particles from a fluid by passing or forcing the fluid through a porous material (the *filter*). In laboratory filtration, filter paper or sintered glass is commonly used.

fine organic chemicals Carbon compounds, such as pesticides, dyes, and drugs, produced only in small quantities. Their main requirement is that they must have a high degree of purity, often higher than 95%. They are manufactured for special purposes, e.g. for use in spectroscopy, pharmacology, and electronics.

fine structure Closely spaced lines seen at high resolution in a spectral line or band. Fine structure may be caused by vibration of the molecules or by electron spin. *Hyperfine structure*, seen at very high resolution, is caused by the atomic nucleus affecting the possible energy levels of the atom.

firedamp Methane occurring in coal mines. Explosions can occur if quantities of methane accumulate. The air left after an explosion is called *blackdamp*. *Afterdamp* is the poisonous carbon monoxide formed.

first-order reaction A reaction in which the rate of reaction is proportional to the concentration of one of the reacting substances. The concentration of the reacting substance is raised to the power one; i.e. rate = k[A]. For example, the decomposition of hydrogen peroxide is a first-order reaction,

rate = $k[H_2O_2]$

Similarly the rate of decay of radioactive material is a first-order reaction,

rate = k[radioactive material]

For a first-order reaction, the time for a definite fraction of the reactant to be consumed is independent of the original concentration. The units of k, the rate constant, are s^{-1}.

Fischer-Tropsch process The catalytic hydrogenation of carbon monoxide in the ratio of 2:1 hydrogen to carbon monoxide at 200°C to produce hydrocarbons, especially motor fuel. The carbon monoxide is derived from *water gas*, which uses coal or natural gas as a source of carbon, and the catalyst is finely divided nickel. The process was extensively used in Germany during World War II and is still used in countries in which crude oil is not an easily available source of motor fuel.

Fittig reaction *See* Wurtz reaction.

Flame test colours

Element	Flame colour
barium	green
calcium	brick red
lithium	crimson
potassium	pale lilac
sodium	yellow
strontium	red

flame test A preliminary test in qualitative analysis in which a small sample of a chemical is introduced into a nonluminous bunsen flame on a clean platinum wire. The flame vaporizes part of the sample and excites some of the atoms, which emit light at wavelengths characteristic of the metallic elements in the sample. Thus the observation of certain colours in the flame indicates the presence of these elements.

The same principles are applied in the modern instrumental method of spectrographic analysis.

flare stack A chimney at the top of which unwanted gases are burnt.

flash photolysis A technique for investigating free radicals in gases. The gas is held at low pressure in a long glass or quartz tube, and an absorption spectrum taken using a beam of light passing down the tube. The gas can be subjected to a very brief intense flash of light from a lamp outside the tube, producing free radicals, which are identified by their spectra. Measurements of the intensity of spectral lines can be made with time using an oscilloscope, and the kinetics of very fast reactions can thus be investigated.

flash point The lowest temperature at which sufficient vapour is given off by a flammable liquid to ignite in the presence of a spark.

fluidization The suspension of a finely-divided solid in an upward-flowing liquid or gas. This suspension mimics many properties of liquids, such as allowing objects to 'float' in it. Fluidized beds so constructed are becoming increasingly important industrially.

fluorescein A fluorescent dye used as an absorption indicator.

fluorescence The absorption of energy by atoms, molecules, etc., followed by immediate emission of electromagnetic radiation as the particles make transitions to lower energy states. *Compare* phosphorescence.

fluoridation The introduction of small quantities of fluoride compounds into the water supply as a public-health measure to reduce the incidence of tooth decay in children.

fluoride *See* halide.

fluorination *See* halogenation.

fluorine A slightly greenish-yellow highly reactive gaseous element belonging to the halogens (group VII of the periodic table). It occurs notably as fluorspar (CaF_2) and cryolite (Na_3AlF_3) but traces are also widely distributed with other minerals. It is slightly more abundant than chlorine, accounting for about 0.065% of the Earth's crust. The high reactivity of the element delayed its isolation. Fluorine is now prepared by electrolysis of molten KF/HF electrolytes, using copper or steel apparatus. Its preparation by chemical methods is impossible.

Fluorine compounds are used in the steel industry, glass and enamels, uranium processing, aluminium production, and in a range of fine organic chemicals. Fluorine is the most reactive and most electronegative element known; in fact it reacts directly with almost all the elements, including the noble gas xenon. As the most powerful oxidizing agent known, it has the pronounced characteristic of bringing out the higher oxidation states when combined with other elements. Ionic fluorides contain the small F^- ion, which is very similar to the O^{2-} ion. With the more electronegative elements, fluorine forms an extensive range of compounds, e.g. SF_6, NF_3, and PF_5, in which the bonding is essentially covalent. A vast number of organic compounds are known in which hydrogen may be replaced by fluorine and in which the C−F bond is characteristically very stable, e.g. $CF_2 = CF_2$, $C_6H_4F_2$, and CF_3COOH; the bonding is again essentially covalent.

Fluorine reacts explosively with hydrogen, even in the dark, to form hydrogen fluoride, HF, which polymerizes as a result of hydrogen bonding and has a boiling point very much *higher* than that of HCl (HF b.p. 19°C; HCl b.p. −85°C).

Unlike the other halogens, fluorine does not form higher oxides or oxyacids; oxygen difluoride in fact reacts with water to give hydrogen fluoride.

Fluorine is strongly electronegative and exhibits a strong electron withdrawing effect on adjacent bonds, thus CF_3COOH is a strong acid (whereas CH_3COOH is not). Although the coordination number seldom exceeds one there are cases in which fluorine bridging occurs, e.g. $(SbF_5)_n$. The element is sufficiently reactive to combine directly with the noble gas xenon at 400°C to form XeF_2 and XeF_4. At 600°C and high pressure it will even form XeF_6.

Fluorine and hydrogen fluoride are extremely dangerous and should only be used in purpose-built apparatus; gloves and face shields should be used when working with hydrofluoric acid and accidental exposure should be treated as a hospital emergency.

Symbol: F; b.p. −188°C; m.p. −220°C; d. 1.696 kg m^{-3}; p.n. 9; r.a.m. 18.99840.

fluorite structure See calcium-fluoride structure.

fluorocarbon A compound derived from a hydrocarbon by replacing hydrogen atoms with fluorine atoms. Fluorocarbons are unreactive and most are stable up to high temperatures. They have a variety of uses — in aerosol propellants, oils and greases, and synthetic polymers such as PTFE. See also chlorofluorocarbon.

fluorspar A naturally occurring form of calcium fluoride, CaF_2.

flux 1. A substance used to keep metal surfaces free of oxide in soldering. See solder. **2.** A substance used in smelting metals to react with silicates and other impurities and form a low-melting slag.

flux See solder.

fluxional molecule A molecule in which the constituent atoms change their relative positions so quickly at room temperature that the normal concept of structure is inadequate; i.e. no specific structure exists for longer than about 10^{-2} second and the relative positions become indistinguishable. For example ClF_3 at −60°C has a distinct 'T' shape but at room temperature the fluorine atoms are visualized as moving rapidly over the surface of the chlorine atom in a state of exchange and are effectively identical.

formaldehyde See methanal.

formalin *See* methanal.

formate *See* methanoate.

formic acid *See* methanoic acid.

formula A representation of a chemical compound using symbols for the atoms and subscript numbers to show the numbers of atoms present. *See* empirical formula, molecular formula, structural formula.

formyl group The group HCO–.

fraction A mixture of liquids with similar boiling points collected by fractional distillation.

fractional crystallization Crystallization of one component from a mixture in solution. When two or more substances are present in a liquid (or in solution), on cooling to a lower temperature one substance will preferentially form crystals, leaving the other substance in the liquid (or dissolved) state. Fractional crystallization can thus be used to purify or separate substances if the correct conditions are known.

fractional distillation (fractionation) A distillation carried out with partial reflux, using a long vertical column (fractionating column). It utilizes the fact that the vapour phase above a liquid mixture is generally richer in the more volatile component. If the region in which refluxing occurs is sufficiently long, fractionation permits the complete separation of two or more volatile liquids. Fractionation is the fundamental process for producing petroleum from crude oil.
Unlike normal reflux, the fractionating column may be insulated to reduce heat loss, and special designs are used to maximize the liquid-vapour interface.

francium A radioactive element of the alkali-metal group. It is found on Earth only as a short-lived product of radioactive decay, occurring in uranium ores in min-

ute quantities. 16 radioisotopes of francium are known.
Symbol: Fr; m.p. 27°C; b.p. 677°C; p.n. 87; stablest isotope ^{223}Fr.

free energy A measure of the ability of a system to do useful work. *See* Gibbs function, Helmholtz function.

free radical An atom or group of atoms with a single unpaired electron. Free radicals are produced by breaking a covalent bond; for example:
$$CH_3Cl \rightarrow CH_3 \cdot + Cl \cdot$$
They are often formed in light-induced reactions. Free radicals are extremely reactive and can only be stabilized and isolated under special conditions.

freezing The process by which a liquid is converted into a solid by cooling; the reverse of melting.

freezing mixtures Two or more substances mixed together to produce a low temperature. A mixture of sodium chloride and ice in water (–20°C) is a common example.

freezing point The temperature at which a liquid is in equilibrium with its solid phase at standard pressure and below which the liquid freezes or solidifies. This temperature is always the same for a particular liquid and is numerically equal to the melting point of the solid.

Friedel-Crafts reaction A method for the substitution of an alkyl or acyl group onto a benzene ring. The aromatic hydrocarbon (e.g. benzene) is refluxed with a haloalkane or acyl halide using aluminium chloride catalyst. The product is an alkylarene, e.g.:
$$C_6H_6 + CH_3I \rightarrow C_6H_5CH_3 + HI$$
or a ketone, for example:
$$C_6H_6 + CH_3COCl \rightarrow C_6H_5COCH_3 + HCl$$
The aluminium chloride accepts a lone pair of electrons on the chlorine, polarizing the chloroalkane to produce positive charge on the carbon. Electrophilic substitution then occurs.

froth flotation An industrial technique for separating the required parts of ores from unwanted gangue. The mineral is pulverized and mixed with water, to which a frothing agent is added. The mixture is aerated and particles of one constituent are carried to the surface by bubbles of air, which adhere preferentially to one of the constituents. Various additives are also used to modify the surface properties of the particles (e.g. to increase the adherence of air bubbles).

fructose ($C_6H_{12}O_6$) A sugar found in fruit juices, honey, and cane sugar. It is a ketohexose, existing in a pyranose form when free. In combination (e.g. in sucrose) it exists in the furanose form.

fuel cell A type of cell in which fuel is converted directly into electricity. In one form, hydrogen gas and oxygen gas are fed to the surfaces of two porous nickel electrodes immersed in potassium hydroxide solution. The oxygen reacts to form hydroxyl (OH^-) ions, which it releases into the solution, leaving a positive charge on the electrode. The hydrogen reacts with the OH^- ions in the solution to form water, giving up electrons to leave a negative charge on the other electrode. Large fuel cells can generate tens of amperes. Usually the e.m.f. is about 0.9 volt and the efficiency around 60%.

fuller's earth A natural clay used as an absorbent and industrial catalyst.

fumaric acid *See* butenedioic acid.

functional group A group of atoms in a compound that is responsible for the characteristic reactions of the type of compound. Examples are:
alcohol $-OH$
aldehyde $-CHO$
amine $-NH_2$
ketone $.CO.$
carboxylic acid $-CO.OH$
acyl halide $-CO.X$ (X = halogen)
nitro compound $-NO_2$
sulphonic acid $-SO_2.OH$
nitrile $-CN$
diazonium salt $-N_2{}^+$
diazo compound $-N=N-$

fundamental units The units of length, mass, and time that form the basis of most systems of units. In SI, the fundamental units are the metre, the kilogram, and the second. *See also* base unit.

furanose A sugar that has a five-membered ring (four carbon atoms and one oxygen atom). *See also* sugar.

fused Describing a solid that has been melted and solidified into a single mass. Fused silica, for example, is produced by melting sand.

fused ring *See* ring.

fusion Melting.

G

gadolinium A ductile malleable silvery element of the lanthanoid series of metals. It occurs in association with other lanthanoids. Gadolinium is used in alloys, magnets, and in the electronics industry.
Symbol: Gd; m.p. 1313°C; b.p. 3266°C; r.d. 7.9; p.n. 64; r.a.m. 157.25.

galactose ($C_6H_{12}O_6$) A sugar found in lactose and many polysaccharides. It is an aldohexose, isomeric with glucose. *See also* sugar.

galena (lead glance) A mineral form of lead(II) sulphide. It is the principal ore of lead.

gallium A soft silvery low-melting metallic element belonging to group III of the periodic table. It is found in minute quantities in several ores, including zinc blende (ZnS) and bauxite ($Al_2O_3.H_2O$). Gallium is used in low-melting alloys, high-temperature thermometers, and as a doping impurity in semiconductors. Gallium arsenide is a semiconductor finding increasing use in electronics, particularly in light emitting diodes.
Symbol: Ga; m.p. 29.78°C; b.p. 2403°C; r.d. 5.90 (solid), 6.09 (liquid); p.n. 31; r.a.m. 69.72.

galvanic cell *See* cell.

galvanizing A process for coating steel with zinc by dipping in a bath of molten zinc, or by electrodeposition. The zinc protects the steel from corrosion.

gamma radiation A form of electromagnetic radiation emitted by changes in the nuclei of atoms. Gamma waves have very high frequency (short wavelength). Gamma radiation shows particle properties much more often than wave properties. The energy of a gamma photon, given by

$$W = h\upsilon$$

(where h is the Planck constant) can be very high. A gamma photon of 10^{24} Hz has an energy of 6.6×10^{-10} J. *See also* electromagnetic radiation.

gamma rays Streams of gamma radiation.

gangue The rock or other undesirable material occurring in an ore.

gas The state of matter in which forces of attraction between the particles of a substance are small. The particles have freedom of movement and gases, therefore, have no fixed shape or volume. The atoms and molecules of a gas are in a continual state of motion and are continually colliding with each other and with the walls of the containing vessel. These collisions with the walls create the pressure of a gas.

gas chromatography A technique widely used for the separation and analysis of mixtures. Gas chromatography employs a column packed with either a solid stationary phase (*gas-solid chromatography* or *GSC*) or a solid coated with a nonvolatile liquid (*gas-liquid chromatography* or *GLC*). The whole column is placed in a thermostatically controlled heating jacket. A volatile sample is introduced into the column using a syringe, and an unreactive carrier gas, such as nitrogen, passed through it. The components of the sample will be carried along in this mobile phase. However, some of the components will cling more readily to the stationary phase than others, either because they become attached to the solid surface or because they dissolve in the liquid. The time taken for different components to pass through the column is characteristic and can be used to identify them. The emergent sample is passed through a detector, which registers the presence of the different components in the carrier gas.
Two types of detector are in common use: the katharometer, which measures

changes in thermal conductivity, and the flame-ionization detector, which turns the volatile components into ions and registers the change in electrical conductivity.

gas constant (universal gas constant) Symbol: R The universal constant 8.314 34 J mol^{-1} K^{-1} appearing in the equation of state for an ideal gas.

gaseous diffusion separation A technique for the separation of gases that relies on their slightly different atomic masses. For example, the isotopes of uranium (^{235}U and ^{238}U) can be separated by first preparing uranium hexafluoride from uranium ore. If this is then made to pass through a series of fine pores, the molecules containing the lighter isotope of uranium will pass through more slowly. As the gases pass through successive 'diaphragms', the proportion of lighter molecules increases and a separation of over 99% is attainable. This method has also been applied to the separation of the isotopes of hydrogen.

gas laws Laws relating the temperature, pressure, and volume of a fixed mass of gas. The main gas laws are Boyle's law and Charles' law. The laws are not obeyed exactly by any real gas, but many common gases obey them under certain conditions, particularly at high temperatures and low pressures. A gas that would obey the laws over all pressures and temperatures is a *perfect* or *ideal gas*.
Boyle's and Charles' laws can be combined into an equation of state for ideal gases:

$$pV_m = RT$$

where V_m is the molar volume and R the molar gas constant. For n moles of gas

$$pV = nRT$$

All real gases deviate to some extent from the gas laws, which are applicable only to idealized systems of particles of negligible volume with no intermolecular forces. There are several modified equations of state that give a better description of the behaviour of real gases, the best known being the van der Waals equation.

gas-liquid chromatography *See* gas chromatography.

gasoline *See* petroleum.

gas-solid chromatography *See* gas chromatography.

gauche conformation *See* conformation.

gauss Symbol: G The unit of magnetic flux density in the c.g.s. system. It is equal to 10^{-4} tesla.

Gay-Lussac's law 1. Gases react in volumes that are in simple ratios to each other and to the products if they are gases (all volumes measured at the same temperature and pressure).
2. *See* Charles' law.

gel A lyophilic colloid that is normally stable but may be induced to coagulate partially under certain conditions (e.g. lowering the temperature). This produces a pseudo-solid or easily deformable jelly-like mass, called a gel, in which intertwining particles enclose the whole dispersing medium. Gels may be further subdivided into elastic gels (e.g. gelatin) and rigid gels (e.g. silica gel).

gel filtration *See* molecular sieve.

gem positions Positions in a molecule on the same atom. For example, 1,1-dichloroethane (CH_3CHCl_2) is a gem dihalide.

geometrical isomerism *See* isomerism.

germanium A hard brittle grey metalloid element belonging to group IV of the periodic table. It is found in sulphide ores such as argyrodite ($4Ag_2S.GeS_2$) and in zinc ores and coal. Germanium was extensively used in early semiconductor devices but has now been largely superseded by silicon. It is used as an alloying agent, catalyst, phosphor, and in infrared equipment.

Symbol: Ge; m.p. 937.4°C; b.p. 2830°C; r.d. 5.32; p.n. 32; r.a.m. 72.59.

germanium(IV) oxide (GeO_2) A compound made by strongly heating germanium in air or by hydrolysing germanium(IV) chloride. It occurs in two forms. One is slightly soluble in water forming an acidic solution; the other is insoluble in water. Germanium oxide dissolves in alkalis to form the anion GeO_3^{2-}

German silver See nickel-silver.

Gibbs function (Gibbs free energy) Symbol: G A thermodynamic function defined by

$$G = H - TS$$

where H is the enthalpy, T the thermodynamic temperature, and S the entropy. It is useful for specifying the conditions of chemical equilibrium for reactions for constant temperature and pressure (G is a minimum). See also free energy.

giga- Symbol: G A prefix denoting 10^9. For example, 1 gigahertz (GHz) = 10^9 hertz (Hz).

glacial ethanoic (acetic) acid Pure water-free ethanoic acid.

glass A hard transparent material made by heating calcium oxide (lime), sodium carbonate, and sand (silicon(IV) oxide). This produces a calcium silicate – the normal type of glass, called *soda glass*. Special types of glass can be obtained by incorporating boron oxide in the glass (borosilicate glass, used for laboratory apparatus) or by including metals other than calcium, e.g. lead or barium.
Glass is an amorphous substance, in the sense that there is no long-range ordering of the atoms on a lattice. It can be regarded as a supercooled liquid, which has not crystallized. Solids with similar non-crystalline structures are also called *glasses*.

Glauber's salt See sodium sulphate.

GLC Gas-liquid chromatography. See gas chromatography.

glucose (dextrose, grape sugar, $C_6H_{12}O_6$) A monosaccharide occurring widely in nature as D-glucose. It occurs as glucose units in sucrose, starch, and cellulose. It is important to metabolism as it participates in energy-storage and energy-release systems. See also sugar.

glutamic acid See amino acids.

glutamine See amino acids.

glyceride An ester formed between glycerol (propane-1,2,3-triol) and one or more carboxylic acids. Glycerol has three alcohol groups, and if all three groups have formed esters, the compound is a *triglyceride*. Naturally occurring fats and oils are triglycerides of long-chain carboxylic acids (hence the name 'fatty acid'). The main carboxylic acids occurring in fats and oils are:
oxtadecanoic acid (stearic acid), a saturated acid $CH_3(CH_2)_{16}COOH$.
hexadecanoic acid (palmitic acid), a saturated acid $CH_3(CH_2)_{14}COOH$.
cis-9-octadecenoic acid (oleic acid), an unsaturated acid

$$CH_3(CH_2)_7CH{:}CH(CH_2)_7COOH$$

glycerine (glycerin) See propane-1,2,3-triol.

glycerol See propane-1,2,3-triol.

glycine See amino acids.

glycogen (animal starch) A polysaccharide that is the main carbohydrate store of animals. It is composed of many glucose units linked in a similar way to starch. Glycogen is readily hydrolysed in a stepwise manner to glucose itself. It is stored largely in the liver and in muscle but is found widely distributed.

glycol See diol.

gold A transition metal that occurs native. It is unreactive and is very ductile and malleable. Gold is used in jewellery, often alloyed with copper. Pure gold is 24 carat; 9

carat indicates that 9 parts in 24 consist of gold.

Symbol: Au; m.p. 1064°C; b.p. 2807°C; r.d. 19.3; p.n. 79; r.a.m. 196.97.

gold(III) chloride (gold trichloride, auric chloride, $AuCl_3$) A compound prepared by dissolving gold in aqua regia. The bright yellow crystals (chloroauric acid) produced on evaporation are heated to form dark red crystals of gold(III) chloride. The chloride decomposes easily (at 175°C) to give gold(I) chloride and chlorine; at higher temperatures it decomposes to give gold and chlorine. Gold(III) chloride is used in photography. It exists as a dimer, Au_2Cl_6.

Goldschmidt process A process for extracting certain metals from their oxides by reduction with aluminium.

Graham's law (of diffusion) The principle that gases diffuse at a rate that is inversely proportional to the square root of their density. Light molecules diffuse faster than heavy molecules. The principle is used in the separation of isotopes.

grain A crystal in a metal that has been prevented from attaining its regular geometrical form.

gram Symbol: g A unit of mass defined as 10^{-3} kilogram.

gram-atom See mole.

gram-equivalent The equivalent weight of a substance in grams.

gram-molecule See mole.

granulation A process for enlarging particles to improve the flow properties of solid reactants and products in chemical processes. The larger a particle, and the freer from fine materials in a solid, the more easily will it flow. Dry granulation produces pellets from dry materials, which are crushed into the desired size. Wet granulation involves the addition of a liquid to the material, and the resulting paste is extruded and dried before cutting into the required size.

grape sugar See dextrose.

graphite An allotrope of carbon. Graphite is a good conductor of heat and electricity. The atoms are arranged in layers which cleave easily. Graphite is used as a solid lubricant. See also carbon.

gravimetric analysis A method of quantitative analysis in which the final analytical measurement is made by weighing. There are many variations in the method but in essence they all consist of:
(1) taking an accurately weighed sample into solution,
(2) precipitation as a known compound by a quantitative reaction,
(3) digestion and coagulation procedures,
(4) filtration and washing,
(5) drying and weighing as a pure compound.
Filtration is a key element in the method and a variety of special filter papers and sinter-glass filters are available.

gray Symbol: Gy The SI unit of absorbed energy dose per unit mass resulting from the passage of ionizing radiation through living tissue. One gray is an energy absorption of one joule per kilogram of mass.

Grignard reagent A type of organometallic compound with the general formula RMgX, where R is an alkyl or aryl group and X is a halogen (e.g. CH_3MgCl). They are prepared by reacting the haloalkane or haloaryl compound with magnesium in dry ether:
$$CH_3Cl + Mg \rightarrow CH_3MgCl$$
Grignard reagents probably have the form $R_2Mg.MgCl_2$. They are used extensively in organic chemistry.
With methanal a primary alcohol is produced:
$$RMgX + HCHO \rightarrow RCH_2OH + Mg(OH)X$$
Other aldehydes give secondary alcohols:
$$RMgX + R'CHO \rightarrow RR'CHOH + Mg(OH)X$$

Alcohols and carboxylic acids give hydrocarbons:

$RMgX + R'OH \rightarrow RR' + Mg(OH)X$

Water also gives a hydrocarbon:

$RMgX + H_2O \rightarrow RH + Mg(OH)X$

Solid carbon dioxide in acid solution gives a carboxylic acid:

$RMgX + CO_2 + H_2O \rightarrow RCOOH + Mg(OH)X$

ground state The lowest energy state of an atom, molecule, or other system. *Compare* excited state.

group 1. In the periodic table, a series of chemically similar elements that have similar electronic configurations. A group is thus a column of the periodic table. For example, the alkali metals, all of which have outer s^1 configurations, are called group I.
2. (functional group) In organic chemistry, an arrangement of atoms that bestows a particular type of property on a molecule and enables it to be placed in a particular class, e.g. the aldehyde group −CHO. *See also* functional group.

group 0 elements *See* noble gases.

group I elements *See* alkali metals.

group II elements *See* alkaline-earth metals.

group III elements A group of elements in the periodic table, often divided into the *main group* (sub-group IIA), boron (B), aluminium (Al), gallium (Ga), indium (In), and thallium (T1); and the B sub-group, scandium (Sc), yttrium (Y), lanthanum (La), and actinium (Ac). The B sub-group division is to some extent not helpful as scandium, yttrium, and lanthanum may be conveniently considered in the transition metals. Further, the elements following lanthanum (the lanthanides) and actinium (the actinoids) are separate groupings with their own characteristic properties. The main group III elements all have three outer electrons (s^2p^1) with no partly filled inner shells. The group is the first group of elements in the p-block. As

with all the groups there is an increase in metallic character as the group is descended.

The boron atom is both small and has a high ionization potential, so bonds to boron are largely covalent with only small degrees of polarization. In fact boron is classed as a metalloid. It has volatile hydrides and a weakly acidic oxide. As the group is descended the ionization potentials decrease and the atomic radii increase leading to increasingly polar interactions and the formation of distinct M^{3+} ions. The increase in metallic character is clearly illustrated by the hydroxides: boric acid $B(OH)_3$ is acidic; aluminium and gallium hydroxides are amphoteric dissolving in acids to give Al^{3+} and Ga^{3+} and in base to give aluminates and gallates; the hydroxides of indium and thallium are distinctly basic. As the elements get heavier the bond energies with other elements become generally smaller. The monovalent state (removal of the p electron only) becomes progressively stable. For example gallium(I) is known in the gas phase and in complex systems sometimes known as 'GaCl$_2$'. Monovalent indium halides are known, for example InX, and thallium(I) is distinctly more stable than thallium(III).

group IV elements A group of elements in the periodic table; divided into the *main group* (sub-group IVA) consisting of carbon (C), silicon (Si), germanium (Ge), tin (Sn), and lead (Pb); and subgroup IVB: titanium (Ti), zirconium (Zr), and hafnium (Hf), which are transition metals. The main-group elements all have electronic structures with four outer electrons (s^2p^2) and no partly filled inner shells. The group shows the common trend towards metallic character with the heavier elements, thus carbon is a typical non-metal, silicon and germanium are metalloids, and tin and lead are characteristically metallic. As observed with other groups the first member of the group is quite different from the rest. The carbon atom is smaller and has a higher ionization potential, both favouring predominance of covalency, but additional factors are:

group V elements

(1) The widespread nature of extensive catenation in carbon compounds.
(2) The possibility of strong overlap of p orbitals or *double bonding*.
(3) The inaccessibility of d orbitals in carbon compounds.
Significant differences are:
(1) Both oxides of carbon are gaseous, CO and CO_2, the other elements form solid oxides, e.g. SiO_2, Pb_3O_4.
(2) The heavier elements have no compounds analogous to aromatic compounds.
(3) The heavier elements have no compounds analogous to ethene or ketones.
(4) The heavier elements readily expand their coordination number, e.g. SiF_6^{2-} and $SnCl_6^{2-}$.

group V elements A group of elements in the periodic table consisting of the *main group* (sub-group VA) consisting of nitrogen (N), phosphorus (P), arsenic (As), antimony (Sb), and bismuth (Bi), and the VB sub-group: vanadium (V), niobium (Nb), and tantalum (Ta), which are transition elements.
The main-group elements all have electronic configurations corresponding to outer s^2p^3 with no vacancies in inner levels.
The ionization potentials are high and the lighter members of the group, N and P, are distinctly electronegative and non-metallic in character. Arsenic and antimony are metalloids and bismuth is weakly metallic (Bi_2O_3 dissolves in acids to give $Bi(OH)_3$).
The general trend to lower energies of formation of covalent bonds as elements become heavier leads to the M(III) state becoming more important as the group is descended (note that the M(V) state is not accessible to nitrogen, which has no outer d levels). The observation of reduced stability of the hydrides MH_3 as the group is descended (NH_3 stable, BiH_3 decomposes at room temperature) is similarly accounted for.
Nitrogen is dissimilar from the other members of the group in that it forms stable multiple bonds, it is limited to coordination number 4, e.g. NH_4^+, it is sufficiently elec-

tronegative to form hydrogen bonds and the nitride ion N^{3-}, and its oxides are irregular with the rest of the group. The group displays a remarkable number of allotropes of its members showing the trend from non-metallic forms through to metallic forms. Thus nitrogen has only the diatomic form; phosphorus has a highly reactive form P_1 (brown), and a tetrahedral form P_4 (white), forms based on broken tetrahedra P_n (red and violet), and a hexagonal layer-type lattice (black). Arsenic and antimony have the As_4 and Sb_4 forms, which are less stable than the layer type in this case; bismuth has the layer lattice form only.

group VI elements A group of elements of the periodic table sometimes called the *chalcogens*. They are oxygen (O), sulphur (S), selenium (Se), tellurium (Te), and polonium (Po). The electronic configurations are all s^2p^4 with no vacant inner orbitals. These configurations are all just two electrons short of an inert gas structure, consequently they have high electron affinities and are almost entirely non-metallic in character.
The group shows the normal property of a trend towards metallic character as it is descended. Selenium, tellurium and polonium have 'metallic' allotropes and polonium has generally metalloid-type properties where they have been studied (Po is very rare). Excluding oxygen, which forms some ionic metal-oxides, the compounds of group VI elements are practically all covalent in their bonding. All the elements combine with a large number of other elements, both metallic and non-metallic, but in contrast to compounds of the halogens they are more generally insoluble in water, and even where soluble they do not ionize readily.
Oxygen behaves like the first members of other groups in having great differences from the rest of the group. For example, oxygen forms *hydrogen bonds* whereas sulphur and the others do not; oxygen cannot expand its outer shell and positive oxidation states are uncommon for oxygen (OF_2, O_2F_2, $O_2^+PtF_6^-$). Oxygen is paramagnetic. Excluding oxygen, group

VI shows similar trends with increasing size to those in group V. For example decreasing stability of H_2X, greater tendency to form complexes such as $SeBr_6^{2-}$, decreasing stability of high formal positive oxidation states, and marginal metallic properties (e.g. Po forms a hydroxide).

The variation of oxidation states, the variety of elements in combination, and in the case of sulphur the existence of many catenated species all combine to make the chemistry of group VI rather complex.

GSC Gas-solid chromatography. *See* gas chromatography.

gun cotton *See* cellulose nitrate.

gunmetal *See* bronze.

gunpowder A powdered mixture of sulphur, charcoal, and potassium nitrate, used as an explosive.

gypsum *See* calcium sulphate.

H

Haber process An important industrial process for the manufacture of ammonia, which is used for fertilizers and for making nitric acid. The reaction is the equilibrium:

$$N_2 + 3H_2 = 3NH_3$$

The nitrogen used is obtained by fractionation of liquid air and the hydrogen by the oxidation of hydrocarbons (from natural gas). The nitrogen and hydrogen are purified and mixed in the correct proportions. The equilibrium amount of ammonia is favoured by low temperatures, but in practice the reaction would never reach equilibrium at normal temperatures. An optimum temperature of about 450°C is therefore used. High pressure also favours the reaction and a pressure of about 250 atmospheres is used. The catalyst is iron with small amounts of potassium and aluminium oxides present. The yield is about 15%. Ammonia is condensed out of the gas mixture at −50°C.

habit *See* crystal habit.

haematite A mineral form of iron(III) oxide. It is the principal ore of iron.

haemoglobin The pigment of the red blood cells that is responsible for the transport of oxygen from the lungs to the tissues. It consists of a basic protein, globin, linked with four haem groups. Haem is a complex compound containing an iron atom. The most important property of haemoglobin is its ability to combine reversibly with one molecule of oxygen per iron atom to form *oxyhaemoglobin*, which has a bright red colour. The iron is present in the divalent state (iron(II)) and this remains unchanged with the binding of oxygen. There are variations in the polypeptide chains, giving rise to different types of haemoglobins in different species. The binding of oxygen depends on the oxygen partial pressure; high pressure favours formation of oxyhaemoglobin and low pressure favours release of oxygen.

hafnium A transition metal found in zirconium ores. Hafnium is difficult to work and can burn in air. It is used in nuclear reactors.
Symbol: Hf; m.p. 2220°C; b.p. 5230°C; r.d. 13.3; p.n. 72; r.a.m. 178.49.

hahmium A transactimide element formed artificially.
Symbol: Ha; p.n. 105.

half cell An electrode in contact with a solution of ions. In general there will be an e.m.f. set up between electrode and solution by transfer of electrons to or from the electrode. The e.m.f. of a half cell cannot be measured directly since setting up a circuit results in the formation of another half cell. *See* electrode potential.

half-life (half-life period) Symbol: $t_{\frac{1}{2}}$ The time taken for half the nuclei of a sample of a radioactive nuclide to decay. The half-life of a nuclide is a measure of its stability. (Stable nuclei can be thought of as having infinitely long half-lives.) If N_0 is the original number of nuclei, the number remaining at the end of one half-life is $N_0/2$, at the end of two half-lives is $N_0/4$, etc.

halide A compound containing a halogen. The inorganic halides of electropositive elements contain ions of the type Na^+Cl^- (sodium chloride) or K^+Br^- (potassium bromide). Transition metal halides often have some covalent bonding. Non-metal halides are covalent compounds, which are usually volatile. Examples are carbon tetrachloride (CCl_4) and silicon tetrachloride ($SiCl_4$). Halides are named as bromides, chlorides, fluorides, or iodides.

haloalkane (alkyl halide) A type of organic compound in which one or more hydrogen atoms of an alkane have been replaced by halogen atoms. Haloalkanes can be made by direct reaction of the alkane with a halogen. Other methods are:

(1) Reaction of an alcohol with the halogen acid (e.g. from $NaBr + H_2SO_4$) or with phosphorus halides (red phosphorus and iodine can be used):

$$ROH + HBr \rightarrow RBr + H_2O$$
$$ROH + PCl_5 \rightarrow RCl + POCl_3 + HCl$$

(2) Addition of an acid to an alkene:

$$RCH{:}CH_2 + HBr \rightarrow RCH_2CH_2Br$$

The haloalkanes are much more reactive than the alkanes, and are useful starting compounds for preparing a wide range of organic chemicals. In particular, they undergo nucleophilic substitutions in which the halogen atom is replaced by some other group (iodine compounds are the most reactive). Some reactions of haloalkanes are:

(1) Refluxing with aqueous potassium hydroxide to give an alcohol:

$$RI + OH^- \rightarrow ROH + I^-$$

(2) Refluxing with potassium cyanide in alcoholic solution to give a nitrile:

$$RI + CN^- \rightarrow RCN + I^-$$

(3) Refluxing with an alkoxide to give an ether:

$$RI + {^-}OR' \rightarrow ROR' + I^-$$

(4) Reaction with alcoholic ammonia solution (100°C in a sealed tube):

$$RI + NH_3 \rightarrow RNH_2 + HI$$

(5) Boiling with alcoholic potassium hydroxide, to eliminate an acid and produce an alkene:

$$RCH_2CH_2I + KOH \rightarrow KI + H_2O + RCH{:}CH_2$$

See also Fittig reaction, Grignard reagent, Wurtz reaction.

haloform Any of the four compounds CHX_3, where X is a halogen atom (F, fluoroform; Cl, chloroform; Br, bromoform; I, iodoform).

halogenation A reaction in which a halogen atom is introduced into a molecule. Halogenations are specified as *chlorinations*, *brominations*, *fluorinations*, etc., according to the element involved. There are several methods.

(1) Direct reaction with the element using high temperature or ultraviolet radiation:

$$CH_4 + Cl_2 \rightarrow CH_3Cl + HCl$$

(2) Addition to a double bond:

$$H_2C{:}CH_2 + HCl \rightarrow C_2H_5Cl$$

(3) Reaction of a hydroxyl group with a halogenating agent, such as PCl_3:

$$C_2H_5OH \rightarrow C_2H_5Cl + OH^-$$

(4) In aromatic compounds direct substitution can occur using aluminium chloride as a catalyst:

$$2C_6H_6 + Cl_2 \rightarrow 2C_6H_5Cl$$

(5) Alternatively in aromatic compounds, the chlorine can be introduced by reacting the diazonium ion with copper(I) chloride:

$$C_6H_5N_2{^+} + Cl^- \rightarrow C_6H_5Cl + N_2$$

halogens A group of elements (group VII of the periodic table) consisting of fluorine, chlorine, bromine, iodine, and the short-lived element astatine. The halogens all have outer valence shells that are one electron short of a noble-gas configuration. Because of this, the halogens are characterized by high electron affinities and high electronegativities, fluorine being the most electronegative element known. The high electronegativities of the halogens favour the formation of both uninegative ions, X, particularly combined with electropositive elements (e.g. NaCl, KBr, CsI, CaF_2), and the single covalent bond −X with elements of moderate to high electronegativity (e.g. HCl, SiF_4, SF_4, $BrCH_3$, Cl_2O). The halogens all form diatomic molecules, X_2, which are characterized by their high reactivities with a wide range of other elements and compounds. As a group, the elements increase in both size and polarizability as the proton number increases and there is an attendant decrease in electronegativity and reactivity. This means that there is no chemistry of positive species apart from a few cationic iodine compounds. The heavier halogens Cl, Br, and I, all form oxospecies with formal positive oxidation numbers $+1$ and $+5$, chlorine and iodine also forming $+3$ and $+7$ species (e.g. HOCl, $HBrO_3$, I_2O_5, HIO_4).

The elements decrease in oxidizing power in the order $F_2 > Cl_2 > Br_2 > I_2$ and the ions X^- may be arranged in order of increasing reducing power $F^- < Cl^- < Br^- < I^-$. Thus any halogen will displace the elements below it from their salts, for example

$$Cl_2 + 2Br^- \rightarrow Br_2 + 2Cl^-$$

A wide range of organic halides is formed in which the C−F bond is characteristically resistant to chemical attack; the C−Cl bond is also fairly stable, particularly in aryl compounds but the alkyl halogen compounds become increasingly susceptible to nucleophilic attack and generally more reactive.

hammer mills Devices used in the chemical industry for crushing and grinding solid materials at high speeds to a specified size. The impact between the particles, grinding plates, and grinding hammers pulverizes the particles. Hammer mills can be used for a greater variety of soft material than other types of grinding equipment. *Compare* ball mill.

hardening (of oils) A process for converting liquid vegetable oils into solid or semi-solid fats. Oils are glycerides of mainly unsaturated carboxylic acids. Hydrogenation of the double carbon-carbon bonds using a nickel catalyst produces saturated glycerides, which have higher melting points. Hydrogenation of vegetable oils is used in the manufacture of margarine.

hardness (of water) Hard water is water that will not readily form a lather with soap owing to the presence of dissolved calcium, iron, and magnesium compounds. Such compounds can react with soap to produce insoluble salts, which collect as solid scum. The effectiveness of the cleansing solution is thus reduced. Hardness of water is of two types, temporary hardness and permanent hardness. Only temporary hardness can be removed by boiling the water.
Hardness of water is usually expressed by assuming that all the hardness is due to dissolved calcium carbonate, which is present as ions. It can be estimated by titration with a standard soap solution or with edta. *See also* soap, water softening.

hard vacuum *See* vacuum.

hard water *See* hardness.

heat Energy transferred as a result of a temperature difference. The term is often loosely used to mean internal energy (i.e. the total kinetic and potential energy of the particles). It is common in chemistry to define such quantities as *heat of combustion*, *heat of neutralization*, etc. These are in fact molar enthalpies for the change, given the symbol $\Delta H_M^{\ominus}$. The superscript symbol denotes standard conditions, whilst the subscript M indicates that the enthalpy change is for one mole. The unit is usually the kilojoule per mole (kJ mol^{-1}). By convention, ΔH is negative for an exothermic reaction. Molar enthalpy changes stated for chemical reactions are changes for standard conditions, which are 298 K (25°C) and 101 325 Pa (1 atmosphere). Thus, the standard molar enthalpy of reaction is the enthalpy change for reaction of substances under these conditions producing reactants under the same conditions. The substances involved must be in their normal equilibrium physical states under these conditions (e.g. carbon as graphite, water as the liquid, etc.). Note that the measured enthalpy change will not usually be the standard change. In addition, it is common to specify the entity involved. For instance $\Delta H_f^{\ominus}(H_2O)$ is the standard molar enthalpy of formation for one mole of H_2O species.

heat engine (thermodynamic engine) A device for converting heat energy into work. Heat engines operate by transferring energy from a high temperature source to a low temperature sink. The theoretical operation of heat engines is useful in the theory of thermodynamics. *See* Carnot cycle.

heat exchangers Devices that enable the heat from a hot fluid to be transferred to a cool fluid without allowing them to come into contact. The normal arrangement is for one of the fluids to flow in a coiled tube through a jacket containing the second fluid. Both the cooling and heating effect may be of benefit in conserving the energy used in a chemical plant and in controlling the process.

heat of atomization The energy required in dissociating one mole of a substance into atoms. *See* heat.

heat of combustion The energy liberated when one mole of a substance burns in excess oxygen. *See* heat.

heat of crystallization The energy liberated when one mole of a substance crystallizes from a saturated solution of this substance.

heat of dissociation The energy required to dissociate one mole of a substance into its constituent elements.

heat of formation The energy change when one mole of a substance is formed from its elements. *See* heat.

heat of neutralization The energy liberated when one mole of an acid or base is neutralized.

heat of reaction The energy change when molar amounts of given substances react completely. *See* heat.

heat of solution The energy change when one mole of a substance is dissolved in a given solvent to infinite dilution (in practice, to form a dilute solution).

heavy hydrogen *Se* deuterium.

heavy-metal pollution *See* pollution.

heavy water Deuterium oxide, D_2O.

hecto- Symbol: h A prefix denoting 10^2. For example, 1 hectometre (hm) $= 10^2$ metres (m).

Heisenberg's uncertainty principle The impossibility of making simultaneous measurements of both the position and the momentum of a subatomic particle (e.g. an electron) with unlimited accuracy. The uncertainty arises because, in order to detect the particle, radiation has to be 'bounced' off it, and this process itself disrupts the particle's position.

Heisenberg's uncertainty principle is not a consequence of 'experimental error'. It represents a fundamental limit to 'objective' scientific observation, and arises from the wave-particle duality of particles and radiation. In one direction, the uncertainty in position Δx and momentum Δp are related by $\Delta x \Delta p \sim h/4\pi$, where h is the Planck constant.

helium A colourless monatomic gas; the first member of the noble gases (group 0 of the periodic table). Helium has the electronic configuration $1s^2$ and consists of a nucleus of two protons and two neutrons (equivalent to an α-particle) with two extranuclear electrons. It has an extremely high ionization potential and is completely resistant to chemical attack of any sort. The gas accounts for only $5.2 \times 10^{-4}\%$ of the atmosphere; up to 7% occurs in some natural gas deposits. Helium is the second most abundant element in the universe, the primary process on the Sun being nuclear fusion of hydrogen to give helium. Helium is recovered commercially from natural gas in both the USA and the USSR and it also forms part of ammonia plant tail gas if natural gas is used as a feedstock. Its applications are in fields in which inertness is required and where the cheaper alternatives, such as nitrogen, are too reactive; for example, high temperature metallurgy, powder technology, and as a coolant in nuclear reactors. Helium is also favoured over nitrogen for diluting oxygen for deep-sea diving (lower solubility in blood) and as a pressurizer for liquefied gas fuels in rockets (total inertness). It is also used as an ideal gas for balloons (no fire risk) and for low-temperature physics research.

Helium is unusual in that it is the only known substance for which there is no triple point (i.e., no combination of pressure and temperature at which all three phases can co-exist). This is because the interatomic forces, which normally participate in the formation of solids, are so weak that they are of the same order as the zero-point energy. At 2.2 K helium undergoes a transition from liquid helium I to liquid helium II, the latter being a true liquid but exhibiting superconductivity and an immeasurably low viscosity. The low viscosity allows the liquid to spread in layers a few atoms thick, described by some as 'flowing uphill'.

Helium also has an isotope. ^{3}He is formed in nuclear reactions and by decay of tritium.

Hell-Volard-Zelinsky reaction A method for the preparation of halogenated carboxylic acids using free halogen in the presence of a phosphorus halide. The halogenation occurs at the carbon atom adjacent to the –COOH group. With Br_2 and PBr_3:

$$RCH_2COOH \rightarrow RCHBrCOOH \rightarrow RCBr_2COOH$$

Helmholtz function (Helmholtz free energy) Symbol: F A thermodynamic function defined by

$$F = U - TS$$

where U is the internal energy, T the thermodynamic temperature, and S the entropy. It is a measure of the ability of a system to do useful work in an isothermal process. *See also* free energy.

hemiacetal *See* acetal.

hemihydrate A crystalline compound with one molecule of water of crystallization per two molecules of compound (e.g. $2CaSO_4.H_2O$).

hemiketal *See* ketal.

henry Symbol: H The SI unit of inductance, equal to the inductance of a closed circuit that has a magnetic flux of one weber per ampere of current in the circuit. 1 H $= 1$ Wb A^{-1}.

Henry's law The concentration (C) of a gas in solution is proportional to the partial pressure (p) of that gas in equilibrium with the solution, i.e. $p = kC$, where k is a proportionality constant. The relationship is similar in form to that for Raoult's law, which deals with ideal solutions.

A consequence of Henry's law is that the 'volume solubility' of a gas is independent of pressure.

heptahydrate A crystalline hydrated compound containing one molecule of compound per seven molecules of water of crystallization.

heptavalent (septavalent) Having a valence of seven.

heptane (C_7H_{16}) A colourless liquid alkane obtained from petroleum refining. It is used as a solvent.

hertz Symbol: Hz The SI unit of frequency, defined as one cycle per second (s^{-1}). Note that the hertz is used for regularly repeated processes, such as vibration or wave motion.

Hesse's law A derivative of the first law of thermodynamics. It states that the total heat change for a given chemical reaction is independent of the route taken.

hetero atom *See* heterocyclic compound.

heterocyclic compound A compound that has a ring containing more than one type of atom. Commonly, heterocyclic compounds are organic compounds with at least one atom in the ring that is not a carbon atom. Pyridine and glucose are examples. The non-carbon atom is called a *hetero atom*. *Compare* homocyclic compound.

heterogeneous Relating to more than one phase. A heterogeneous mixture, for instance, contains two or more distinct phases.

heterolytic fission (heterolysis) The breaking of a covalent bond so that both electrons of the bond remain with one fragment. A positive ion and a negative ion are produced:

$$RX \rightarrow R^+ + X^-$$

Compare homolytic fission.

hexadecanoate (palmitate) A salt or ester of hexadecanoic acid.

hexadecanoic acid (palmitic acid) A crystalline carboxylic acid:
$$CH_3(CH_2)_{14}COOH$$
It is present as glycerides in fats and oils. *See* glyceride.

hexagonal close-packed crystal A crystal structure in which layers of close-packed atoms are stacked in an ABABAB arrangement. The second layer B fits into the holes of the first layer A, with the third layer over the first. The coordination number is 12. Zinc and magnesium have hexagonal close-packed structures.
The unit cell is based on a hexagon with atoms occupying corner and mid positions. *See* face-centred cubic crystal.

hexagonal crystal *See* crystal system.

hexane (C_6H_{14}) A liquid alkane obtained from the light fraction of crude oil. The principal use of hexane is as a solvent.

hexanoic acid (caproic acid, $CH_3(CH_2)_4COOH$) An oily carboxylic acid found (as glycerides) in cow's milk and some vegetable oils.

hexose A sugar that has six carbon atoms in its molecules. *See* sugar.

hexyl group The group $C_5H_{11}CH_2-$, having a straight chain of carbon atoms.

histidine *See* amino acids.

Hofmann degradation A method of preparing primary amines from acid amides. The amide is refluxed with aqueous sodium hydroxide and bromine:
$$RCONH_2 + NaOH + Br_2 \rightarrow$$
$$RCONHBr + NaBr + H_2O$$
$$RCONHBr + OH^- \rightarrow RCON^-Br + H_2O$$
$$RCON^-Br \rightarrow R-N = C = O + Br^-$$
$$RNCO + 2OH^- \rightarrow RNH_2 + CO_2{}^{2-}$$
The reaction is a 'degradation' in that a carbon atom is removed from the amide chain.

Hofmann's method A rather outmoded method for determining the vapour density of volatile liquids. A known weight of sample is introduced into a mercury barometer tube, which is surrounded by a heating jacket. The volume of vapour can thus be read off directly, the temperature is known, and the pressure is obtained by taking the atmospheric pressure minus the mercury height in the barometer (with corrections for the density of mercury at higher temperatures). The method's only advantage is that it may be used for samples that decompose at their normal boiling point. *See also* Dumas' method, Victor Meyer's method.

holmium A soft malleable silvery element of the lanthanoid series of metals. It occurs in association with other lanthanoids. It has few applications.
Symbol: Ho; m.p. 1474°C; b.p. 2695°C; r.d. 8.80; p.n. 67; r.a.m. 164.93.

homocyclic compound A compound containing a ring made up of the same atoms. Benzene is an example of a homocyclic compound. *Compare* heterocyclic compound.

homogeneous Relating to a single phase. A homogeneous mixture, for instance, consists of only one phase.

homologous series A group of organic compounds possessing the same functional group and having a regular structural pattern so that each member of the series differs from the next one by a fixed number of atoms. The members of a homologous series can be represented by a general formula. For example, the homologous series of alkane alcohols CH_3OH, C_2H_5OH, C_3H_7OH, ..., has a general formula $C_nH_{2n+1}OH$. Each member differs by CH_2 from the next. Any two successive members of a series are called *homologues*.

homologues *See* homologous series.

homolytic fission (homolysis) The breaking of a covalent bond so that one electron from the bond is left on each fragment. Two free radicals result:

$$RR' \rightarrow R\cdot + R'\cdot$$

Compare heterolytic fission.

HPLC High-performance liquid chromatography; a sensitive analytical technique, similar to gas-liquid chromatography but using a liquid carrier. The carrier is specifically choosen for the particular substance to be detected.

Hückel rule *See* aromatic compound.

Hund's rule A rule that states that the electronic configuration in degenerate orbitals will have the minimum number of paired electrons.

hybrid orbital *See* orbital.

hydrate A compound coordinated with water molecules. When water is bound up in compound it is known as the *water of crystallization.*

hydration The solvation of such species as ions in water.

hydrazine (N_2H_4) A colourless liquid that can be prepared by the oxidation of ammonia with sodium chlorate(I) or by the gas phase reaction of ammonia with chlorine. Hydrazine is a weak base, forming salts (e.g. $N_2H_4.HCl$) with strong acids. It is a powerful reducing agent, reducing salts of the noble metals to the metal. Anhydrous hydrazine ignites spontaneously in oxygen and reacts violently with oxidizing agents. The aqueous solution, *hydrazine hydrate*, has been used as a fuel for jet engines and for rockets.

hydrazoic acid (hydrogen azide, azoimide, HN_3) A colourless liquid with a nauseating smell. It is highly poisonous and explodes in the presence of oxygen and oxidizing agents. It can be made by distilling a mixture of sodium azide (NaN_3) and a dilute acid. It is usually used as an aqueous solution. The salts of hydrazoic acid (*azides*), especially lead azide ($Pb(N_3)_2$), are used in detonators because of their ability to explode when given a mechanical shock.

hydrazone A type of organic compound containing the $C:NNH_2$ group, formed by the reaction between an aldehyde or ketone and hydrazine (N_2H_4). Derivatives of hydrazine are often used to produce crystalline products, which have sharp melting points that can be used to characterize the original aldehyde or ketone. Phenylhydrazine ($C_6H_5NH.NH_2$), for instance, produces *phenylhydrazones.*

hydride A compound of hydrogen. Ionic hydrides are formed with highly electropositive elements and contain the H^- ion (hydride ion). Non-metals form covalent hydrides, as in methane (CH_4) or silane (SiH_4). The boron hydrides are electron-deficient covalent compounds. Many transition metals absorb hydrogen to form interstitial hydrides.

hydrobromic acid (HBr) A colourless liquid produced by adding hydrogen bromide to water. It shows the typical properties of a strong acid and it is a strong reducing agent. A convenient way of producing hydrobromic acid is to bubble hydrogen sulphide through bromine water. Although it is not as strong as hydrochloric acid it dissociates extensively in water and is a good proton donor.

hydrocarbon Any compound containing only the elements carbon and hydrogen. Examples are the alkanes, alkenes, alkynes, and aromatics such as benzene and naphthalene.

hydrochloric acid (HCl) A colourless fuming liquid made by adding hydrogen chloride to water:

$$HCl(g) + H_2O(l) \rightarrow H_3O^+(aq) + Cl^-(aq)$$

Dissociation into ions is extensive and hydrochloric acid shows the typical properties of a strong acid. It reacts with carbonates to give carbon dioxide and yields hydrogen when reacted with all but the most unreactive metals. Hydrochloric acid is used in the manufacture of dyes, drugs, and photographic materials. It is also used to pickle metals, i.e. clean the surface prior to electroplating. Hydrochloric acid donates protons with ease and is the strongest of the hydrohalic acids. The concentrated acid is oxidized to chlorine by such agents as potassium permanganate and manganese(IV) oxide.

hydrofluoric acid (HF) A colourless liquid produced by dissolving hydrogen fluoride in water. It is a weak acid, but will dissolve most silicates and hence can be used to etch glass. As the interatomic distance in HF is relatively small, the H–F bond energy is very high, therefore — unable to donate protons easily — hydrofluoric acid is a weak acid. *See* bond energy.

hydrogen A colourless gaseous element. Hydrogen has some similarities to both the alkali metals (group I) and the halogens (group VII), but is not normally classified in any particular group of the periodic table. It is the most abundant element in the Universe and the ninth most abundant element in the Earth's crust and atmosphere (by mass). It occurs principally in the form of water and petroleum products; traces of molecular hydrogen are found in some natural gases and in the upper atmosphere.

The gas may be prepared in the laboratory by the reaction of dilute hydrochloric acid with a metal that lies above hydrogen in the electrochemical series, magnesium and zinc being commonly used.

$$Zn + 2HCl \rightarrow H_2 + ZnCl_2$$

Reactions of the amphoteric metals zinc and aluminium with dilute aqueous alkali, to form zincates or aluminates, are also a convenient source of hydrogen. Electrolysis of dilute mineral acids may be used to obtain hydrogen but care must be taken to avoid mixing with the oxygen released at the anode as this leads to explosive mixtures. Industrially, hydrogen is obtained as a by-product of the electrolytic cells used in the production of sodium hydroxide (reaction of the Na/Hg amalgam with water), or by the water-gas route in which steam is decomposed by hot coke.

The main use of hydrogen is as a chemical feedstock for the manufacture of ammonia and of a range of organic compounds. Small-scale uses include reducing atmospheres for metallurgy, hardening edible oils, and pharmaceutical manufacture. Hydrogen also has a potential future use as a fuel.

Hydrogen occupies a unique position among the elements as hydrogen atoms are the simplest of all atoms. The hydrogen atom consists of a proton (positive charge) with one extranuclear electron $(1s^1)$. The chemistry of hydrogen depends on one of three processes:

(1) Loss of the electron to form H^+.
(2) Gain of an electron to form H^-.
(3) Sharing of electrons by covalent bond formation as in H_2 or HCl.

The hydrogen atom with the 1s electron removed would have an extremely small ionic radius and the positive hydrogen ion occurs only in association with other species as in H^+NH_3 or H^+FH. The ion commonly written H^+ in solution is in fact the solvated proton (hydroxonium or hydronium) H_3O^+, which is formed by ionization of acids:

$$H_2O + HCl \rightarrow H_3O^+ + Cl^-$$

It is believed that the lifetime of any one H_3O^+ ion is extremely short as the protons appear to undergo very rapid exchange between water molecules.

Hydrogen also forms a number of compounds in which it is regarded as gaining an electron and becoming H^- (the hydride ion). It can form ionic hydrides only with the most electropositive elements (groups IA and IIA). The ionic nature of these compounds is indicated by the fact that the melts are good conductors and that electrolysis liberates hydrogen at the anode. These hydrides are prepared by heating the element in a stream of hydrogen:

$$H_2 + 2M \rightarrow 2MH$$

Examples are CH_4, NH_3, and HCl. These compounds are generally low-boiling. General methods of preparation are: (1) The hydrolysis of the appropriate '-ide' compound; e.g. silicides give silane, nitrides give ammonia, sulphides give H_2S. (2) Reduction of a chloride; for example:

$$SiCl_4 + LiAlH_4 \rightarrow SiH_4 + LiCl + AlCl_3$$

A third class of hydride is that of metallic hydrides formed between hydrogen and many transition metals. Palladium in particular is renowned for its ability to absorb hydrogen as PdH_x, where x takes values up to 1.8. Titanium and zirconium behave similarly but the exact nature of the compounds formed and of the bond type remains uncertain. There are changes in the magnetic properties of the metal (indicating some electron interaction) and in the lattice dimensions but discrete phases of the type MH or MH_2 are not isolated. The class is sometimes referred to as the *interstitial hydrides*.

Atomic nuclei possess the property of 'spin' and for diatomic molecules there exists the possibility of having the spins of adjacent nuclei aligned (*ortho*) or opposed (*para*). Because of the small mass of hydrogen, these forms are more important in hydrogen molecules than in other diatomic molecules. The two forms are in equilibrium with parahydrogen dominant at low temperatures, rising to 75% orthohydrogen at room temperatures. Although chemically identical the melting point and boiling point of the para form are both about $0.1°$ lower than the 3:1 equilibrium mixture.

Natural hydrogen in molecular or combined forms contains about one part in 2000 of deuterium, D; an isotope of hydrogen that contains one proton and one neutron in its nucleus. Although the effect of isotopes on chemical properties is normally small, in the case of hydrogen the difference in mass number leads to a lowering of some reaction rates known as the 'deuterium isotope effect'. Hydrogen also exhibits two less common forms of bonding. Boron hydrides form a wide variety of compounds in which the hydrogen acts as a bridging species involving 'three-centre two-electron' bonds. Such species are said to be 'electron deficient' as they do not have sufficient electrons for conventional two-electron covalent bonds. The second, less common, form is that of the coordinated hydrides in which the H^- ion acts as a ligand bound to a transition metal atom. Symbol: H; m.p. $-259°C$; b.p. $-252°C$; d. 0.0899 kg m^{-3}; p.n. 1; r.a.m. 1.0079.

hydrogenation The reaction of a compound with hydrogen. An example is the hydrogenation of nitrogen to form ammonia in the Haber process. In organic chemistry, hydrogenation refers to the addition of hydrogen to multiple bonds, usually with the aid of a catalyst. Unsaturated natural liquid vegetable oils can be hydrogenated to form saturated semisolid fats — a reaction used in making types of margarine.

hydrogen azide *See* hydrazoic acid.

hydrogen bond (H-bonding) An intermolecular bond between molecules in which hydrogen is bound to a strongly electronegative element. Bond polarization by the electronegative element X leads to a positive charge on hydrogen $X^{\delta-} \rightarrow H^{\delta+}$; this hydrogen can then interact directly with electronegative elements of adjacent molecules. The H-bond is represented as a dotted line:
$$X^{\delta-} - H^{\delta+} .... X^{\delta-} - H^{\delta+} ...$$
The length of a hydrogen bond is characteristically $0.15-0.2$ nm.
Hydrogen bonding may lead to the formation of dimers (for example, in carboxylic acids) and is used to explain the anomalously high boiling points of H_2O and HF.

hydrogen bromide (HBr) A colourless sharp-smelling gas that is very soluble in water. It is produced by direct combination of hydrogen and bromine in the presence of a platinum catalyst or by the reaction of phosphorus tribromide with water. It dissolves in water to give hydrobromic acid. Hydrogen bromide is rather inactive chemically. It will not conduct electricity in the molten state, indicating that it is a molecular compound.

hydrogencarbonate (bicarbonate) A salt containing the ion $^-HCO_3$.

hydrogen chloride (HCl) A colourless gas that has a strong irritating odour and fumes strongly in moist air. It is prepared by the action of concentrated sulphuric acid on sodium chloride. The gas is made industrially by burning a stream of hydrogen in chlorine. It is not particularly reactive but will form dense white clouds of ammonium chloride when mixed with ammonia. It is very soluble in water and ionizes almost completely to give hydrochloric acid. It will also dissolve in many non-aqueous solvents, including toluene, but will not ionize and the resultant solution shows no acidic properties. Hydrogen chloride is used in the manufacture of organic chlorine compounds, such as polyvinyl chloride (PVC).

Unlike the other hydrogen halides hydrogen chloride will not dissociate on heating, indicating a strong H−Cl bond. *See* hydrochloric acid.

hydrogen electrode A type of half cell based on hydrogen, and assigned zero electrode potential, so that other elements may be compared with it. It is also called the standard hydrogen half cell. The hydrogen is bubbled over a platinum electrode, coated in 'platinum black', in a 1 M acid solution. Hydrogen is adsorbed on the platinum black, which has a high surface area, enabling the equilibrium

$$H(g) = H^+(aq) + e^-$$

to be set up. The platinum is inert and has no tendency to form platinum ions in solution. *See also* electrode potential.

hydrogen fluoride (HF) A colourless liquid produced by the reaction of concentrated sulphuric acid on calcium fluoride:

$$CaF_2(s) + H_2SO_4(aq) \rightarrow CaSO_4(aq) + 2HF(l)$$

It produces toxic corrosive fumes and dissolves readily in water to give hydrofluoric acid. Hydrogen fluoride is atypical of the hydrogen halides as the individual H−F units are associated into much larger units, forming zigzag chains and rings. This is caused by hydrogen bonds that form between the hydrogen and the highly electronegative fluoride ions. Hydrogen fluoride is used extensively as a catalyst in the petroleum industry. *See* hydrofluoric acid.

hydrogen peroxide (H_2O_2) A colourless syrupy liquid, usually used in solution in water. Although it is stable when pure, on contact with bases such as manganese(IV) oxide it gives off oxygen, the manganese(IV) oxide acting as a catalyst:

$$2H_2O_2 \rightarrow 2H_2O + O_2$$

Hydrogen peroxide can act as an oxidizing agent, converting iron(II) ions to iron(III) ions, or as a reducing agent with potassium manganate(VII). It is used as a bleach and in rocket fuel. The strength of solutions is usually given as *volume strength* — the volume of oxygen (dm^3) at STP given by decomposition of 1 dm^3 of the solution.

hydrogen sulphide (sulphuretted hydrogen, H_2S) A colourless very poisonous gas with an odour of bad eggs. Hydrogen sulphide is prepared by reacting hydrochloric acid with iron(II) sulphide. It is tested for by mixing with lead nitrate, with which it gives a black precipitate. Its aqueous solution is weakly acidic. Hydrogen sulphide reduces iron(III) chloride to iron(II) chloride, forming hydrochloric acid and a yellow precipitate of sulphur. Hydrogen sulphide precipitates insoluble sulphides, and is used in qualitative analysis. It burns with a blue flame in oxygen to form sulphur(IV) oxide and water. Natural gas contains some hydrogen sulphide, which is removed before supply to the consumer.

hydrolysis A reaction between a compound and water. Some examples are:
Salts of weak acids
$$Na_2CO_3 + 2H_2O \rightarrow 2NaOH + H_2CO_3$$
Esters
$$CH_3COOC_2H_5 + H_2O = CH_3COOH + C_2H_5OH$$
Certain inorganic halides
$$SiCl_4 + 4H_2O \rightarrow Si(OH)_4 + 4HCl$$

hydrophilic Water attracting. *See* lyophilic.

hydrophobic Water repelling. *See* lyophobic.

hydroxide A compound containing the ion OH^- or the group $-OH$.

2-hydroxypropanoic acid (lactic acid) A colourless liquid carboxylic acid:
$$CH_3CH(OH)COOH$$

hygroscopic

See optical activity.

hygroscopic Describing a substance that absorbs moisture from the atmosphere. *See also* deliquescent.

hyperfine structure *See* fine structure.

hypobromous acid *See* bromic(I) acid.

hypochlorous acid *See* chloric(I) acid.

hypophosphorous acid *See* phosphinic acid.

I

ideal gas See gas laws, kinetic theory.

imine A compound containing the group
−NH−, where the nitrogen atom is not
joined to carbonyl groups or to other hydrogen atoms. The group is the *imino
group*.

imino group See imine.

indicator A compound that reversibly
changes colour depending on the pH of
the solution in which it is dissolved. The
visual observation of this change is therefore a guide to the pH of the solution and
it follows that careful choice of indicators
permits a wide range of end points to be
detected in acid−base titrations.
Redox titrations require either specific indicators, which detect one of the components of the reaction (e.g. starch for
iodine, potassium thiocyanate for Fe^{3+}) or
true redox indicators in which the transition potential of the indicator between oxidized and reduced forms is important. The
transition potential of a redox indicator is
analogous to the transition pH in acid−
base systems.
Complexometric titrations require indicators that complex with metal ions and
change colour between the free state and
the complex state. See also absorption indicator.

indium A soft silvery metallic element
belonging to group III of the periodic table. It is found in minute quantities, primarily in zinc ores and is used in alloys, in
several electronic devices, and in electroplating.
Symbol: In; m.p. 156.61°C; b.p. 2080°C;
r.d. 7.31; p.n. 49; r.a.m. 114.82.

inductive effect The effect in which
substituent atoms or groups in an organic
compound can attract (−I) or push away
electrons (+I), forming polar bonds.

inert gases See noble gases.

infrared (IR) Electromagnetic radiation
with longer wavelengths than visible radiation. The wavelength range is approximately 0.7 μm to 1 mm. Many materials
transparent to visible light are opaque to
infrared, including glass. Rock salt, quartz,
germanium, or polyethylene prisms and
lenses are suitable for use with infrared.
Infrared radiation is produced by movement of charges on the molecular scale;
i.e. by vibrational or rotational motion of
molecules. Of particular importance in
chemistry is the absorption spectrum of
compounds in the infrared region. Certain
bonds between pairs of atoms (C−C,
C=C, C=O, etc.) have characteristic vibrational frequencies, which correspond
to bands in the infrared spectrum. Infrared
spectra are thus used in finding the structures of new organic compounds. They
are also used to 'fingerprint' and thus
identify known compounds. At shorter
wavelengths, infrared absorption corresponds to transitions between rotational
energy levels, and can be used to find the
dimensions of molecules (by their moment
of inertia). See also electromagnetic radiation.

inhibitor A substance that slows down
the rate of reaction. Hydrogen sulphide,
hydrogen cyanide, mercury salts, and arsenic compounds readily inhibit heterogeneous catalysts by adsorption. For
example, arsenic compounds inhibit platinum catalysts in the oxidation of
sulphur(IV) oxide to sulphur(VI) oxide. Inhibitors must not be confused with negative catalysts; inhibitors do not change the
pathway of a reaction.

inorganic chemistry The branch of
chemistry concerned with elements other
than carbon and with the preparation,
properties, and reactions of their compounds. Certain simple carbon compounds are treated in inorganic chemistry,
including the oxides, carbon disulphide,
the halides, hydrogen cyanide, and salts,

such as the cyanides, cyanates, carbonates, and hydrogencarbonates.

insoluble Describing a compound that has a very low solubility (in a specified solvent).

instrumentation The measurement of the conditions and the control of processes within a chemical plant. The instruments can be classified into three groups: those for current information using mercury thermometers, weighing scales, and pressure gauges; those for recording viscosity, fluid flow, pressure, and temperature; and those instruments that control and maintain the desired conditions including pH and the flow of materials.

intercallation compound A compound that has a structure based on layers and in which there are layers of a different character interleaved in the basic structural units. For example, the micas phlogopite ($KMg_3(OH)_2Si_3AlO_{10}$) and muscovite ($KAl_2(OH)_2Si_3AlO_{10}$) are formed by interleaving K^+ ions replacing a quarter of the silicon layers in talc and pyrophyllite respectively. *See also* lamellar compound.

intermediate bonding A form of covalent bond that also has an ionic or electrovalent character. *See* polar bond.

intermolecular forces Forces of attraction between molecules rather than forces within the molecule (chemical bonding). If these intermolecular forces are weak the material will be gaseous and as their strength progressively increases materials become progressively liquids and solids. The intermolecular forces are divided into H-bonding forces and van der Waals forces, and the major component is the electrostatic interaction of dipoles. *See* hydrogen bond, van der Waals force.

internal energy Symbol: U The energy of a system that is the total of the kinetic and potential energies of its constituent particles (e.g. atoms and molecules). If the temperature of a substance is raised, by transferring energy to it, the internal energy increases (the particles move faster). Similarly, work done on or by a system results in an increase or decrease in the internal energy. The relationship between heat, work, and internal energy is given by the first law of thermodynamics. Sometimes the internal energy of a system is loosely spoken of as 'heat' or 'heat energy'. Strictly, this is incorrect; heat is the transfer of energy as a result of a temperature difference.

internal resistance Resistance of a source of electricity. In the case of a cell, when a current is supplied, the potential difference between the terminals is lower than the e.m.f. The difference (i.e. e.m.f. − p.d.) is proportional to the current supplied. The internal resistance (r) is given by:
$$r = (E - V)/I$$
where E is the e.m.f., V the potential difference between the terminals, and I the current.

interstitial *See* defect.

interstitial compound A crystalline compound in which atoms of a non-metal (e.g. carbon, hydrogen, or boron) occupy interstitial positions in the crystal lattice of a metal (usually a transition metal). Interstitial compounds are often non-stoichiometric. Their physical properties are often similar to those of metals; e.g. they have a metallic lustre and are electrical conductors.

inversion A change from one optical isomer to the other. *See* Walden inversion.

invert sugar *See* sucrose.

iodic(V) acid (iodic acid, HIO_3) A colourless deliquescent crystalline solid produced by the reaction of concentrated nitric acid with iodine. Iodic(V) acid is a strong oxidizing agent. It will liberate iodine from solutions containing iodide ions and it reacts vigorously with organic materials, often producing flames. It dissociates extensively in water and hence is a strong acid.

iodic(VII) acid (periodic acid, H_5IO_6) A white crystalline solid made by low-temperature electrolysis of concentrated iodic(V) acid. It exists in a number of forms, the most common of which is paraiodic(VII) acid. Iodic(VII) acid is a powerful oxidizing agent. It is a weak acid, which – by cautious heating under vacuum – can be converted to dimesoiodic(VII) acid ($H_4I_2O_9$), and metaiodic(VII) acid (HIO_4). All three will oxidize manganese to permanganate and this reaction can be used to determine small amounts of manganese in steel.

iodic acid See iodic(V) acid.

iodide See halide.

iodine A dark-violet volatile solid element belonging to the halogens (group VII of the periodic table). It occurs in seawater and is concentrated by various marine organisms in the form of iodides. Significant deposits also occur in the form of iodates. The element is conveniently prepared by the oxidation of iodides in acid solution (MnO_2). Industrial methods similarly use oxidation of iodides or reduction of iodates to iodides by sulphur dioxide followed by oxidation, depending on the source of the raw materials. Iodine and its compounds are used in chemical synthesis, photography, pharmaceuticals, and dye-stuffs manufacture.

Iodine has the lowest electronegativity of the stable halogens and consequently is the least reactive. It combines only slowly with hydrogen to form hydroiodic acid, HI. Iodine also combines directly with many electropositive elements, but does so much more slowly than does bromine or chlorine. Because of the larger size of the iodine ion and the consequent low lattice energies, the iodides are generally more soluble than related bromides or chlorides. As with the other halides, iodides of Ag(I), Cu(I), Hg(I), and Pb(II) are insoluble unless complexing ions are present.

Iodine also forms a range of covalent iodides with the metalloids and non-metallic elements (this includes a vast range of organic iodides) but these are generally less thermodynamically stable and are more readily hydrolysed than chlorine or bromine analogues.

Four oxides are known of which iodine(V) oxide, I_2O_5, is the most important. The other oxides, I_2O_4, I_4O_9, and I_2O_7 are much less stable and of uncertain structure. Like chlorine (but *not* bromine) iodine forms oxo-species based on IO^-, IO_2^-, IO_3^-, and IO_4^-. The chemistry of the other oxo-species in solution is complex. Elemental iodine reacts with alkalis in a similar way to bromine and chlorine.

The ionization potential of iodine is sufficiently low for it to form a number of compounds in which it is electropositive. It forms I^+ cations, for example, by reaction of solid silver nitrate with iodine solution, and such cations are sufficiently electrophilic to substitute aromatic compounds such as phenol. Iodine also exhibits the interesting property of forming solutions that are violet coloured in non-donor type solvents, such as carbon tetrachloride, but in donor solvents, such as ethanol or dioxan, there is a strong iodine–solvent interaction, which gives the solution a deep brown colour. Even though iodine solutions were commonly used as antiseptic agents, the element is classified as toxic, and care should be taken to avoid eye intrusions or excessive skin contact.

Symbol: I; m.p. 113.5°C; b.p. 184°C; r.d. 4.95; p.n. 53; r.a.m. 126.9045.

iodine monochloride (ICl) A dark red liquid made by passing chlorine over iodine. It has properties similar to those of its constituent halogens. Iodine monochloride is used as a nonaqueous solvent and as an iodating agent in organic reactions.

iodine(V) oxide (iodine pentoxide, I_2O_5) A white crystalline solid made by heating iodic(V) acid to a temperature of 200°C. It is a very strong oxidizing agent. Iodine oxide is the acid anhydride of iodic(V) acid, which is reformed when water is added to the oxide. Its main use is in titration work, measuring traces of carbon monoxide in the air.

iodine pentoxide See iodine(V) oxide.

iodine trichloride See diiodine hexachloride.

iodoethane (ethyl iodide, C_2H_5I) A colourless liquid alkyl halide made by reaction of ethanol with iodine in the presence of red phosphorus.

iodoform See triiodomethane.

iodoform reaction See triiodomethane.

iodomethane (methyl iodide, CH_3I) A liquid alkyl halide made by reaction of methanol with iodine in the presence of red phosphorus.

ion exchange A process that takes place in certain insoluble materials, which contain ions capable of exchanging with ions in the surrounding medium. Zeolites, the first ion exchange materials, were used for water softening. These have largely been replaced by synthetic resins made of an inert backbone material, such as polyphenylethene, to which ionic groups are weakly attached. If the ions exchanged are positive, the resin is a cationic resin. An anionic resin exchanges negative ions. When all available ions have been exchanged (e.g. sodium ions replacing calcium ions) the material can be regenerated by passing concentrated solutions (e.g. sodium chloride) through it. The calcium ions are then replaced by sodium ions. Ion-exchange techniques are used for a vast range of purification and analytical purposes.
For example, solutions with ions that can be exchanged for OH^- and H^+ can be estimated by titrating the resulting solution with an acid or a base.

ionic bond See electrovalent bond.

ionic crystal A crystal composed of ions of two or more elements. The positive and negative ions are arranged in definite patterns and are held together by electrostatic attraction. Common examples are sodium chloride and caesium chloride.

ionic product The product of concentrations:
$$K_W = [H^+][OH^-]$$
in water as a result of a small amount of self-ionization.

ionic radius A measure of the effective radius of an ion in a compound. For an isolated ion, the concept is not very meaningful, since the ion is a nucleus surrounded by an 'electron cloud'. Values of ionic radii can be assigned, however, based on the distances between ions in crystals. Different methods exist for determining ionic radii and often different values are quoted for the same ion. The two main methods are those of Goldschmidt and of Pauling. The Goldschmidt radii are determined by substituting data from one compound to another, to produce a set of ionic radii for different ions. The Pauling radii are assigned by a more theoretical treatment for apportioning the distance between an anion and a cation.

ionic strength For an ionic solution a quantity can be introduced which emphasizes the charges of the ions present:
$$I = \frac{1}{2}\Sigma_i m_i z^2_i$$
where m is the molality and z the ionic charge. The summation is continued over all the different ions in the solution, i.

ionization The process of producing ions. There are several ways in which ions may be formed from atoms or molecules. In certain chemical reactions ionization occurs by transfer of electrons; for example, sodium atoms and chlorine atoms react to form sodium chloride, which consists of sodium ions (Na^+) and chloride ions (Cl^-). Certain molecules can ionize in solution; acids, for example, form hydrogen ions as in the reaction
$$H_2SO_4 \rightarrow 2H^+ + SO_4^{2-}$$

The 'driving force' for ionization in a solution is solvation of the ions by molecules of the solvent. H^+, for example, is solvated as a hydronium ion, H_3O^+.

Ions can also be produced by ionizing radiation; i.e. by the impact of particles or photons with sufficient energy to break up molecules or detach electrons from atoms: $A \rightarrow A^+ + e^-$. Negative ions can be formed by capture of electrons by atoms or molecules: $A + e^- \rightarrow A^-$.

ionization energy *See* ionization potential.

ionization potential (IP) Symbol: I
The energy required to remove an electron from an atom (or molecule or group) in the gas phase, i.e. the energy required for the process:
$$M \rightarrow M^+ + e^-$$
It gives a measure of the ability of metals to form positive ions. The second ionization potential is the energy required to remove two electrons and form a doubly charged ion:
$$M \rightarrow M^{2+} + e^-$$
Ionization potentials stated in this way are positive; often they are given in electronvolts. *Ionization energy* is the energy required to ionize one mole of the substance, and is usually stated in kilojoules per mole (kJ mol^{-1}).

In chemistry, the terms 'second', 'third', etc., ionization potentials are usually used for the formation of doubly, triply, etc., charged ions. However, in spectroscopy and physics, they are often used with a different meaning. The second ionization potential is the energy to remove the second least strongly bound electron in forming a singly charge ion. For lithium ($1s^2 2s^1$) it would refer to removal of a 1s electron to produce an excited ion with the configuration $1s^1 2s^1$. Note also that ionization potentials are now stated as energies. Originally they were the potential through which an electron had to be accelerated to cause ionization by electron impact:
$$M + e^- \rightarrow M^{2+} + 2e^-$$
They were thus stated in volts.

IP *See* ionization potential.

IR *See* infrared.

iridium A white transition metal that is highly resistant to corrosion. It is used to harden platinum for electrical contacts and in jewellery.
Symbol: Ir; m.p. 2410°C; b.p. 4130°C; r.d. 22.4; p.n. 77; r.a.m. 192.22.

iron A transition element occurring in many ores, especially the oxides (haematite and magnetite) and carbonate. It is extracted in a blast furnace using coke, limestone, and hot air. The coke and air form carbon monoxide, which then reduces the iron ore to iron. The limestone removes acidic impurities and forms a layer of slag above the molten iron at the base of the furnace. Steel is formed by reducing the carbon content to between 0.1 and 1.5%. Iron is used as a catalyst in the Haber process for ammonia production. It corrodes in air and moisture to hydrated iron(III) oxide (rust).
The most stable oxidation state is +3, which is yellow, but +2 (green) and +6 (easily reduced) also exist. Solutions of iron(II) ions give a green precipitate with sodium hydroxide solution, whereas iron(III) ions give a brown precipitate. The concentration of iron(II) ions can be estimated in acid solution by titration with standard potassium manganate(VII) solution. Iron(III) ions must first be reduced to iron(II) ions with sulphur dioxide.
Symbol: Fe; m.p. 1539°C; b.p. 2887°C; r.d. 7.9; p.n. 26; r.a.m. 55.85.

iron(II) chloride (ferrous chloride, $FeCl_2$) A compound prepared by passing dry hydrogen chloride gas over heated iron. White feathery anhydrous crystals are produced. Hydrated iron(II) chloride ($FeCl_2.6H_2O$) is prepared by reacting excess iron with dilute or concentrated hydrochloric acid. Green crystals of the hexahydrate are obtained on crystallization from solution. Iron(II) chloride is readily soluble in water, producing an acidic solution due to salt hydrolysis.

iron(III) chloride (ferric chloride, $FeCl_3$) A compound prepared in the anhydrous state as dark red crystals by passing dry chlorine over heated iron. The product sublimes and is collected in a cooled receiver. The hydrated salt ($FeCl_3.6H_2O$) is prepared by adding excess iron(III) oxide to concentrated hydrochloric acid. On crystallization yellow-brown crystals of the hexahydrate are formed. Iron(III) chloride is very soluble in water and undergoes salt hydrolysis. At temperatures below 400°C the anhydrous salt exists as a dimer, Fe_2Cl_6.

iron(II) oxide (ferrous oxide, FeO) A black powder formed by the careful reduction of iron(III) oxide using either carbon monoxide or hydrogen. It can also be prepared by heating iron(II) oxalate in the absence of air. It is only really stable at high temperatures and disproportionates slowly on cooling to give iron(III) oxide and iron. Iron(II) oxide can be reduced by heating in a stream of hydrogen. When exposed to air, it is oxidized to iron(III) oxide. It is a basic oxide, dissolving readily in dilute acids to form iron(II) salt solutions. If heated to a high temperature in an inert atmosphere, iron(II) oxide disproportionates to give iron and triiron tetroxide.

iron(III) oxide (ferric oxide, Fe_2O_3) A rusty-brown solid prepared by the action of heat on iron(III) hydroxide or iron(II) sulphate. It occurs in nature as the mineral haematite. Industrially it is obtained by roasting iron pyrites. Iron(III) oxide dissolves in dilute acids to produce solutions of iron(III) salts. It is stable at red heat, decomposes around 1300°C to give triiron tetroxide, and can be reduced to iron by hydrogen at 1000°C. Iron(III) oxide is not ionic in character but has a structure similar to that of aluminium(III) oxide.

iron(II) sulphate (ferrous sulphate, green vitriol, $FeSO_4.7H_2O$) A compound that occurs in nature as the mineral melanterite (or copperas). It is made industrially from iron pyrites. In the laboratory iron(II) sulphate is prepared by dissolving excess iron in dilute sulphuric acid. On crystallization, green crystals of the heptahydrate are obtained. Careful heat-

ing of the hydrated salt yields anhydrous iron(II) sulphate; on further heating the sulphate decomposes to give iron(III) oxide, sulphur(IV) oxide, and sulphur(VI) oxide. The hydrated crystals oxidize easily on exposure to air owing to the formation of basic iron(III) sulphate. A freshly prepared solution of iron(II) sulphate absorbs nitrogen(II) oxide (brown-ring test). Iron(II) sulphate crystals are isomorphous with the sulphates of zinc, magnesium, nickel, and cobalt.

iron(III) sulphate (ferric sulphate, $Fe_2(SO_4)_3$) A compound prepared in the hydrated state by the oxidation of iron(II) sulphate dissolved in dilute sulphuric acid, using an oxidizing agent, such as hydrogen peroxide or concentrated nitric acid. On crystallization, the solution deposits a white mass of the nonahydrate ($Fe_2(SO_4)_3.9H_2O$). The anhydrous salt can be prepared by gently heating the hydrated salt. Iron(III) sulphate decomposes on heating to give iron(III) oxide and sulphur(VI) oxide. It forms alums with the sulphates of the alkali metals.

irreversible change *See* reversible change.

irreversible reaction A reaction in which conversion to products is complete; i.e. there is little or no back reaction.

isobars 1. Two or more nuclides that have the same nucleon numbers but different proton numbers.
2. Lines joining points of equal pressure.

isocyanide *See* isonitrile.

isocyanide test (carbylamine reaction) A test for the primary amine group. The suspected primary amine is warmed with trichloromethane in an alcoholic solution of potassium hydroxide. The resulting isocyanide (RNC) has a characteristic smell of bad onions and is very toxic:

$$CHCl_3 + 3KOH + RNH_2 \rightarrow RNC + 3KCl + 3H_2O$$

isoelectronic Describing compounds that have the same number of electrons. For example, carbon monoxide (CO) and nitrogen (N_2) are isoelectronic.

isoleucine *See* amino acids.

isomer Any one of the possible compounds arising from the different ways of grouping the atoms in a given molecular formula. For example, methoxymethane and ethanol are isomers of C_2H_6O. Types of isomers are classified according to the type of isomerism exhibited. *See* isomerism.

isomerism The existence of two or more chemical compounds with the same molecular formulae but different structural formulae or different spatial arrangements of atoms. The different forms are known as *isomers*. For example, the compound C_4H_{10} may be butane (with a straight chain of carbon atoms) or 2-methyl propane ($CH_3CH(CH_3)CH_3$, with a branched chain).

Structural isomerism is the type of isomerism in which the structural formulae of the compounds differ. There are two main types. In one the isomers are different types of compound. An example is ethanol (C_2H_5OH) and methoxymethane (CH_3OCH_3), both having the formula C_2H_6O. In the other type of structural isomerism, the isomers differ because of the position of a functional group. For example, the primary alcohol propan-1-ol (C_3H_7OH) and the secondary alcohol propan-2-ol ($CH_3CH(OH)CH_3$) are isomers.

Stereoisomerism occurs when two compounds with the same molecular formulae and the same groups differ only in the arrangement of the groups in space. There are two types of stereoisomerism.

Cis-trans isomerism occurs when there is restricted rotation about a bond between two atoms. Groups attached to each atom may be on the same side of the bond (the *cis isomer*) or opposite sides (the *trans isomer*). *Cis-trans* isomerism also occurs in some inorganic complex compounds, when two groups may be at adjacent (*cis*)

or opposite (*trans*) positions. *Cis-trans* isomerism was formerly called *geometrical isomerism*. Sometimes the terms *syn* and *anti* are used, which correspond to *cis* and *trans*, respectively.

Optical isomerism occurs when the compound has no plane of symmetry and can exist in left- and right-handed forms that are mirror images of each other. Such molecules have an asymmetric atom – i.e. one attached to four different groups – called a *chiral centre*. *See also* optical activity.

isomorphism The existence of compounds with the same crystal structure.

isomorphism, law of *See* Mitscherlich's law.

isonitrile (isocyanide) An organic compound of the formula $R-NC$.

isoprene *See* methylbuta-1,3-diene.

isotherm A line on a chart or graph joining points of equal temperature. *See also* isothermal change.

isothermal change A process that takes place at a constant temperature. Throughout an isothermal process, the system is in thermal equilibrium with its surroundings. For example, a cylinder of gas in contact with a constant-temperature box may be compressed slowly by a piston. The work done appears as energy, which flows into the reservoir to keep the gas at the same temperature. Isothermal changes are contrasted with *adiabatic changes*, in which no energy enters or leaves the system, and the temperature changes. In practice no process is perfectly isothermal and none is perfectly adiabatic, although some can approximate in behaviour to one of these ideals.

isotones Two or more nuclides that have the same neutron numbers but different proton numbers.

isotonic Describing solutions that have the same osmotic pressure.

butane   2 – methylpropane

structural isomers of the hydrocarbon C_4H_{10}

ethanol   dimethyl ether

structural isomers differing in functional groups

propan-1-ol   propan-2-ol

structural isomers differing in the position of the functional group

Structural isomerism

118

cis – 1, 2 – dichloroethane   trans – 1, 2 – dichloroethane

cis-trans isomers differing in arrangement about a double bond

cis

trans

cis-trans isomers differing in arrangement about a single bond
with restricted rotation

cis

trans

cis-trans isomers in a planar metal complex

Cis-trans isomerism

119

isotopes Two or more species of the same element differing in their mass numbers because of differing numbers of neutrons in their nuclei. The nuclei must have the same number of protons (an element is characterized by its proton number). Isotopes of the same element have very similar properties because they have the same electron configuration, but differ slightly in their physical properties. An unstable isotope is termed a *radioactive isotope* or *radioisotope*.

Isotopes of elements are useful in chemistry for studies of the mechanisms of chemical reactions. A standard technique is to *label* one of the atoms in a molecule by using an isotope of the element. It is then possible to trace the way in which this atom behaves throughout the course of the reaction. For example, in the esterification reaction:

$$ROH + R'COOH = H_2O + R'COOR$$

it is possible to find which bonds are broken by using a labelled oxygen atom. If the reaction is performed using ^{18}O in the alcohol it is found that this nuclide appears in the ester, showing that the C–OH bond of the acid is broken in the reaction. In labelling, radioisotopes are detected by counters; stable isotopes can also be used, and detected by a mass spectrum.

Isotopes are also used in kinetic studies. For example, if the bond between two atoms X–Y is broken in the rate-determining step, and Y is replaced by a heavier isotope of the element, Y^*, then the reaction rate will be slightly lower with the Y^* present. This *kinetic isotope effect* is particularly noticeable with hydrogen and deuterium compounds, because of the large difference in mass.

isotope effect *See* isotopes.

isotopic mass (isotopic weight) The mass number of a given isotope of an element.

istopic number The difference between the number of neutrons in an atom and the number of protons.

isotopic weight *See* isotopic mass.

J

jeweller's rouge Iron(III) oxide (Fe_2O_3), used as a mild abrasive.

joule Symbol: J The SI unit of energy and work, equal to the work done when the point of application of a force of one newton moves one metre in the direction of action of the force. 1 J = 1 N m. The joule is the unit of all forms of energy.

Kekulé structure A structure of benzene in which there is a hexagonal ring with alternate double and single bonds. Two possible Kekulé structures contribute to the resonance hybrid of benzene. *See* benzene, resonance.

kelvin Symbol: K The SI base unit of thermodynamic temperature. It is defined as the fraction $1/273.16$ of the thermodynamic temperature of the triple point of water. Zero kelvin (0 K) is absolute zero. One kelvin is the same as one degree on the Celsius scale of temperature.

kerosine *See* paraffin.

ketal A type of organic compound formed by addition of an alcohol to a ketone. Addition of one molecule gives a *hemiketal*; two molecules give a full ketal. *See* acetal.

keto-enol tautomerism An equilibrium between two isomers in which one isomer is a ketone (the *keto form*) and the other an enol. The equilibrium is by transfer of a hydrogen atom between the oxygen atom of the carbonyl group and an adjacent carbon atom.

ketohexose A ketose sugar with six carbon atoms. *See* sugar.

ketone A type of organic compound with the general formula RCOR, having two alkyl or aryl groups bound to a carbonyl

Ketone

group. They are made by oxidizing secondary alcohols (just as aldehydes are made from primary alcohols). Simple examples are propanone (acetone, CH_3COCH_3) and butanone (methylethyl ketone, $CH_3COC_2H_5$).

The chemical reactions of ketones are similar in many ways to those of aldehydes. The carbonyl group is polarized, with positive charge on the carbon and negative charge on the oxygen. Thus nucleophilic addition can occur at the carbonyl group. Ketones thus:

(1) Undergo addition reactions with hydrogen cyanide and hydrogensulphite ions.

(2) Undergo condensation reactions with hydroxylamine, hydrazine, and their derivatives.

(3) Are reduced to (secondary) alcohols. They are not, however, easily oxidized. Strong oxidizing agents give a mixture of carboxylic acids. They do not react with Fehling's solution or Tollen's reagent, and do not easily polymerize. *See also* aldehyde.

ketopentose A ketose sugar with five carbon atoms. *See* sugar.

ketose A sugar containing a ketone ($=CO$) or potential ketone group. *See* sugar.

Keto-enol tautomerism

kieselguhr A naturally occurring very porous form of silica (SiO_2) used as an absorbent material and catalyst support.

killed spirits Zinc(II) chloride solution used as a flux for solder, so called because it is made by adding zinc to hydrochloric acid ('spirits of salt').

kilo- Symbol: k A prefix denoting 10^3. For example, 1 kilometre (km) = 10^3 metres (m).

kilogram (kilogramme) Symbol: kg The SI base unit of mass, equal to the mass of the international prototype of the kilogram, which is a piece of platinum-iridium kept at Sèvres in France.

kilogramme An alternative spelling of *kilogram*.

kilowatt-hour Symbol: kWh A unit of energy, usually electrical, equal to the energy transferred by one kilowatt of power in one hour. It has a value of 3.6×10^6 joules.

kinetic energy *See* energy.

kinetic isotope effect *See* isotopes.

kinetics A branch of physical chemistry concerned with the study of rates of chemical reactions and the effect of physical conditions that influence the rate of reaction, e.g. temperature, light, concentration, etc. The measurement of these rates under different conditions gives information on the mechanism of the reaction; i.e. on the sequence of processes by which reactants are converted into products.

kinetic theory A theory explaining physical properties in terms of the motion of particles. The kinetic theory of gases assumes that the molecules or atoms of a gas are in continuous random motion and the pressure (p) exerted on the walls of a containing vessel arises from the bombardment by these fast moving particles. When the temperature is raised the speeds increase; so consequently does the pressure. If more particles are introduced or the volume is reduced there are more particles to bombard unit area of the walls and the pressure also increases. When a particle collides with the wall it experiences a rate of change of momentum, which is proportional to the force exerted. For a large number of particles this provides a steady force per unit area (or pressure) on the wall.

Following certain additional assumptions, the kinetic theory leads to an expression for the pressure exerted by an ideal gas. The assumptions are:

(1) The particles behave as if they are hard smooth perfectly elastic points.

(2) They do not exert any appreciable force on each other except during collisions.

(3) The volume occupied by the particles themselves is a negligible fraction of the volume of the gas.

(4) The duration of each collision is negligible compared with the time between collisions.

By considering the change in momentum on impact with the walls it can be shown that

$$p = \rho c^{2/3}$$

where ρ is the density of the gas and c is the root-mean-square speed of the molecules. The mean-square speed of the molecules is proportional to the absolute temperature:

$$Nmc^2 = RT$$

See also degrees of freedom.

Kipp's apparatus An apparatus for the production of a gas from the reaction of a liquid on a solid. It consists of three globes, the upper globe being connected via a wide tube to the lower globe. The upper globe is the liquid reservoir. The middle globe contains the solid and also has a tap at which the gas may be drawn off. When the gas is drawn off the liquid rises from the lower globe to enter the middle globe and reacts with the solid, thereby releasing more gas. When turned off the gas released forces the liquid back down into the lower globe and up into the reservoir, thus stopping the reaction.

Kjeldahl's method

Kjeldahl's method A method used for the determination of nitrogen in organic compounds. The nitrogenous substance is converted to ammonium sulphate by boiling with concentrated sulphuric acid (often with a catalyst such as $CuSO_4$) in a specially designed long-necked *Kjeldahl flask*. The mixture is then made alkaline and the ammonia distilled off into standard acid for measurement by titration.

Kolbe electrolysis The electrolysis of sodium salts of carboxylic acids to prepare alkanes. The alkane is produced at the anode after discharge of the carboxylate anion and decomposition of the radical:

$$RCOO^- \rightarrow RCOO\cdot + e^-$$
$$RCOO\cdot \rightarrow R\cdot + CO_2$$
$$R\cdot + RCOO\cdot \rightarrow R - R + CO_2$$

and

$$2R\cdot \rightarrow R - R$$

As the reaction is a coupling reaction, only alkanes with an even number of carbon atoms in the chain can be prepared in this way.

Kroll process A method of obtaining certain metals by reducing the metal chloride with magnesium. Titanium can be obtained in this way.

$$TiCl_4 + 2Mg \rightarrow 2MgCl_2 + Ti$$

krypton A colourless odourless monatomic element of the rare-gas group, now known to form unstable compounds. It occurs in minute quantities (0.001% by volume) in air. Krypton is used in fluorescent lights. Symbol: Kr; m.p. $-157.3°C$; b.p. $-153.4°C$; d. 3.73 kg m^{-3}; p.n. 36; r.a.m. 83.80.

kurchatovium *See* element 104.

L

label A stable or radioactive nuclide used to investigate some process, such as a chemical reaction. *See* isotopes.

labile Describing a complex with ligands that can easily be replaced by more strongly bonded ligands.

Lactam

lactam A type of organic compound containing the −NH.CO.− group as part of a ring in the molecule. They can be regarded as formed from a straight-chain compound that has an amine group (−NH₂) at one end of the molecule and a carboxylic acid group (−COOH) at the other; i.e. from an amino acid. The reaction of the amine group with the carboxylic acid group, with elimination of water, leads to the cyclic lactam, which is thus an internal amide.

lactate A salt or ester of lactic acid.

lactic acid *See* 2-hydroxypropanoic acid.

lactone A type of organic compound containing the group −O.CO.− as part of a ring in the molecule. A lactone can be regarded as formed from a compound with an alcohol (−OH) group on one end of the chain and a carboxylic acid (−COOH) group on the other. The lactone then results from reaction of the −OH group with the −COOH group; i.e. it is an internal ester.

lactose (milk sugar, $C_{12}H_{22}O_{11}$) A sugar found in milk. It is a disaccharide composed of glucose and galactose units.

laevo-form *See* optical activity.

laevorotatory *See* optical activity.

lake A pigment made by combining an organic dyestuff with an inorganic compound (e.g. an oxide).

lamellar compound A compound with a crystal structure composed of thin plates or layers. Silicates form many compounds with distinct layers. Examples are talc $(Mg_3(OH)_2Si_4O_{10})$ and pyrophyllite $(Al_2(OH)_2Si_4O_{10})$. *See also* intercallation compound.

lamp black A black pigment; a finely divided form of carbon formed by incomplete combustion of an organic compound.

lanthanides *See* lanthanoids.

lanthanoids (lanthanides, lanthanons) A group of elements whose electronic configurations display back filling of the 4f-level. There is a maximum of 14 electrons in an f-orbital. The element lanthanum itself has no f-electrons (La $[Xe]5d^16s^2$) and is thus strictly not a lanthanoid, but it is included by convention, thus giving a closely related series of 15 elements. Excluding lanthanum, the elements all have $4f^x6s^2$

Lactone

125

configurations but gadolinium and luteci-um have an additional $5d^1$ electron.

The characteristic oxidation state is M^{3+} and the great similarity in the size of the ions leads to a very close similarity of chemical properties and hence to great difficulties of separation using convention-al methods. Chromatographic and sol-vent-extraction methods have been specially developed for the lanthanoids.

lanthanons See lanthanoids.

lanthanum A soft ductile malleable sil-very metallic element that is the first mem-ber of the lanthanoid series. It is found associated with other lanthanoids in many minerals, including monazite and bastnasite. Lanthanum is used in several alloys (especially for lighter flints), as a cat-alyst, and in compound form in carbon-arc searchlights, etc., and in the glass industry. Symbol: La; m.p. 920°C; b.p. 3457°C; r.d. 6.15; p.n. 57; r.a.m. 138.91.

laser An acronym for Light Amplification by Stimulated Emission of Radiation. A la-ser device produces high-intensity, mono-chromatic, coherent beams of light. In the laser process the molecules of a sample (such as ruby doped with Cr^{3+} ions) are promoted to an excited state. As the sam-ple is in a cavity between two reflective surfaces, when a molecule emits spontaneously, the photon so generated ricochets backwards and forwards. In this way other molecules are stimulated to emit photons of the same energy. If one of the reflective surfaces is partially transmit-ting this radiation can be tapped.

latent heat The heat evolved or ab-sorbed when a substance changes its phys-ical state, e.g. the latent heat of fusion is the heat absorbed when a substance changes from a solid to a liquid.

lattice A regular three-dimensional ar-rangement of points. A lattice is used to describe the positions of the particles (at-oms, ions, or molecules) in a crystalline solid. The lattice structure can be exam-ined by X-ray diffraction techniques.

lattice energy The energy released when ions of opposite charge are brought together from infinity to form one mole of a given crystal. The lattice energy is a mea-sure of the stability of a solid ionic sub-stance, with respect to ions in the gas. See also Born-Haber cycle.

laughing gas See dinitrogen monox-ide.

lauric acid See dodecanoic acid.

law of conservation of energy See conservation of energy, law of.

law of conservation of mass See conservation of mass, law of.

law of constant composition See constant composition, law of.

law of constant proportions See constant proportions, law of.

law of definite proportions See definite proportions, law of.

law of isomorphism See Mitscher-lich's law.

law of mass action See mass action, law of.

law of octaves See Newlands' law.

lawrencium A radioactive transuranic element of the actinoid series, not found naturally on Earth. Several very short-lived isotopes have been synthesized. Symbol: Lr; p.n. 103.

laws of chemical combination See chemical combination, laws of.

leaching The washing out of a soluble material from an insoluble solid using a solvent. This is often carried out in batch tanks or by dispersing the crushed solid in a liquid.

lead A dense, dull, grey, soft metallic ele-ment; the fifth member of group IV of the periodic table and the end product of radioactive decay series. It occurs in small quantities in a wide variety of minerals but only a few are economically important. The most important one is galena (PbS), found in Australia, Mexico, the USA, and Canada. Other minerals are anglesite (Pb-SO$_4$), litharge (PbO), and cerussite (PbCO$_3$).

Galena is often associated with zinc ores and the smelting operations of both lead and zinc industries are closely integrated. The ore is concentrated by froth flotation and the concentrate roasted then reduced,

$$2PbS + 3O_2 \rightarrow 2PbO + 2SO_2$$

$$PbS + 2O_2 \rightarrow PbSO_4$$

$$2PbO + 2C \rightarrow 2Pb + 2CO$$

$$2PbO + PbS \rightarrow 3Pb + SO_2$$

$$PbSO_4 + 2C \rightarrow Pb + 2CO + SO_2$$

Silver is often recovered in economic quantities from crude lead.

The outer s^2p^2 configurations of tin and lead give rise to similar properties for the two metals. There is however a much greater predominance of the divalent state for lead. Both oxides are amphoteric, lead(II) oxide (PbO) leading to plumbites and lead(IV) oxide (PbO$_2$) to plumbates on dissolution in alkalis. Lead also forms mixed oxides Pb$_2$O$_3$ (a yellow solid, better written as Pb(II)Pb(IV)O$_3$) and Pb$_3$O$_4$ (red lead, Pb$_2$(II)Pb(IV)O$_4$). Both metals have low melting points and neither is attacked by dilute acids; they differ however in their reaction with concentrated nitric acid

$$3Pb + 8HNO_3 \rightarrow 3Pb(NO_3)_2 + 2NO + 4H_2O$$

Tin gives hydrated Sn(IV) oxide. Concentrated hydrochloric acid will not attack lead.

Lead, like tin, forms halides, PbX$_2$, with all halogens. PbCl$_4$ is the only known lead(IV) halide (all halides of SnX$_4$ are known). The great stability of the Pb(II) state leads to Pb(IV) compounds being powerful oxidizing agents.

Lead is used in accumulators (lead–acid), alloys, radiation shielding, and water and sound proofing. It is also used in the petrochemical, paint, and glass industries.

Symbol: Pb; m.p. 328°C; b.p. 1755°C; r.d. 11.34; p.n. 82; r.a.m. 207.19.

lead(II) acetate *See* lead(II) ethanoate.

lead–acid accumulator A type of electrical accumulator used in vehicle batteries. It has two sets of plates: spongy lead plates connected in series to the negative terminal and lead oxide plates connected to the positive terminal. The material of the electrodes is held in a hard lead-alloy grid. The plates are interleaved. The electrolyte is dilute sulphuric acid. The e.m.f. when fully charged is about 2.2 V. This falls to a steady 2 V when current is drawn. As the accumulator begins to run down, the e.m.f. falls further. During discharge the electrolyte becomes more dilute and its relative density falls. To recharge the accumulator, charge is passed through it in the opposite direction to the direction of current supply. This reverses the cell reactions and increases the relative density of the electrolyte (c. 1.25 for a fully charged accumulator).

The electrolyte contains hydrogen ions (H$^+$) and sulphate ions (SO$_4^{2-}$). During discharge, H$^+$ ions react with the lead(IV) oxide to give lead(II) oxide and water

$$PbO_2 + 2H^+ + 2e^- \rightarrow PbO + H_2O$$

This reaction takes electrons from the plate, causing the positive charge. There is a further reaction to yield soft lead sulphate:

$$PbO + SO_4^{2-} + 2H^+ \rightarrow PbSO_4 + H_2O + 2e^-$$

Electrons are released to the electrode, producing the negative charge.

During charging the reactions are reversed:

$$PbSO_4 + 2e^- \rightarrow Pb + SO_4^{2-}$$

$$PbSO_4 + 2H_2O \rightarrow PbO_2 + 4H^+ + SO_4^{2-} + 2e^-$$

lead(II) carbonate (PbCO$_3$) A white poisonous powder that occurs naturally as the mineral cerussite. It forms rhombic crystals and can be precipitated by reacting a cold aqueous solution of a soluble lead salt (e.g. lead(II) nitrate) with ammonium carbonate.

lead(II) carbonate hydroxide (white lead, 2PbCO$_3$.Pb(OH)$_2$) The most important basic lead carbonate. It can be manufactured electrolytically. It has been used as a pigment in white and coloured paints but it has the considerable drawback that it is poisonous.

lead-chamber process A former process for the manufacture of sulphuric acid by oxidizing sulphur(IV) oxide with nitrogen monoxide. The reaction, which was carried out in large lead chambers, has now been replaced by the contact process.

lead dioxide See lead(IV) oxide.

lead(II) ethanoate (lead(II) acetate, $Pb(CH_3CO_2)_2$) A compound usually occurring as the hydrate $Pb(CH_3CO_2)_2.3H_2O$, forming monoclinic crystals. At $100°C$ it loses ethanoic acid and water, forming a basic lead(II) ethanoate. Its solubility in water is 50 g per 100 g of water at $25°C$. Lead(II) ethanoate can be obtained as an anhydrous salt. Its chief asset is that it is one of the few common lead(II) salts that are soluble in water.

lead(IV) hydride See plumbane.

lead monoxide See lead(II) oxide.

lead(II) oxide (lead monoxide, PbO) A yellow crystalline powder formed by roasting molten lead in air. *Litharge*, the most common form, is obtained when lead(II) oxide is heated above its melting point. If prepared below its melting point, another form, called *massicot*, is obtained. Litharge is used in the rubber industry, in the manufacture of paints and varnishes, and in the manufacture of lead glazes for pottery.

lead(IV) oxide (lead dioxide, PbO_2) A dark-brown solid forming hexagonal crystals. On heating to $310°C$ it decomposes into lead(II) oxide and oxygen. It can be prepared electrolytically or by reacting lead(II) oxide with potassium chlorate. Lead(IV) oxide has been used in the manufacture of matches.

lead(II) sulphate ($PbSO_4$) A white crystalline solid that occurs naturally as the mineral anglesite. It is almost insoluble in water and can be precipitated by reacting an aqueous solution containing sulphate ions with a solution of a soluble lead(II) salt (e.g. lead(II) ethanoate). It forms basic lead(II) sulphates when shaken together with lead(II) hydroxide and water. Basic lead(II) sulphates are useful as pigments.

lead(II) sulphide (PbS) A black solid that occurs naturally as the mineral galena. It can be prepared as a precipitate by reacting a solution of a soluble lead(II) salt with hydrogen sulphide or with a solution of a soluble sulphide. Lead(II) sulphide is useful as a rectifier in electrical components.

lead tetraethyl (tetraethyl lead, $Pb(CH_2CH_3)_4$) A poisonous liquid that is insoluble in water but soluble in organic solvents. It is manufactured by the reaction of an alloy of sodium and lead with 1-chloroethane. The product is obtained by steam distillation. Lead tetraethyl is used as an additive in internal-combustion engine fuel to increase its octane number and thus prevent preignition (knocking).

Leblanc process An obsolete process for the manufacture of sodium carbonate. Sodium chloride is converted to sodium sulphate by heating with sulphuric acid. This sulphate is then roasted in a rotary furnace where it is first reduced to the sulphide using carbon, and then immediately converted to the carbonate by the action of limestone. Sodium carbonate solution is obtained by leaching with water and this is subsequently dried and calcined to obtain the solid.

Le Chatelier's principle If a system is at equilibrium and a change is made in the conditions, the equilibrium adjusts so as to oppose the change. The principle can be applied to the effect of temperature and pressure on chemical reactions. A good example is the Haber process for synthesis of ammonia:
$$N_2 + 3H_2 = 2NH_3$$
The 'forward' reaction
$$N_2 + 3H_2 \rightarrow 2NH_3$$

is exothermic. Thus, reducing the temperature displaces the equilibrium towards production of NH_3 (as this tends to increase temperature). Increasing the pressure also favours formation of NH_3, because this leads to a reduction in the total number of molecules (and hence pressure).

Leclanché cell A primary voltaic cell consisting, in its 'wet' form, of a carbon rod anode and a zinc cathode, with a 10–20% solution of ammonium chloride as electrolyte. Manganese(IV) oxide mixed with crushed carbon in a porous bag or pot surrounding the anode acts as a depolarizing agent. The dry form (*dry cell*) is widely used for torch batteries, transistor radios, etc. It has a mixture of ammonium chloride, zinc chloride, flour, and gum forming an electrolyte paste. Sometimes the dry cell is arranged in layers to form a rectangular battery, which has a longer life than the cylinder type.

leucine *See* amino acids.

Lewis acid A substance that can accept an electron pair to form a coordinate bond; a *Lewis base* is an electron-pair donor. In this model, the neutralization reaction is seen as the acquisition of a stable octet by the acid, for example:
$$Cl_3B + :NH_3 \rightarrow Cl_3B:NH_3$$
Metal ions in coordination compounds are also electron-pair acceptors and therefore Lewis acids. The definition includes the 'traditional' Bronsted acids since H^+ is an electron acceptor, but in common usage the terms Lewis acid and Lewis base are reserved for systems without acidic hydrogen atoms.

Lewis base *See* Lewis acid.

Lewis octet theory *See* octet.

L-form *See* optical activity.

***l*-form** *See* optical activity.

Liebig condenser A simple type of laboratory condenser. It consists of a straight glass tube, in which the vapour is condensed, with a surrounding glass jacket through which cooling water flows.

ligand A molecule or ion that forms a coordinate bond to a metal atom or ion in a complex. *See also* complex, chelate.

light (visible radiation) A form of electromagnetic radiation able to be detected by the human eye. Its wavelength range is between about 400 nm (far 'red') and about 700 nm (far 'violet'). The boundaries are not precise as individuals vary in their ability to detect extreme wavelengths; this ability also declines with age.

lime *See* calcium oxide, calcium hydroxide.

limestone A natural form of calcium carbonate. It is used in making calcium compounds, carbon dioxide, and cement.

lime water A solution of calcium hydroxide in water. If carbon dioxide is bubbled through lime water a milky precipitate of calcium carbonate forms. Prolonged bubbling of carbon dioxide turns the solution clear again as a result of the formation of calcium hydrogencarbonate $(Ca(HCO_3)_2)$.

line spectrum A spectrum composed of a number of discrete lines corresponding to single wavelengths of emitted or absorbed radiation. Line spectra are produced by atoms or simple (monatomic) ions in gases. Each line corresponds to a change in electron orbit, with emission or absorption of radiation. *See also* spectrum.

liquid The state of matter in which the particles of a substance are loosely bound by intermolecular forces. The weakness of these forces permits movement of the particles and consequently liquids can change their shape within a fixed volume. The liquid state lacks the order of the solid state. Thus, amorphous materials, such as glass, in which the particles are disordered and can move relative to each other, can be classed as liquids.

liquid air A pale blue liquid. It is a mixture of liquid oxygen (boiling at $-183°C$) and liquid nitrogen (boiling at $-196°C$).

litharge See lead(II) oxide.

lithia See lithium oxide.

lithia water See lithium hydrogen carbonate.

lithium A light silvery moderately reactive metal; the first member of the alkali metals (group I of the periodic table). It occurs in a number of complex silicates, such as spodumene, lepidolite, and petalite, and a mixed phosphate, tryphilite. It is a rare element accounting for 0.0065% of the Earth's crust. Lithium ores are treated with concentrated sulphuric acid and lithium sulphate subsequently separated by crystallization. The element is obtained by conversion to the chloride and electrolysis of the fused chloride or of solutions of LiCl in pyridine.

Lithium has the electronic configuration $1s^2 2s^1$ and is entirely monovalent in its chemistry. The lithium ion is, however, much smaller than the ions of the other alkali metals; consequently it is polarizing and a certain degree of covalency occurs in its bonds. Lithium also has the highest ionization potential of the alkali metals.

The element reacts with hydrogen to form lithium hydride, LiH, a colourless high-melting solid, which releases hydrogen at the anode during electrolysis (confirming the ionic nature Li^+H^-). The compound reacts with water to release hydrogen and is also frequently used as a reducing agent in organometallic synthesis. Lithium reacts with oxygen to give Li_2O (sodium gives the peroxide) and with nitrogen to form Li_3N on fairly gentle warming. The metal itself reacts only slowly with water, giving the hydroxide (LiOH) but lithium oxide reacts much more vigorously to give again the hydroxide; the nitride is hydrolysed to ammonia. The metal reacts with halogens to form halides (LiX). Apart from the fluoride the halides are readily soluble both in water and in oxygen-containing organic solvents. In this property lithium partly resembles magnesium which has a similar charge/size ratio. Compared to the other carbonates of group I, lithium carbonate is thermally unstable decomposing to Li_2O and CO_2. This is because the small Li^+ ion leads to particularly high lattice energies favouring the formation of Li_2O.

Lithium also forms a wide range of alkyl- and aryl-compounds with organic compounds, which are particularly useful in organic synthesis.

Lithium compounds impart a characteristic purple colour to flames.

Symbol: Li; m.p. $179°C$; b.p. $1336°C$; r.d. 0.82; p.n. 3; r.a.m. 6.9.

lithium aluminium hydride See lithium tetrahydridoaluminate(III).

lithium carbonate (Li_2CO_3) A white solid obtained by the addition of excess sodium carbonate solution to a solution of a lithium salt. It is soluble in the presence of excess carbon dioxide, forming lithium hydrogen carbonate. If heated (to $780°C$) in a stream of hydrogen it undergoes thermal decomposition to give lithium oxide and carbon dioxide. Lithium carbonate forms monoclinic crystals and differs from the other group I carbonates in being only sparingly soluble in water. It has a close resemblance to the carbonates of magnesium and calcium — an example of a diagonal relationship in the periodic table.

lithium chloride (LiCl) A white solid produced by dissolving lithium carbonate or oxide in dilute hydrochloric acid and crystallizing the product. Below $19°C$ the dihydrate ($LiCl.2H_2O$) is obtained; at $19°C$ the chloride loses one molecule of water and becomes anhydrous at $93.5°C$. Lithium chloride is used as a flux in welding aluminium. The compound forms cubic crystals; the anhydrous salt is isomorphous with sodium chloride. It dissolves in such organic solvents as alcohols, ketones, and esters. Lithium chloride is probably the most deliquescent substance known and forms hydrates with 1, 2, and 3 molecules of water.

lithium hydride (LiH) A white crystalline solid produced by direct combination of the elements at a temperature above 500°C. Electrolysis of the fused salt yields hydrogen at the anode. Lithium hydride reacts violently with water to give lithium hydroxide and hydrogen; the reaction is highly exothermic. The compound is used as a reducing agent and in the preparation of other hydrides. Lithium hydride has a cubic structure and is more stable than the group I hydrides.

lithium hydrogencarbonate ($LiHCO_3$) A compound known only in solution, produced by the action of carbon dioxide, water, and lithium carbonate. When a solution of the hydrogencarbonate is heated, carbon dioxide and lithium carbonate are produced. Solutions of the hydrogencarbonate are sold and used in medicine under the name of *lithia water*.

lithium hydroxide (LiOH) A white solid made industrially as the monohydrate ($LiOH.H_2O$) by reacting lime with a lithium ore or with a salt made from the ore. Lithium hydroxide has a closer resemblance to the group II hydroxides than to the group I hydroxides.

lithium oxide (lithia, Li_2O) A white solid produced by burning metallic lithium in air above its melting point. It can also be produced by the thermal decomposition of the carbonate or hydroxide. Lithium oxide reacts slowly with water forming a solution of lithium hydroxide; the reaction is exothermic. The oxide has a calcium-fluoride structure.

lithium sulphate (Li_2SO_4) A white solid prepared by the addition of excess lithium oxide or carbonate to a solution of sulphuric acid. It is readily soluble in water from which it crystallizes as the monohydrate. In contrast to the other group I sulphates it does not form alums and it is not isomorphous with these other sulphates.

lithium tetrahydridoaluminate (III) (lithium aluminium hydride, $LiAlH_4$) A white solid produced by action of lithium hydride on aluminium chloride, the hydride being in excess. Lithium tetrahydridoaluminate reacts violently with water. It is a powerful reducing agent, reducing ketones and carboxylic acids to their corresponding alcohols. In inorganic chemistry it is used in the preparation of hydrides.

litmus A natural pigment that changes colour when in contact with acids and alkalis; above a pH of 8.3 it is blue and below a pH of 4.5 it is red. Thus it gives a rough indication of the acidity or basicity of a solution; because of its rather broad range of colour change it is not used for precise work. Litmus is used both in solution and as litmus paper.

litre Symbol: l A unit of volume now defined as 10^{-3} metre^3; i.e. 1000 cm^3. The millilitre (ml) is thus the same as the cubic centimetre (cm^3). However, the name is not recommended for precise measurements since the litre was formerly defined as the volume of one kilogram of pure water at 4°C and standard pressure. On this definition, one litre is the same as 1000.028 cubic centimetres.

lixiviation The process of separating soluble components from a mixture by washing them out with water.

localized bond A bond in which the electrons contributing to the bond remain between the two atoms concerned, i.e. the bonding orbital is localized. The vast majority of bonds are of this type. *Compare* delocalized bond.

lone pair A pair of valence electrons having opposite spin that are located together on one atom, i.e. are not shared as in a covalent bond. Lone pairs occupy similar positions in space to bond pairs and account for the shapes of molecules. A molecule with a lone pair can donate this to an electron acceptor, such as H^+ or a metal ion, to form coordinate bonds. *See* complex, Lewis base.

long period *See* period.

lowering of vapour pressure A colligative property of solutions in which the vapour pressure of a solvent is lowered as a solute is introduced. When both solvent and solute are volatile the effect of increasing the solute concentration is to lower the partial vapour pressure of each component. When the solute is a solid of negligible vapour pressure the lowering of the vapour pressure of the solution is directly proportional to the number of species introduced rather than to their nature and the proportionality constant is regarded as a general solvent property. Thus the introduction of the same number of moles of any solute causes the same lowering of vapour pressure, if dissociation does not occur. If the solute dissociates into two species on dissolution the effect is doubled. The kinetic model for the lowering of vapour pressure treats the solute molecules as occupying part of the surface of the liquid phase and thereby restricting the escape of solvent molecules. The effect can be used in the measurement of relative molecular masses, particularly for large molecules, such as polymers. *See also* Raoult's law.

Lowry-Brønsted theory *See* acid.

lumen Symbol: lm The SI unit of luminous flux, equal to the luminous flux emitted by a point source of one candela in a solid angle of one steradian. 1 lm = 1 cd sr.

luminescence The emission of radiation from a substance in which the particles have absorbed energy and gone into excited states. They then return to lower energy states with the emission of electromagnetic radiation. If the luminescence persists after the source of excitation is removed it is called *phosphorescence*: if not, it is called *fluorescence*.

lutetium A silvery element of the lanthanoid series of metals. It occurs in association with other lanthanoids. One of its few uses is as a catalyst.
Symbol: Lu; m.p. 1663°C; b.p. 3395°C; r.d. 9.84; p.n. 71; r.a.m. 174.97.

lux Symbol: lx The SI unit of illumination, equal to the illumination produced by a luminous flux of one lumen falling on a surface of one square metre. 1 lx = 1 lm m^{-2}.

Lyman series A series of lines in the ultraviolet spectrum emitted by excited hydrogen atoms. They correspond to the atomic electrons falling into the lowest energy level and emitting energy as radiation. The wavelength (λ) of the radiation in the Lyman series is given by
$$1/\lambda = R(1/1^2 - 1/n^2)$$
where n is an integer and R is the Rydberg constant. *See also* spectral series.

lyophilic Solvent attracting. When the solvent is water, the word *hydrophilic* is often used. The terms are applied to: (1) Ions or groups on a molecule. In aqueous or other polar solutions ions or polar groups are lyophilic. For example, the $-COO^-$ group on a soap is the lyophilic (hydrophilic) part of the molecule.
(2) The disperse phase in colloids. In lyophilic colloids the dispersed particles have an affinity for the solvent, and the colloids are generally stable. *Compare* lyophobic.

lyophobic Solvent repelling. When the solvent is water, the word *hydrophobic* is used. The terms are applied to:
(1) Ions or groups on a molecule. In aqueous or other polar solvents, the lyophobic group will be non-polar. For example, the hydrocarbon group on a soap molecule is the lyophobic (hydrophobic) part.
(2) The disperse phase in colloids. In lyophobic colloids the dispersed particles are not solvated and the colloid is easily solvated. Gold and sulphur sols are examples. *Compare* lyophilic.

lysine *See* amino acids.

M

macromolecule A large molecule; e.g. a natural or synthetic polymer.

macromolecular crystal A crystal composed of atoms joined together by covalent bonds, which form giant three-dimensional or two-dimensional networks. Diamond is an example of a macromolecular crystal.

magnesia See magnesium oxide.

magnesite A mineral form of magnesium carbonate, $MgCO_3$.

magnesium A light metallic element; the second member of group II of the periodic table (the alkaline earths). It has the electronic configuration of neon with two additional outer 3s electrons. The element accounts for 2.09% of the Earth's crust and is eighth in order of abundance. It occurs in a wide variety of minerals such as brucite ($Mg(OH)_2$), carnallite ($MgCl_2.KCl.$ $6H_2O$), epsomite ($MgSO_4.7H_2O$), magnesite ($MgCO_3$), and dolomite ($MgCO_3.$ $CaCO_3$); and also as $MgCl_2$ in sea water. The metal is obtained by several routes, depending on the mineral used, but all leading to the chloride, followed by electrolysis of the fused chloride.

The element has a fairly low ionization potential and is an electropositive element. It reacts directly with oxygen, nitrogen, and sulphur on heating, to form the oxide, MgO, the nitride, Mg_3N_2, and the sulphide MgS, all of which are hydrolysed by water to give the hydroxide.

Magnesium halides are also formed by direct reaction with the halogen but show a slight deviation in the general behaviour by dissolving normally in water and only undergoing significant hydrolysis when solutions are evaporated:

$$MgX_2 + 2H_2O \rightarrow Mg(OH)_2 + 2HX$$

Magnesium hydrodixe is considerably less soluble than the hydroxides of the heavier elements in group II and is a weaker base.

It forms a soluble sulphate, chlorate, and nitrate. Like the heavier alkaline earths, magnesium forms an insoluble carbonate, but $MgCO_3$ is the least thermally stable of the group. Unlike calcium, it does not form a carbide.

Magnesium metal is industrially important as a major component in lightweight alloys (with aluminium and zinc). The metal surfaces develop an impervious oxide film which protects them from progressive deterioration.

Because of its ionic nature magnesium forms very few coordination compounds, Donor-acceptor species in aqueous solutions are short-lived, but magnesium bromide and iodide have sufficient acceptor properties to dissolve in donor solvents such as alcohols and ketones. Magnesium appears to be an essential element for life, occurring in chlorophyll.

The organometallic compounds of magnesium are important in synthetic organic chemistry; magnesium forms Grignard reagents, $RMgX$, and related di-alkyls, R_2Mg, by reaction of the metal directly with the alkyl bromide or iodide. The element crystallizes with the close-packed hexagonal structure.

Symbol: Mg; m.p. 651°C; b.p. 1107°C; r.d. 1.74; p.n. 12; r.a.m. 24.3.

magnesium bicarbonate See magnesium hydrogencarbonate.

magnesium carbonate ($MgCO_3$) A white solid that occurs naturally as the mineral magnesite and in association with calcium carbonate in dolomite. Magnesium carbonate is sparingly soluble in water but reacts with dilute acids to give the salt, carbon dioxide, and water. Because of this it is used in medicine as a mild antacid. Magnesium carbonate also decomposes easily to give magnesium oxide, which is an important refractory material. It is also mixed with asbestos and used to cover hot pipes.

magnesium chloride ($MgCl_2$) A solid that exists in a variety of hydrated forms, most commonly as the hexahydrate, $MgCl_2.6H_2O$. When heated, this hydrated form is hydrolysed by its water of crystallization, with the evolution of hydrogen chloride and the formation of the oxide:

$$MgCl_2 + H_2O \rightarrow MgO + 2HCl$$

The anhydrous chloride must therefore be prepared by evaporating an aqueous solution in an atmosphere of hydrogen chloride. It is used in the preparation of cotton fabrics.

magnesium hydrogencarbonate (magnesium bicarbonate, $Mg(HCO_3)_2$) The solid hydrogencarbonate is unknown at room temperature. The compound is produced in solution when water containing carbon dioxide dissolves magnesium carbonate:

$$MgCO_3 + H_2O + CO_2 \rightarrow Mg(HCO_3)_2$$

It is a cause of temporary hardness in water.

magnesium hydroxide ($Mg(OH)_2$) A white solid that dissolves sparingly in water to give an alkaline solution. It can be prepared by adding an alkali to a soluble magnesium compound. Magnesium hydroxide is used in medicine as an antacid.

magnesium oxide (magnesia, MgO) A white solid that occurs naturally as the mineral periclase. It can be prepared by heating magnesium in oxygen or by decomposition of magnesium hydroxide, carbonate, or nitrate. Magnesium oxide is weakly basic because of the attraction of the oxide ions for protons from water molecules:

$$O^{2-} + H_2O \rightarrow 2OH^-$$

Magnesium oxide melts at 2800°C and is widely used as a refractory lining for metal, glass, and cement furnaces.

magnesium peroxide (Mg_2O_2) A white insoluble solid prepared by reacting sodium or barium peroxide with a concentrated solution of a magnesium salt. Magnesium peroxide is used as a bleach for dyestuffs and silks.

magnesium sulphate ($MgSO_4$) A solid that occurs naturally in many salt deposits in combination with other minerals. One hydrated form, $MgSO_4.7H_2O$, is known as *Epsom salt*. Epsom salt is made on a commercial scale by reacting magnesium carbonate with sulphuric acid. Magnesium sulphate is used as a purgative drug and as an antidote for barium and barbiturate poisoning. It is also used in the dyeing and sizing of textiles, in man-made fibre production, and as a source of magnesium in fertilizers.

magnetic quantum number *See* atom.

magnetic separation A method of separating crushed mineral mixtures using the magnetic properties which a component may possess, such as ferromagnetism or diamagnetism.

magnetism The study of the nature and cause of magnetic force fields, and how different substances are affected by them. Magnetic fields are produced by moving charge – on a large scale (as with a current in a coil, forming an *electromagnet*), or on the small scale of the moving charges in the atoms. It is generally assumed that the Earth's magnetism and that of other planets, stars, and galaxies have the same cause.

Substances may be classified on the basis of how samples interact with fields. Different types of magnetic behaviour result from the type of atom. *Diamagnetism*, which is common to all substances, is due to the orbital motion of electrons. *Paramagnetism* is due to electron spin, and a property of materials containing unpaired electrons. It is particularly important in transition-metal chemistry, in which the complexes often contain unpaired electrons. Magnetic measurements can give information about the bonding in these complexes. *Ferromagnetism*, the strongest effect, also involves electron spin and the alignment of magnetic moments in domains.

malachite A green mineral consisting of copper(II) carbonate and hydroxide ($CuCO_3.Cu(OH)_2$). It is used as an ore and a pigment.

maleic acid See butenedioic acid.

maltose ($C_{12}H_{22}O_{11}$) A sugar found in germinating cereal seeds. It is a disaccharide composed of two glucose units. Maltose is an important intermediate in the enzyme hydrolysis of starch. It is further hydrolysed to glucose.

manganate(VI) A salt containing the ion MnO_4^{2-}.

manganate(VII) (permanganate) A salt containing the ion MnO_4^-. Manganate(VII) salts are purple, and strong oxidizing agents. In basic solutions the dark green manganate(VI) ion is formed.

manganese A transition metal occurring naturally as oxides, e.g. pyrolusite (MnO_2). Its main use is in alloy steels made by adding pyrolusite to iron ore in an electric furnace. Manganese decomposes cold water and dilute acids to give hydrogen and reacts with oxygen and nitrogen when heated. The oxidation states are $+7$, $+6$, $+4$, and (the most stable) $+2$. Manganese(II) salts are pale pink and with alkali the solutions precipitate manganese(II) hydroxide, which rapidly oxidizes in air to brown manganese(III) oxide.
Symbol: Mn; m.p. 1244°C; b.p. 2040°C; r.d. 7.4; p.n. 25; r.a.m. 54.94.

manganese dioxide See manganese(IV) oxide.

manganese(II) oxide (manganous oxide, MnO) A green powder prepared by heating manganese(II) carbonate or oxalate in the absence of air. Alternatively it may be prepared by heating the higher manganese oxides in a stream of hydrogen. Manganese(II) oxide is a basic oxide and is almost insoluble in water. At high temperatures it is reduced by hydrogen to manganese. On exposure to air, manganese(II) oxide rapidly oxidizes. Manganese(II) oxide has a crystal lattice similar to that of sodium(I) chloride.

manganese(III) oxide (manganic oxide, manganese sesquioxide, Mn_2O_3) A black powder obtained by igniting manganese(IV) oxide or a manganese(II) salt in air at 800°C. It reacts slowly with cold dilute acids to form manganese(III) salts. Manganese(III) oxide occurs in nature as braunite ($3Mn_2O_3.MnSiO_3$) and the monohydrate ($Mn_2O_3.H_2O$) occurs in the mineral manganite. The manganese(III) oxide crystal lattice contains Mn^{3+} and O^{2-} ions. With concentrated alkalis, it undergoes disproportionation to give manganese(II) and manganese(IV) ions.

manganese(IV) oxide (manganese dioxide, MnO_2) A black powder prepared by the action of heat on manganese(II) nitrate. Hydrated forms occur naturally as the minerals pyrolusite and wad (a less pure form). Manganese(IV) oxide is insoluble in water. It is a powerful oxidizing agent: it reacts with hot concentrated hydrochloric acid to produce chlorine and with warm sulphuric acid to give oxygen. The dihydrate ($MnO_2.2H_2O$) is formed when potassium permanganate is reduced in alkaline solution. It is used as a catalyst in the laboratory preparation of chlorine, as a depolarizer in electric dry cells, and in the glass industry. At 500–600°C, manganese(IV) oxide decomposes to give manganese(III) oxide and trimanganese tetroxide. It has a good electrical conductivity.

manganese sesquioxide See manganese(III) oxide.

manganic oxide See manganese(III) oxide.

manganous oxide See manganese(II) oxide.

mannose ($C_6H_{12}O_6$) A simple sugar found in many polysaccharides. It is an aldohexose, isomeric with glucose.

manometer A device for measuring pressure. A simple type is a U-shaped glass tube containing mercury or other liquid. The pressure difference between the arms of the tube is indicated by the difference in heights of the liquid.

marble A dense mineral form of calcium carbonate ($CaCO_3$), formed from limestone.

Markovnikoff's rule A rule that predicts the quantities of the products formed when an acid (HA) adds to the double bond in an alkene. If the alkene is not symmetrical two products may result; for instance $(CH_3)_2C:CH_2$ can yield either $(CH_3)_2HCCH_2A$ or $(CH_3)_2ACCH_3$. The rule states that the major product will be the one in which the hydrogen atom attaches itself to the carbon atom with the larger number of hydrogen atoms. In the example above, therefore, the major product is $(CH_3)_2ACCH_3$.

The Markovnikoff rule is explainable if the mechanism is ionic. The first step is addition of H^+ to one side of the double bond, forming a carbonium ion. The more stable form of carbonium ion will be the form in which the positive charge appears on the carbon atom with the largest number of alkyl groups – thus the hydrogen tends to attach itself to the other carbon. The positive charge is partially stabilized by the electron-releasing (inductive) effect of the alkyl groups.

Additions of this type do not always follow the Markovnikoff rule. Under certain conditions the reaction may involve the free radicals H· and A·, in which case the opposite (anti-Markovnikoff) effect occurs.

marsh gas Methane produced in marshes by decomposing vegetation.

maser An acronym for Microwave Amplification by Stimulated Emission of Radiation. A process similar to the laser principle, but using microwaves instead of visible radiation.

mass Symbol: m A measure of the quantity of matter in an object. Mass is determined in two ways: the *inertial mass* of a body determines its tendency to resist change in motion; the *gravitational mass* determines its gravitational attraction for other masses. The SI unit of mass is the kilogram.

mass action, law of At constant temperature, the rate of a chemical reaction is directly proportional to the active mass of the reactants, the active mass being taken as the concentration in mol dm^{-3}. In general, the rate of reaction decreases steadily as the reaction proceeds; a measure of the concentration of any of the reactants will give a measure of the rate of reaction. For the reaction $A + B \rightarrow$ products, the law of mass action states that

$$rate = k[A][B]$$

where [A] represents the concentration of A in mol dm^{-3} and k is a constant dependent on the reaction. The term *active mass* can only equal the concentration in mol dm^{-3} if there is no interaction or interference between the reacting molecules. Many systems exhibit interactions and interference and consequently the concentration in mol dm^{-3} has to be multiplied by an activity coefficient in order to obtain the effective active mass.

mass-energy equation The equation $E = mc^2$, where E is the total energy (rest-mass energy + kinetic energy + potential energy) of a mass m, c being the velocity of light. The equation is a consequence of Einstein's Special theory of relativity; mass is a form of energy and energy also has mass. Conversion of rest-mass energy into kinetic energy is the source of power in radioactive substances and the basis of nuclear-power generation.

mass number *See* nucleon number.

mass spectrometer An instrument for producing ions in a gas and analysing them according to their charge/mass ratio. The earliest experiments by J. J. Thomson used a stream of positive ions from a discharge tube, which were deflected by parallel electric and magnetic fields at right angles to the beam. Each type of ion formed a parabolic trace on a photographic plate (a *mass spectrograph*).

In modern instruments, the ions are produced by ionizing the gas with electrons. The positive ions are accelerated out of this ion source into a high-vacuum region. Here, the stream of ions is deflected and focused by a combination of electric and magnetic fields, which can be varied so that different types of ion fall on a detector. In this way, the ions can be analysed according to their mass, giving a *mass spectrum* of the material. Mass spectrometers are used for accurate measurements of relative atomic mass and for analysis of isotope abundance. They can also be used to identify compounds and analyse mixtures. An organic compound bombarded with electrons forms a number of fragment ions. (Ethane (C_2H_6), for instance, might form CH^+, $C_2H_5^+$, CH_2^+, etc.) The relative proportions of different types of ions is used to find the structure of new compounds. The characteristic spectrum can also identify compounds by comparison with standard spectra.

mass spectrum *See* mass spectrometer.

matrix A continuous solid phase in which particles of a different solid phase are embedded.

matte A mixture of iron and copper sulphides obtained at an intermediate stage in smelting copper ores.

maxwell Symbol: Mx A unit of magnetic flux used in the c.g.s. system. It is equal to 10^{-8} Wb.

mechanism A step-by-step description of the events taking place in a chemical reaction. It is a theoretical framework accounting for the fate of bonding electrons and illustrates which bonds are broken and which are formed. For example, in the chlorination of methane to give chloromethane:

step 1
$$Cl:Cl \rightarrow 2Cl\cdot$$
step 2

$$Cl\cdot + CH_4 \rightarrow HCl + CH_3\cdot$$
step 3
$$CH_3\cdot + Cl:Cl \rightarrow CH_3Cl + Cl\cdot$$

mega- Symbol: M A prefix denoting 10^6. For example, 1 megahertz (MHz) $= 10^6$ hertz (Hz).

melting (fusion) The process by which a solid is converted into a liquid by heat or pressure.

melting point The temperature at which a solid is in equilibrium with its liquid phase at standard pressure and above which the solid melts. This temperature is always the same for a particular solid. Ionically bonded solids generally have much higher melting points than those in which the forces are covalent or intermolecular.

membrane A thin pliable sheet of tissue or other material acting as a boundary. The membrane may be either natural (as in cells, skin, etc.) or synthetic modifications of natural materials (cellulose derivatives or rubbers). In many physicochemical studies membranes are supported on porous materials, such as porcelain, to provide mechanical strength. Membranes are generally permeable to some degree. Membranes can be prepared to permit the passage of other molecules and micromolecular material. Because of permeability effects, concentration diffferences at a membrane give rise to a whole range of membrane-equilibrium studies, of which osmosis, dialysis, and ultra-filtration are examples.

See also semipermeable membrane.

mendelevium A radioactive transuranic element of the actinoid series, not found naturally on Earth. Several short-lived isotopes have been synthesized. Symbol: Md; p.n. 101; stablest isotope ^{258}Md (half-life 60 days).

Mendius reaction The reduction of the cyanide group to a primary amine group using sodium in alcohol:
$$RCN + 2H_2 \rightarrow RCH_2NH_2$$

It is a method of increasing the chain length of compounds in ascending a homologous series.

mercuric chloride *See* mercury(II) chloride.

mercuric oxide *See* mercury(II) oxide.

mercuric sulphide *See* mercury(II) sulphide.

mercurous chloride *See* mercury(I) chloride.

mercurous oxide *See* mercury(I) oxide.

mercurous sulphide *See* mercury(I) sulphide.

mercury A transition metal occurring naturally as cinnabar (mercury(II) sulphide). The vapour is very poisonous. Mercury is used in thermometers, special amalgams for dentistry, and in scientific apparatus. Mercury compounds are used as fungicides, timber preservatives, and explosives.
Symbol: Hg; m.p. −39°C; b.p. 357°C; r.d. 13.5; p.n. 80; r.a.m. 200.59. *See* amalgam, zinc group.

mercury cell A voltaic or electrolytic cell in which one or both of the electrodes consists of mercury or an amalgam. Amalgam electrodes are used in the Daniell cell and the Weston cadmium cell. Flowing mercury electrodes are also used in electrolytic cells. In the production of chlorine, sodium chloride solution is electrolysed in a cell with carbon anodes and a flowing mercury cathode. At the cathode, sodium metal is formed, which forms an amalgam with the mercury.

mercury(I) chloride (mercurous chloride, calomel, Hg_2Cl_2) A white precipitate prepared by adding dilute hydrochloric acid to a mercury(I) salt solution or by subliming mercury(II) chloride with mercury. Mercury(I) chloride is sparingly soluble in water and is blackened by both ammonia

gas, which it absorbs, and by alkalis. It is used as a purgative.

mercury(II) chloride (mercuric chloride, $HgCl_2$) A colourless crystalline compound prepared by direct combination of mercury with cold dry chlorine. Mercury(II) chloride is soluble in water; it also dissolves in concentrated hydrochloric acid due to the formation of complex ions, $HgCl_4^{2-}$ and $HgCl_3^{-}$. On heating, mercury(II) chloride sublimes forming a white translucent mass. It is extremely poisonous, but in dilute solution (1:1000) it is used as an antiseptic.

mercury(I) oxide (mercurous oxide, Hg_2O) (There is some doubt as to the existence of this compound.) The black precipitate formed on addition of sodium hydroxide solution to a solution of mercury(I) nitrate is thought by some to be mercury(I) oxide; others, who doubt its existence, think that the blackness of the precipitate is due to some free mercury. X-ray examination of this black compound has shown it to be an intimate mixture of mercury(II) oxide and mercury.

mercury(II) oxide (mercuric oxide, HgO) A poisonous compound formed as a yellow powder by the addition of sodium(I) hydroxide to a solution of mercury(II) nitrate or as a red solid by heating mercury at 350°C for a long time. The difference in colour is simply one of particle size. If the oxide is strongly heated it decomposes to give mercury and oxygen.

mercury(I) sulphide (mercurous sulphide, Hg_2S) (There is some doubt as to the existence of this compound.) The brownish black precipitate formed when mercury is treated with cold concentrated sulphuric acid for a long time is thought to be mercury(I) sulphide. Alternatively it may be prepared by the action of hydrogen sulphide or an alkaline sulphide on a mercury(I) salt solution. As soon as the mercury(I) sulphide has been formed it disproportionates to give mercury(II) sulphide and mercury.

mercury(II) sulphide (mercuric sulphide, vermilion, HgS) A compound that occurs in nature as the minerals cinnabar (a red solid) and metacinnabar (a black solid). It is prepared as a black precipitate by the action of hydrogen sulphide on a soluble mercury(II) salt solution. On heating, mercury(II) sulphide sublimes and becomes red. It is insoluble in dilute hydrochloric and nitric acids but will dissolve in concentrated nitric acid and aqua regia. Mercury(II) sulphide is used as the pigment vermilion.

meso-form See optical activity.

meta- 1. Designating a benzene compound with substituents in the 1,3 positions. The position on a benzene ring that is two carbon atoms away from a substituent is the meta position. This was used in the systematic naming of benzene derivatives. For example, meta-dinitrobenzene (or *m*-dinitrobenzene) is 1,3-dinitrobenzene.
2. Certain acids regarded as formed from an anhydride and water are named meta acids to distinguish them from the more hydrated ortho acids. For example, H_2SiO_3 ($SiO_2 + H_2O$) is metasilicic acid; H_4SiO_4 ($SiO_2 + 2H_2O$) is orthosilicic acid. See also ortho-, para-.

metabolism The chemical reactions that take place in cells. The molecules taking part in these reactions are termed *metabolites*. Metabolic reactions characteristically occur in small steps that together make up a *metabolic pathway*. They involve the breaking down of molecules to provide energy (catabolism) and the building up of more complex molecules and structures from simpler molecules (anabolism).

metabolite A substance that takes part in a metabolic reaction, either as reactant or product. Some are synthesized within the organism itself, while others have to be taken in as food.

metal See metals.

metal carbonyl A coordination compound formed between a metal and carbonyl groups. Transition metals form many such compounds.

metaldehyde See ethanal.

metallic bond A bond formed between atoms of a metallic element in its zero oxidation state and in an array of similar atoms. The outer electrons of each atom are regarded as contributing to an 'electron gas', which occupies the whole crystal of the metal. It is the attraction of the positive atomic cores for the negative electron gas that provides the strength of the metallic bond.
Quantum mechanical treatment of the electron gas restricts its energy to a series of 'bands'. It is the behaviour of electrons in these bands that gives rise to semiconductors.

metallic crystal A crystal formed by metal atoms in the solid state. Each atom contributes its valence (outer) electrons to a 'sea' of electrons, which are free to migrate through the solid, and the ions remaining are arranged in a lattice. The ability of electrons to move through the lattice accounts for the electrical and thermal conductivity of metals.

metallocene A sandwich compound in which a metal atom is coordinated to two cyclopentadienyl ions. Ferrocene ($Fe(C_5H_5)_2$) is the commonest example.

metalloid Any of a class of chemical elements that are intermediate in properties between metals and non-metals. Examples are germanium, arsenic, and tellurium.
There is, in fact, no clear-cut distinction between metals and non-metals. In the periodic table, there is a change from metallic to non-metallic properties across the table, and an increase in metallic properties down a group. Consequently there is a diagonal region around the centre of the table (B, Si, As, Te) in which there is a borderline between metals and non-metals, and the metalloids are the borderline

cases. Elements such as arsenic, germanium, and tellurium are semiconductors, but other elements are often said to be metalloids according to their chemical properties. Tin, for instance, forms salts with acids but also forms stannates with alkalis. Its oxide is amphoteric. Note also that tin has metallic (white tin) and non-metallic (grey tin) allotropes.

metallurgy The study of metals, especially methods of extracting metals from their ores and the formation and properties of alloys.

metals Any of a class of chemical elements with certain characteristic properties. In everyday usage, metals are elements (and alloys) such as iron, aluminium, and copper, which are lustrous malleable solids – usually good conductors of heat and electricity. Note that this is not a strict definition – some metals are poor conductors, mercury is a liquid, etc.
In chemistry, metals are distinguished by their chemical properties, and there are two main groups. Reactive metals, such as the alkali metals and alkaline-earth metals, are electropositive elements. They are high in the electrochemical series and tend to form compounds by losing electrons to give positive ions. They have basic oxides and hydroxides. This typical metallic behaviour decreases across the periodic table and increases down a group in the table. The other type of metals are the transition elements, which are less reactive, have variable valences, and tend to form complexes. In the solid and liquid states metals have metallic bonds, formed by positive ions with free electrons.
See also metalloid, non-metal.

metastable species An excited state of an atom, ion, or molecule, that has a relatively long lifetime before reverting to the ground state. Metastable species are intermediates in some reactions.

metastable state A condition of a system or body in which it appears to be in stable equilibrium but, if disturbed, can settle into a lower energy state. For exam-

ple, supercooled water is liquid below $0°C$ (at standard pressure). When a small crystal of ice or dust (for example) is introduced, rapid freezing occurs.

metathesis Double decomposition.

Methanal trimer

methanal (formaldehyde, HCOH) A colourless gaseous aldehyde. It is manufactured by the oxidation of methanol ($500°C$ and a silver catalyst)

$$2CH_3OH + O_2 \rightarrow 2HCOH + 2H_2O$$

The compound is used in the manufacture of urea-formaldehyde resins. A solution of methanal (40%) in water is called *formalin*. It is extensively used as a preservative for biological specimens.
If an aqueous solution of methanal is evaporated a polymer – *polymethanal* – is formed:

$$-O-CH_2-O-CH_2-O-CH_2-$$

This was formerly called *paraformaldehyde*. If methanal is distilled from acidic solutions a cyclic *methanal trimer* ($C_3O_3H_6$) is produced.

methane (CH_4) A gaseous alkane. Natural gas is about 99% methane and this provides an important starting material for the organic-chemicals industry. Methane can be chlorinated directly to produce the more reactive chloromethanes, or it can be 'reformed' by partial oxidation or using steam to give mixtures of carbon oxides and hydrogen. Methane is the first member of the homologous series of alkanes.

methanoate (formate) A salt or ester of methanoic acid.

methanoic acid (formic acid, HCOOH) A liquid carboxylic acid made by the action of sulphuric acid on sodium methanoate (NaOOCH). It is a strong reducing agent. Methanoic acid occurs naturally in ants and nettles.

methanol (methyl alcohol, wood alcohol, CH_3OH) A colourless liquid alcohol, which is used as a solvent and in the manufacture of methanal (formaldehyde) for the plastics and drugs industries. Methanol was originally produced from the distillation of wood. Now it is manufactured by the catalytic oxidation of methane from natural gas.

methionine See amino acids.

methoxy group The group CH_3O-.

methyl acetate See methyl ethanoate.

methyl alcohol See methanol.

methylated spirits Ethanol to which is added methanol (about 9.5%), pyridine (about 0.5%), and a blue dye. The ethanol is denatured in this way so that it can be sold without excise duty for use as a fuel and solvent.

methylation A reaction in which a methyl group (CH_3-) is introduced. Friedel–Crafts reactions with halo-methanes are examples of methylations.

methylbenzene (toluene, $C_6H_5CH_3$) A colourless liquid hydrocarbon, similar to benzene both in structure and properties. As methylbenzene is much less toxic than benzene it is more widely used, especially as a solvent. Large quantities are required for the manufacture of TNT (trinitrotoluene).
Methylbenzene is itself produced either from the fractional distillation of coal tar or from methylcyclohexane (a constituent of some crude oils):
$$C_6H_{11}CH_3 \rightarrow C_6H_5CH_3 + 3H_2$$
A catalyst of aluminium and molybdenum oxides is employed at high temperatures and pressures.

methyl bromide See bromomethane.

methyl chloride See chloromethane.

methylene See carbene.

methylene group The group $:CH_2$.

methyl ethanoate (methyl acetate, CH_3COOCH_3) A colourless liquid ester with a fragrant odour, used as a solvent.

methyl ethyl ketone See butanone.

methyl group The group CH_3-.

methyl iodide See iodomethane.

methyl orange An acid-base indicator that is red in solutions below a pH of 3 and yellow above a pH of 4.4. As the transition range is clearly on the acid side, methyl orange is suitable for the titration of an acid with a moderately weak base, such as sodium carbonate.

methylphenols (cresols, $HOC_6H_4CH_3$) Compounds with both methyl and hydroxyl groups substituted onto the benzene ring. There are three isomers, with the methyl group in the 2-, 3-, and 4- positions, respectively. A mixture of the isomers can be obtained from coal tar. It is used as a germicide.

methyl red An acid-base indicator that is red in solutions below a pH of 4.2 and yellow above a pH of 6.3. It is often used for the same types of titration as methyl orange but the transition range of methyl red is nearer neutral (pH7) than that of methyl orange. The molecules are structurally similar.

metre Symbol: m The SI base unit of length, defined as the length equal to 1 650 763.73 wavelengths in vacuum corresponding to the transition between the levels $2p_{10}$ and $5d_5$ of the ^{86}Kr atom.

metric system A system of units based on the metre and the kilogram and using multiples and submultiples of 10. SI units, c.g.s. units, and m.k.s. units are all scientific metric systems of units.

metric ton *See* tonne.

mho *See* siemens.

micro- Symbol: μ A prefix denoting 10^{-6}. For example, 1 micrometre (μm) = 10^{-6} metre (m).

micron Symbol: μ A unit of length equal to 10^{-6} metre.

microwaves A form of electromagnetic radiation, ranging in wavelength from about 1 mm (where it merges with infrared) to about 120 mm (bordering on radio waves). Microwaves are produced by various electronic devices including the klyston; they are often carried over short distances in tubes of rectangular section called *wave guides*.
Spectra in the microwave region can give information on the rotational energy levels of certain molecules.
See also electromagnetic radiation.

milk sugar *See* lactose.

milli- Symbol: m A prefix denoting 10^{-3}. For example, 1 millimetre (mm) = 10^{-3} metre (m).

millimetre of mercury *See* mmHg.

mirror image A shape that is identical to another except that its structure is reversed as if viewed in a mirror. If an object is not symmetrical it cannot be superimposed on its mirror image. For example, the left hand is the mirror image of the right hand.

Misch metal A pyrophoric alloy of cerium and other lanthanoids, used in lighter flints.

miscible Denoting combinations of substances that, when mixed, give rise to only one phase; i.e. substances that dissolve in each other.

Mitscherlich's law (law of isomorphism) The law stating that substances that crystallize in isomorphous forms (i.e.

have identical crystalline forms and form mixed crystals) have similar chemical compositions. The law can be used to indicate the formulae of compounds. For instance, the fact that chromium oxide is isomorphous with Fe_2O_3 and Al_2O_3 implies that its formula is Cr_2O_3.

mixed indicator A mixture of two or more indicators so as to decrease the pH range or heighten the colour change, etc.

mixture Two or more substances forming a system in which there is no chemical bonding between the two. In homogeneous mixtures (e.g. solutions or mixtures of gases) the molecules of the substances are mixed, and there is only one phase. In heterogeneous mixtures (e.g. gunpowder or certain alloys) different phases can be distinguished. Mixtures differ from chemical compounds in that:
(1) The chemical properties of the components of a mixture are the same as those of the pure substances.
(2) The mixture can be separated by physical means (e.g. distillation or crystallization) or mechanically.
(3) The proportions of the components can vary. Some mixtures (e.g. certain solutions) can only vary in proportions between definite limits.

m.k.s. system A system of units based on the metre, the kilogram, and the second. It formed the basis for SI units.

mmHg (millimetre of mercury) A former unit of pressure defined as the pressure that will support a column of mercury one millimetre high under specified conditions. It is equal to 133.322 4 Pa, and is almost identical to the torr.

molal concentration *See* concentration.

molar 1. Denoting a physical quantity divided by the amount of substance. In almost all cases the amount of substance will be in moles. For example, volume (V) divided by the number of moles (n) is molar volume $V_m = v/n$.

2. A *molar solution* contains one mole of solute per cubic decimeter of solvent.

molarity A measure of the concentration of solutions based upon the number of molecules or ions present, rather than on the mass of solute, in any particular volume of solution. The molarity (M) is the number of moles of solute in one cubic decimetre (litre). Thus a 0.5M solution of hydrochloric acid contains $0.5 \times (1 + 35.5)$ g HCl per dm^3 of solution.

mole Symbol: mol The SI base unit of amount of substance, defined as the amount of substance that contains as many elementary entities as there are atoms in 0.012 kilogram of ^{12}C. The elementary entities may be atoms, molecules, ions, electrons, photons, etc., and they must be specified. The amount of substance is proportional to the number of entities, the constant of proportionality being the Avogadro number. One mole contains $6.022\ 045 \times 10^{23}$ entities. One mole of an element with relative atomic mass A has a mass of A grams (this mass was formerly called one *gram-atom* of the element). One mole of a compound with relative molecular mass M has a mass of M grams (this was formerly called one *gram-molecule* of the compound).

molecular crystal A crystal in which molecules, as opposed to atoms, occupy lattice points. Examples include iodine and solid carbon dioxide (dry ice). Because the forces holding the molecules together are weak, molecular crystals have low melting points. When the molecules are small, the crystal structure approximates to a close-packed arrangement. *See* close packing.

molecular formula The formula of a compound showing the number and types of the atoms present in a molecule, but not the arrangement of the atoms. For example, C_2H_6O represents the molecular formula both of ethanol (C_2H_5OH) and methoxymethane (CH_3OCH_3). The molecular formula can only be determined if the molecular weight is known. *Compare* empirical formula, structural formula.

molecularity The total number of reacting molecules in the individual steps of a chemical reaction. Thus, a unimolecular step has molecularity 1, a bimolecular step etc. Molecularity is always an integer, whereas the order of a reaction need not necessarily be so. The molecularity of a reaction gives no information about the mechanism by which it takes place.

molecular orbital *See* orbital.

molecular sieve A substance through which molecules of a limited range of sizes can pass, enabling volatile mixtures to be separated. Zeolites and other metal aluminium silicates can be manufactured with pores of constant dimensions in their molecular structure. When a sample is passed through a column packed with granules of this material, some of the molecules enter these pores and become trapped. The remainder of the mixture passes through the interstices in the column. The trapped molecules can be recovered by heating. Molecular-sieve chromatography is widely used in chemistry and biochemistry laboratories.

A modified form of molecular sieve is used in *gel filtration*. The sieve is a continuous gel made from a polysaccharide. In this case, molecules larger than the largest pore size are totally excluded from the column.

molecular spectrum The absorption or emission spectrum that is characteristic of a molecule. Molecular spectra are usually band spectra.

molecular weight *See* relative molecular mass.

molecule A particle formed by the combination of atoms in a whole-number ratio. A molecule of an element (combining atoms are the same, e.g. O_2) or of a compound (different combining atoms, e.g. HCl) retains the properties of that element or compound. Thus, any quantity of a compound is a collection of many identical molecules. Molecular sizes are characteristically 10^{-10} to 10^{-9} m.

143

Many molecules of natural products are so large that they are regarded as giant molecules (macromolecules); they may contain thousands of atoms and have complex structural formulae that require very advanced techniques to identify. *See also* formula, relative molecular mass.

mole fraction The number of moles of a given component in a mixture divided by the total number of moles present of all the components. The mole fraction of component A is

$$n_A/(n_A + n_B + n_C + \ldots)$$

where n_A is the number of moles of A, etc.

molybdenum A transition element that occurs naturally in molybdenite (MoS_2). It is used in alloy steels and in lamp bulbs. The compound ammonium molybdate, dissolved in nitric acid, is used as a test for phosphates(V).
Symbol: Mo; m.p. 2620°C; b.p. 4800°C; r.d. 10.2; p.n. 42; r.a.m. 95.94.

monatomic Denoting a molecule, radical, or ion consisting of only one atom. For example, helium is a monatomic gas and H· is a monatomic radical.

monobasic acid An acid which has only one active proton, such as hydrochloric acid.

monochlorobenzene *See* chlorobenzene.

monoclinic crystal *See* crystal system.

monosaccharide A sugar that cannot be hydrolysed to simpler carbohydrates of smaller carbon content. Glucose and fructose are examples.

monomer The molecule, group, (or compound) from which a dimer, trimer, or polymer is formed.

monotropy The existence of a single allotrope of an element that is always more stable than the other(s) regardless of temperature; phosphorus, for example, exhibits monotropy. The phase diagram for a monotropic element shows that one allotrope always has a lower vapour pressure than the other(s) at all temperatures; this is the stable allotrope.

mordant An inorganic compound used to fix dye in cloth. The mordant (e.g. aluminium hydroxide or chromium salts) is precipitated in the fibres of the cloth, and the dye then absorbs in the particles.

Moseley's law Lines in the X-ray spectra of elements have frequencies that depend on the proton number of the element. For a set of elements, a graph of the square root of the frequency of X-ray emission against proton number is a straight line (for spectral lines corresponding to the same transition).

mother liquor The solution remaining after the formation of crystals.

multicentre bond A two-electron bond formed by the overlap of orbitals from more than two atoms (usually 3). The bridging in diborane is believed to take place by overlap of an sp^3 hybrid orbital from each boron atom with the 1s orbital on the hydrogen atom. This multicentre bond is called a two-electron three-centre bond. The molecule is electron deficient. *See also* electron-deficient compounds.

multidentate ligand A ligand which possesses at least two sites at which it can coordinate.

multiple bond A bond between two atoms involving more than one pair of electrons, e.g. a double bond or a triple bond. This additional bonding arises from overlap of atomic orbitals that are perpendicular to the inter-nuclear axis and gives rise to an increase in electron density above and below the inter-nuclear axis. Such bonds are called pi bonds. If the sigma bond axis is taken as the z-axis (i.e. the inter-nuclear axis) then overlap of orbitals along the x-axis gives rise to a pi bond in the xz plane. Similarly orbitals on the y-axis form a pi bond in the yz plane.

multiple proportions, law of Proposed by Dalton in 1804, the principle that when two elements A and B combine to form more than one compound, the weights of B that combine with a fixed weight of A are in small whole-number ratios. For example, in nitrous oxide, N_2O, nitric oxide, NO, and dinitrogen tetroxide, N_2O_4, the amounts of nitrogen combined with a fixed weight of oxygen are in the ratio 4:2:1.

multiple-range indicator *See* universal indicator.

Mumetal (*Trademark*) A magnetic alloy containing about 75% nickel, the remainder being iron, copper, and chromium and sometimes molybdenum. It is easily magnetized and demagnetized, has a high permeability, and is used in transformer cores and electromechanical equipment.

muriate An obsolete name for a chloride; e.g. muriate of potash (KCl).

mutarotation A change in the optical rotation of a solution with time.

nano- Symbol: n A prefix denoting 10^{-9}. For example, 1 nanometre (nm) = 10^{-9} metre (m).

Representations of naphthalene

naphthalene ($C_{10}H_8$) A white crystalline solid with a distinctive smell of mothballs. Naphthalene is found in both the middle- and heavy-oil fractions of crude oil and is obtained by fractional crystallization. It is used in the manufacture of benzene-1,2-dicarboxylic anhydride and thence in the production of plastics and dyes.

The structure of naphthalene is 'benzene-like', having two six-membered rings fused together. The reactions are characteristic of aromatic compounds. *See also* aromatic compound.

nascent hydrogen A particularly reactive form of hydrogen, which is believed to exist briefly between its generation (e.g. by the action of dilute acid on magnesium) and its appearance as bubbles of normal hydrogen gas. It is thought that part of the free energy of the production reaction remains with the hydrogen molecules for a short time. Nascent hydrogen may be used to produce the hydrides of phosphorus, arsenic, and antimony, which are not readily formed from ordinary hydrogen.

Natta process A method for the manufacture of isotactic polypropene using Ziegler catalysts. *See* addition polymerization, Ziegler process.

natural abundance *See* abundance.

natural gas Gas obtained from underground deposits and often associated with sources of crude oil. It contains a high proportion of methane and other volatile hydrocarbons.

neodymium A toxic silvery element having two allotropic forms and belonging to the lanthanoid series of metals. It occurs in association with other lanthanoids. Neodymium is used in various alloys, as a catalyst, and in compound form in carbon-arc searchlights, etc., and in the glass industry. Symbol: Nd; m.p. 1010°C; b.p. 3127°C; r.d. 6.8 or 7.0; p.n. 60; r.a.m. 144.24.

neon An inert colourless odourless monatomic element of the rare-gas group. It occurs in minute quantities (0.0018% by volume) in air. It is used in neon signs and lights, electrical equipment, and gas lasers. Symbol: Ne; m.p. −248.67°C; b.p. −246.05°C; d. 0.9 kg m^{-3}; p.n. 10; r.a.m. 20.18.

neoprene A type of synthetic rubber made by polymerization of 2-chlorobuta-1,2,-diene ($H_2C{:}CHCCl{:}CH_2$). It is more resistant to oil, solvents, and temperature than natural rubbers.

neptunium A toxic radioactive silvery element of the actinoid series of metals that was the first transuranic element to be synthesized (1940). Found on Earth only in minute quantities in uranium ores, it is obtained as a by-product in ^{239}Pu production.

Symbol: Np; m.p. 640°C; b.p. 3902°C (est.); r.d. 20.25; p.n. 93; stablest isotope ^{237}Np (half-life 2.2×10^6 years).

neutralization The stoichiometric reaction of an acid and a base in volumetric analysis. The neutralization point or end point is detected with indicators.

neutron number Symbol: N The number of neutrons in the nucleus of an atom; i.e. the nucleon number (A) minus the proton number (Z).

Newlands' law (law of octaves) The observation that, when the elements are arranged in order of increasing atomic weight, there is a similarity between members that are eight elements apart (inclusive). For example, lithium has similarities to sodium, beryllium to magnesium, and so on. Newlands discovered this relationship in 1863 before the announcement of Mendeléev's famous periodic law (1869). It is now recognized that Newlands' law arises from there being eight members in each short period.

newton Symbol: N The SI unit of force, equal to the force needed to accelerate one kilogram by one metre second^{-2}. 1 N = 1 kg m s^{-2}.

Nichrome (*Trademark*) Any of a group of nickel-chromium-iron alloys, containing 60–80% nickel and about 16% chromium; small amounts of other elements, such as carbon or silicon, may be added. They can withstand very high temperatures and their high electrical resistivity makes them suitable for use in heating elements.

nickel A transition metal that occurs naturally as the sulphide and silicate. It is extracted by the Mond process, which involves reduction of nickel oxide using carbon monoxide followed by the formation and subsequent decomposition of volatile nickel carbonyl. Nickel is used as a catalyst in the hydrogenation of alkenes, e.g. margarine manufacture, and in coinage alloys. Its main oxidation state is +2 and these compounds are usually green.

Symbol: Ni; m.p. 1450°C; b.p. 2840°C; r.d. 8.9; p.n. 28; r.a.m. 58.70.

nickel carbonyl ($Ni(CO)_4$) A colourless liquid with a highly toxic vapour. It is prepared by passing carbon monoxide over finely divided nickel at a temperature of approximately 60°C. The product is purified by fractional distillation. Nickel carbonyl decomposes on heating to give pure nickel. It is used as a catalyst and in the preparation of nickel (Mond process).

nickelic oxide See nickel(III) oxide.

nickel-iron accumulator (Edison cell, Nife or NIFE cell) A type of secondary cell in which the electrodes are formed by steel grids. The positive electrode is impregnated with a nickel-nickel hydride mixture and the negative electrode is impregnated with iron oxide. Potassium hydroxide solution forms the electrolyte. The nickel-iron cell is lighter and more durable than the lead accumulator, and it can work with higher currents. Its e.m.f. is approximately 1.3 volts.

nickelous oxide See nickel(II) oxide.

nickel(II) oxide (nickelous oxide, NiO) A pale green powder formed by heating nickel(II) hydroxide, nitrate, or carbonate in the absence of air. It is a basic oxide dissolving in dilute acids to give green solutions of nickel(II) salts. Nickel(II) oxide may be reduced by carbon, carbon monoxide, or hydrogen.

nickel(III) oxide (nickelic oxide, Ni_2O_3) A black powder formed by the action of heat on nickel(II) oxide in air. It can also be prepared by igniting nickel(II) nitrate or carbonate. It exists as the dihydrate, $Ni_2O_3.2H_2O$.

nickel-silver (German silver) Any of a group of alloys containing nickel, copper, and zinc in various proportions (but no silver). They are white or silvery in colour, can be highly polished, and have good corrosion-resistance. They are used, for example, in silver plating, chromium plating, and enamelling.

NIFE cell *See* nickel-iron accumulator.

ninhydrin A colourless organic compound that gives a blue coloration with amino acids. It is used as a test for amino acids; in particular, to show the positions of spots of amino acids in paper chromatography.

niobium A transition element used in welding, special steels, and nuclear reactor work.
Symbol: Nb; m.p. 2468°C; b.p. 4742°C; r.d. 8.6; p.n. 41; r.a.m. 92.91.

nitrate A salt or ester of nitric acid.

nitration A reaction introducing the nitro ($-NO_2$) group into an organic compound. Nitration of aromatic compounds is usually carried out using a mixture of concentrated nitric and sulphuric acids, although the precise conditions will differ from compound to compound. The attacking species is NO_2^+ (the nitryl cation), and the reaction is an example of electrophilic substitution.

nitre *See* potassium nitrate.

nitric acid (HNO_3) A colourless fuming corrosive liquid that is a strong acid. Nitric acid can be made in a laboratory by the distillation of a mixture of an alkali metal nitrate and concentrated sulphuric acid. Commercially it is prepared by the catalytic oxidation of ammonia and is supplied as concentrated nitric acid, which contains 68% of the acid and is often coloured yellow by dissolved oxides of nitrogen. Nitric acid is a strong oxidizing agent. Most metals are converted to their nitrates with the evolution of oxides of nitrogen (the composition of the mixture of the oxides depends on the temperature and on the concentration of the nitric acid used). Some non-metals (e.g. sulphur and phosphorus) react to produce oxyacids. Organic substances (e.g. sawdust and ethanol) react violently, but the more stable aromatic compounds, such as benzene and toluene, can be converted to nitro compounds in controllable reactions.

nitric oxide *See* nitrogen monoxide.

nitriding *See* case hardening.

nitrile (cyanide) A type of organic compound containing the $-CN$ group. Nitriles are colourless liquids with pleasant smells. They can be prepared by either refluxing an organic halogen compound with an alcoholic solution of potassium cyanide, or by dehydration of an amide with phosphorus(V) oxide. Characteristic reactions are hydrolysis (to the carboxylic acid and ammonia) and reduction with hydrogen gas (to an amine).

nitrile rubber A copolymer of butadiene and acrylonitrile ($CH_2=CHCN$). It is an important rubber due to its oil- and solvent-resistance.

nitrite A salt or ester of nitrous acid.

nitrobenzene ($C_6H_5NO_2$) A yellow organic oil obtained by refluxing benzene with a mixture of concentrated nitric and sulphuric acids. The reaction is a typical electrophilic substitution on the benzene ring by the nitryl cation (NO_2^+).

nitrocellulose *See* cellulose nitrate.

Nitro compound

nitro compound A type of organic compound containing the nitro ($-NO_2$) group attached to an aromatic ring. Nitro compounds can be prepared by nitration using a mixture of concentrated nitric and sulphuric acids. They can be reduced to aromatic amines:

$$RNO_2 + 3H_2 \rightarrow RNH_2 + 2H_2O$$

They can also undergo further substitution on the benzene ring. The nitro group directs substituents into the 3 position.

nitrogen The first element of group V of the periodic table; a very electronegative element existing in the uncombined state as gaseous diatomic N_2 molecules. Nitrogen has the electronic configuration $[He]2s^22p^3$. It is typically non-metallic and its bonding is primarily by polarized covalent bonds. With electropositive elements the nitride ion N^{3-} may be formed.

Nitrogen accounts for about 78% of the atmosphere (by volume) and it also occurs as sodium nitrate in various mineral deposits. It is separated for industrial use by the fractionation of liquid air. Pure nitrogen is prepared by thermal decomposition of azides;

$$2NaN_3 \rightarrow 2Na + 3N_2$$

The 'active nitrogen' gas obtained by passing an electric discharge through nitrogen is a mixture of ordinary nitrogen molecules and excited single nitrogen atoms.

Nitrogen is a diatomic molecule, which is effectively triple-bonded and has a high dissociation energy (940 kJ mol^{-1}). It is therefore inert and it only reacts readily with lithium and the highly electropositive elements of group II. The direct combination of nitrogen and hydrogen occurs at elevated temperatures and pressures ($400-600°C$, 100 atmospheres) and is the basis of the industrially important Haber process for the manufacture of ammonia.

Nitrogen forms at least eight oxides:
Dinitrogen monoxide (nitrous oxide, N_2O) – a neutral oxide.
Nitrogen monoxide (nitric oxide, NO) – a neutral oxide.
Nitrogen(III) oxide (nitrogen sesquioxide, N_2O_3) – an unstable oxide that is the anhydride of nitrous acid.
Nitrogen dioxide (NO_2), which gives a mixture of nitrous and nitric acids in water.
Dinitrogen tetroxide (N_2O_4), which is in equilibrium with nitrogen dioxide.
Dinitrogen pentoxide (N_2O_5) – the anhydride of nitric acid.
Nitrogen trioxide (NO_3).

Nitrogen also forms a number of binary halides such as NF_3, NCl_3, and halogen azides such as ClN_3. Except for NF_3 these are all very explosive, and even NF_3, which is reputedly stable, has been known to explode.

Transition metals form nitrides that are different from the ionic or covalent nitrides, the stoichiometries are, for example, ZrN, W_2N, Mn_4N. They are often called 'interstitial' nitrides. They are very hard and their formation is the basis of surface hardening by exposing hot metal to ammonia gas.

Nitrogen has two isotopes; ^{14}N, the common isotope and ^{15}N, about 0.3% abundant. ^{15}N is used as a marker in mass spectrometric studies.

Symbol: N; m.p. $-210°C$; b.p. $-196°C$; d. 1.2506 kg m^{-3}; p.n. 7; r.a.m. 14.

nitrogen dioxide (NO_2) A brown gas produced by the dissociation of dinitrogen tetroxide (with which it is in equilibrium), the dissociation being complete at $140°C$. Further heating causes dissociation to colourless nitrogen monoxide and oxygen:

$$2NO_2(g) = 2NO(g) + O_2(g)$$

Nitrogen dioxide can also be made by the action of heat on metal nitrates (not the nitrates of the alkali metals or some of the alkaline-earth metals).

nitrogen fixation A reaction in which atmospheric nitrogen is converted into nitrogen compounds. Nitrogen fixation occurs naturally by certain bacteria (in the soil and in the roots of leguminous plants). Small amounts of nitrogen(II) oxide (NO) are also formed in thunderstorms by direct combination of the elements in the electrical discharge. Fixation of nitrogen is important because nitrogen is an essential element for plant growth, and vast amounts of nitrogenous fertilizers are also required for agriculture. Former processes for fixing nitrogen were the Birkeland-Eyde process (reacting N_2 and O_2 in an electric arc – the equivalent of the thunderstorm in nature), and the cyanamide process (heating calcium dicarbide with nitrogen to give $CaCN_2$). Now the major method is the Haber process for making ammonia.

nitrogen monoxide (nitric oxide, NO) A colourless gas that is insoluble in water but dissolves in a solution containing iron(II) ions owing to the formation of the complex ion $(FeNO)^{2+}$: the nitrogen monoxide can be released by heating. Nitrogen monoxide is prepared by the action of nitric acid on copper turnings; the impure product can be purified by using a solution of iron(II) ions to absorb the product. Commercially, nitrogen monoxide is prepared by the catalytic oxidation of ammonia or by the direct union of nitrogen and oxygen in an electric arc. Nitrogen monoxide is the most heat-stable of the oxides of nitrogen, only decomposing above 1000°C. At ordinary temperatures it combines immediately with oxygen to give nitrogen dioxide:

$$2NO(g) + O_2(g) \rightarrow 2NO_2(g)$$

nitrogenous base See quaternary ammonium compound.

nitroglycerine A highly explosive substance used in dynamite. It is obtained by treating glycerol (1,2,3-trihydroxypropane) with a mixture of concentrated nitric and sulphuric acids. It is not a nitro compound, but a nitrate ester

$$CH_2(NO_3)CH(NO_3)CH_2(NO_3)$$

nitro group The group $-NO_2$; the functional group of nitro compounds.

nitronium ion See nitryl ion.

nitrophenols ($C_6H_4(OH)NO_2$) Organic compounds formed directly or indirectly by the nitration of phenol. Three isomeric forms are possible. The 2 and 4 isomers are produced by the direct nitration of phenol and can be separated by steam distillation, the 2 isomer being steam volatile. The 3 isomer is produced from nitrobenzene by formation of 1,3-dinitrobenzene, conversion to 3-nitrophenylamine, and thence by diazotization to 3-nitrophenol.

nitrous acid (HNO_2) A weak acid known only in solution, obtained by acidifying a solution of a nitrite. It readily decomposes on warming or shaking to nitrogen monoxide and nitric acid. The use of nitrous acid is very important in the dyestuffs industry in the diazo reaction: nitrous acid is liberated by acidifying a solution of a nitrite (usually sodium nitrite) in the presence of the compound to be diazotized. Nitrous acid and the nitrites are normally reducing agents but in certain circumstances they can behave as oxidizing agents, e.g. with sulphur dioxide and hydrogen sulphide.

nitrous oxide See dinitrogen oxide.

nitryl ion (nitronium ion) The positive ion NO_2^+. See nitration.

NMR See nuclear magnetic resonance.

nobelium A radioactive transuranic element of the actinoid series, not found naturally on Earth. Several very short-lived isotopes have been synthesized. Symbol: No; p.n. 102.

noble gases (rare gases, inert gases, group 0 elements) A group of monatomic, low-boiling gases classified as belonging to group 0 of the periodic table and occupying a position between the highly electronegative group VII elements and the highly electropositive group I elements. They are: helium (He), neon (Ne), argon (Ar), krypton (Kr), xenon (Xe), and radon (Rn). This complete filling of the s and p levels gives a closed-shell structure that is associated with a general lack of chemical reactivity and very high ionization energies. It is the acquisition of a stable noble-gas configuration that is associated with determining the valence of many covalent and ionic compounds.

Prior to 1962 the noble gases were frequently called *inert gases* as no chemical compounds were known (there were a few clathrates and 'hydrates'), but the realization that the ionization potential of xenon was sufficiently low to be accessible to chemical reaction led to the preparation of several fluorides, oxides, oxyfluorides, and a hexafluoroplatinate of xenon. Several unstable krypton and radon compounds have been synthesized.

Argon forms about 1% of the atmosphere but the other gases are present in only minute traces (excluding radon). They are obtained by fractionation of liquid air or (in the US) of natural gas. Considerable amounts of argon are produced as a by-product from ammonia-plant tail gas. Their applications depend heavily on their inertness; for example, welding (Ar), filling light bulbs (Ar), gas-stirring in high temperature metallurgy (Ar), powder technology (Ar), discharge tubes (Ne), and chemical research as inert carriers (He, Ar). Neon has been proposed as an oxygen diluent for deep-sea diving because of its lower solubility in blood. Liquid helium is extensively used in low-temperature work and radon is used therapeutically as a source of α-particles.

The gases helium and argon are formed by radioactive decay in various minerals, for example argon by electron capture of ^{40}K. This mechanism may be applied to age-determination of minerals. Radon is generally obtained as ^{222}Rn from decay of radium. The half-life is only about 3.8 days.

non-benzenoid aromatic *See* aromatic compound.

non-localized bond *See* delocalized bond.

non-metal Any of a class of chemical elements. Non-metals lie in the top right-hand region of the periodic table. They are electronegative elements with a tendency to form covalent compounds or negative ions. They have acidic oxides and hydroxides. In the solid state, non-metals are either covalent volatile crystals or macromolecular (giant-covalent) crystals. *See also* metal, metalloid.

non-polar compound A compound that has molecules with no permanent dipole moment. Examples of non-polar

compounds are hydrogen, tetrachloromethane, and carbon dioxide.

non-polar solvent *See* solvent.

normality The number of gram equivalents per cubic decimetre of a given solution.

normal solution A solution that contains one gram equivalent weight per litre of solution. Values are designated by the symbol N, e.g. 0.2N, N/10, etc. Because there is not a clear definition of equivalent weight suitable for all reactions, a solution may have one value of normality for one reaction and another value in a different reaction. Because of this many workers prefer the molar solution notation.

NTP *See* STP.

nuclear magnetic resonance (NMR) A method of investigating nuclear spin. In an external magnetic field the nucleus can have certain quantized energy states, corresponding to certain orientations of the spin magnetic moment. Hydrogen nuclei, for instance, can have two energy states, and transitions between the two occur by absorption of radiofrequency radiation. In chemistry, this is the basis of a spectroscopic technique for investigating the structure of molecules. Radiofrequency radiation is fed to a sample and the magnetic field is changed slowly. Absorption of the radiation is detected when the difference between the nuclear levels corresponds to absorption of a quantum of radiation. This difference depends slightly on the electrons around the nucleus – i.e. the position of the atom in the molecule. Thus a different absorption frequency is seen for each type of hydrogen atom. In ethanol, for example, there are three frequencies, corresponding to hydrogen atoms on the CH_3, the CH_2, and the OH. The intensity of absorption also depends on the number of hydrogen atoms (3:2:1). NMR spectroscopy is a powerful method of finding the structures of organic compounds.

nucleon number (mass number) Symbol: A The number of nucleons (protons plus neutrons) in an atomic nucleus.

nucleophile An electron-rich ion or molecule that takes part in an organic reaction. The nucleophile can be a negative ion (Br^-, CN^-) or a molecule with a lone pair of electrons (NH_3, H_2O). The nucleophile attacks positively charged parts of molecules, which usually arise from the presence of an electronegative atom elsewhere in the molecule. *Compare* electrophile.

nucleophilic addition A class of reaction involving the addition of a small molecule to an unsaturated organic compound. Since these reactions are initiated by nucleophiles, the unsaturated bond must contain an electronegative atom, which creates an electron-deficient area in the molecule. Nucleophilic addition is a characteristic reaction of aldehydes and ketones and is often followed by the subsequent elimination of a different small molecule, particularly water. The addition of hydrogen cyanide to propanone is an example:

$$CH_3COCH_3 + HCN \rightarrow$$
$$CH_3C(CN)(OH)CH_3$$

See also condensation reaction.

nucleophilic substitution A reaction involving the substitution of an atom or group of atoms in an organic compound with a nucleophile as the attacking substituent. Since nucleophiles are electron-rich species, nucleophilic substitution occurs in compounds in which a strongly electronegative atom or group leads to a dipolar bond. The electron-deficient centre can then be attacked by the electron-rich nucleophile causing the electronegative atom or group to be displaced. In general terms:

$$R-Le + Nu^- \rightarrow R-Nu + Le^-$$

where Nu^- represents the incoming nucleophile and Le^- represents the leaving group.

There are two possible mechanisms for nucleophilic substitution. In the S_N1 (sub-

stitution, nucleophilic, monomolecular) reaction the molecule first forms a carbonium ion; for example:

$$RCH_2Cl \rightarrow RCH_2^+ + Cl^-$$

The nucleophile then attaches itself to this carbonium ion:

$$RCH_2^+ + OH^- \rightarrow RCH_2OH$$

In the S_N2 (substitution, nucleophilic, bimolecular) reaction the nucleophile approaches as the other group leaves, forming a transition state in which the carbon has five attached groups.

The preferred mechanism depends on several factors:

(1) The stability of the intermediate in the S_N1 mechanism.

(2) Steric factors affecting the formation of the transition state in the S_N2 mechanism.

(3) The solvent in which the reaction occurs: polar solvents will stabilize polar intermediates and so favour the S_N1 mechanism.

In practice, however, both routes or some intermediate mechanism is thought to operate in many cases.

See also substitution reaction.

nucleus The compact positively charged centre of an atom made up of one or more nucleons (protons and neutrons) around which is a cloud of electrons. The density of nuclei is about 10^{15} kg m^{-3}. The number of protons in the nucleus defines the element, being its proton number (or atomic number). The nucleon number, or atomic mass number, is the sum of the protons and neutrons. The simplest nucleus is that of a hydrogen atom, ^1H, being simply one proton (mass 1.67×10^{-27} kg). The most massive naturally occurring nucleus is ^{238}U of 92 protons and 146 protons (mass 4×10^{-25} kg, radius 9.54×10^{-15} m). Only certain combinations of protons and neutrons form stable nuclei. Others undergo spontaneous decay.

A nucleus is depicted by a symbol indicating nucleon number (mass number), proton number (atomic number), and element name. For example, $^{23}_{11}Na$ represents a nucleus of sodium having 11 protons and mass 23, hence there are $(23 - 11) = 12$ neutrons.

S_N1 mechanism: note that the entering ion may enter either side of the intermediate. If the reactant has a chiral centre, a mixture of optical isomers is formed.

S_N2 mechanism: note that the entering ion enters from the opposite side to the leaving group, and inversion occurs.

Nucleophilic substitution

nuclide A nuclear species with a given number of protons and neutrons; for example, ^{23}Na, ^{24}Na, and ^{24}Mg are all different nuclides. Thus:

$^{23}_{11}Na$ has 11 protons and 12 neutrons

$^{24}_{11}Na$ has 11 protons and 13 neutrons

$^{24}_{12}Mg$ has 12 protons and 12 neutrons

The term is applied to the nucleus and often also to the atom.

nylon A type of synthetic polymer linked by amide groups $-NH.CO-$. Nylon polymers can be made by copolymerization of a molecule containing two amine groups with one containing two carboxylic acid groups.

occlusion 1. The process in which small amounts of one substance are trapped in the crystals of another; for example, pockets of liquid occluded during crystallization from a solution.
2. Absorption of a gas by a solid; for example, the occlusion of hydrogen by palladium.

octadecanoic acid (stearic acid, $CH_3(CH_2)_{16}COOH$) A solid carboxylic acid present in fats and oils as the glyceride.

octadecenoic acid cis-9-octadecenoic acid (or oleic acid) is a naturally occurring carboxylic acid present (as glycerides) in fats and oils:

$$CH_3(CH_2)_7CH:CH(CH_2)_7COOH$$

octahedral complex See complex.

octahydrate A crystalline compound containing eight molecules of water of crystallization per molecule of compound.

octavalent Having a valence of eight.

octane (C_8H_{18}) A liquid alkane obtained from the light fraction of crude oil. Octane and its isomers are the principal constituents of petrol, which is obtained as the re-fined light fraction from crude oil. See also octane rating.

octane rating A rating for the performance of petrol in internal-combustion engines. The octane rating measures the freedom from 'knocking' – i.e. preignition of the fuel in the engine. This depends on the relative proportions of branched-chain and straight-chain hydrocarbons present. High proportions of branched-chain alkanes are better in high-performance engines. In rating fuels, 2,2,4-trimethylpentane (isooctane) is given a value of 100 and heptane is given a value 0. The performance of a fuel is compared with a mixture of these hydrocarbons.

octaves, law of See Newlands' law.

octet A stable shell of eight electrons in an atom. The completion of the octet gives rise to particular stability and this is the basis of the Lewis octet theory, thus:
(1) the noble gases have complete octets and are chemically inert,
(2) the bonding in small covalent molecules is frequently achieved by the central atom completing its octet by sharing electrons with surrounding atoms, e.g. CH_4, H_2O,

1,6-diaminohexane

hexanedioic acid

Formation of nylon

155

(3) the ions formed by electropositive and electronegative elements are generally those with a complete octet, e.g. Na^+, Ca^{2+}, O^{2-}, Cl^-.

oersted Symbol: Oe A unit of magnetic field strength in the c.g.s. system. It is equal to $10^3/4\pi$ A m^{-1}.

ohm Symbol: Ω The SI unit of electrical resistance, equal to a resistance that passes a current of one ampere when there is an electric potential difference of one volt across it. $1\,\Omega = 1\,V\,A^{-1}$. Formerly, it was defined in terms of the resistance of a column of mercury under specified conditions.

oil Any of various viscous liquids. *See* petroleum, glyceride.

oil of vitriol Sulphuric(VI) acid.

oleate A salt or ester of oleic acid.

olefin *See* alkene.

oleic acid *See* octadecenoic acid.

oleum (pyrosulphuric acid, $H_2S_2O_7$) A colourless fuming liquid formed by dissolving sulphur(VI) oxide in concentrated sulphuric acid. It gives sulphuric acid on diluting with water. *See* contact process.

onium ion An ion formed by addition of a proton (H^+) to a molecule. The hydronium ion (H_3O^+) and ammonium ion (NH_4^+) are examples.

open-hearth process A method of making steel by heating pig iron in an open-hearth furnace by burning a mixture of gas and air. Oxygen is injected through pipes to oxidize impurities and burn off carbon, and lime is used to form a slag.

optical activity The ability of certain compounds to rotate the plane of polarization of plane-polarized light when the light is passed through them. Optical activity can be observed in crystals, gases, liquids, and solutions. The amount of rotation depends on the concentration of the compound.

Optical activity is caused by the interaction of the varying electric field of the light with the electrons in the molecule. It occurs when the molecules are asymmetric – i.e. they have no plane of symmetry. Such molecules have a mirror image that cannot be superimposed on the original molecule. In organic compounds this usually means that they contain a carbon atom attached to four different groups, forming a chiral centre. The two mirror-image forms of an asymmetric molecule are optical isomers. One isomer will rotate the polarized light in one sense and the other by the same amount in the opposite sense. Such isomers are described as *dextrorotatory* or *laevorotatory*, according to whether they rotate the plane to the 'right' or 'left' respectively (rotation to the left is clockwise to an observer viewing the light coming towards him). Dextrorotatory compounds are given the symbol *d* or (+) and laevorotatory compounds *l* or (−). A mixture of the two isomers in equal amounts does not show optical activity. Such a mixture is sometimes called the (±) or *dl*-form, a *racemate*, or a *racemic mixture*.

Optical isomers have identical physical properties (apart from optical activity) and cannot be separated by fractional crystallization or distillation. Their general chemical behaviour is also the same, although they do differ in reactions involving other optical isomers. Many naturally occurring substances are optically active (only one optical isomer exists naturally) and biochemical reactions occur only with the natural isomer. For instance, the natural form of glucose is *d*-glucose and living organisms cannot metabolize the *l*-form.

The terms 'dextrorotatory' and 'laevorotatory' refer to the effect on polarized light. A more common method of distinguishing two optical isomers is by their *D*-form (*dextro-form*) or *L*-form (*laevo-form*). This convention refers to the absolute structure of the isomer according to specific rules. Sugars are related to a particular configuration of glyceraldehyde (2,3-dihydroxypropanal). For alpha amino

D – lactic acid   L – lactic acid

Optical isomers of lactic acid

D – form   L – form   meso – form

Isomers of tartaric acid

Optical isomers

acids the 'corn rule' is used: the structure of the acid $RC(NH_2)(COOH)$ H is drawn with H at the top; viewed from the top the groups spell CORN in a clockwise direction for all D-amino acids (i.e. the clockwise order is $-COOH,R,NH_2$). The opposite is true for L-amino acids. Note that this convention does not refer to optical activity: D-alanine is dextrorotatory but D-cystine is laevorotatory.

An alternative is the R/S system for showing configuration. There is an order of priority of attached groups based on the proton number of the attached atom:

I, Br, Cl, SO_3H, $OCOCH_3$, OCH_3, OH, NO_2, NH_2, $COOCH_3$, $CONH_2$, $COCH_3$, CHO, CH_2OH, C_6H_5, C_2H_5, CH_3, H

Hydrogen has the lowest priority. The chiral carbon is viewed such that the group of lowest priority is hidden behind it. If the other three groups are in descending priority in a clockwise path, the compound is R-. If descending priority is anticlockwise it is S-.

The existence of a carbon atom bound to four different groups is not the strict condition for optical activity. The essential point is that the molecule should be asymmetric. Inorganic octahedral complexes, for example, can show optical isomerism. It is also possible for a molecule to contain asymmetric carbon atoms and still have a plane of symmetry. One structure of tartaric acid has two parts of the molecule that are mirror images, thus having a plane of symmetry. This (called the *meso-form*) is not optically active. *See also* resolution (of racemates).

optical isomerism *See* isomerism, optical activity.

optical rotary dispersion (ORD) The phenomenon in which the amount of

157

The corn rule for absolute configuration of alpha amino acids

"R" – configuration   "S" – configuration

The R/S system

Absolute configuration

rotation of plane-polarized light by an opti-cally active substance depends on the wavelength of the light. Plots of rotation against wavelength can be used to give in-formation about the molecular structure of optically active compounds.

optical rotation Rotation of the plane of polarization of plane-polarized light by an optically active substance.

orbital A region around an atomic nu-cleus in which there is a high probability of finding an electron. The modern picture of the atom according to quantum mechan-ics does not have electrons moving in fixed elliptical orbits. Instead, there is a finite probability that the electron will be found in any small region at all possible distances from the nucleus. In the hydrogen atom the probability is low near the nucleus, in-creases to a maximum, and falls off to in-

finity. It is useful to think of a region in space around the nucleus — in the case of hydrogen the region within a sphere — within which there is a high chance of find-ing the electron. Each of these, called an *atomic orbital*, corresponds to a sub-shell and can 'contain' two electrons with oppo-site spins. Another way of visualizing an orbital is as a cloud of electron charge (the average distribution with time).

Similarly, in molecules the electrons move in the combined field of the nuclei and can be assigned to *molecular orbitals*. In con-sidering bonding between atoms it is use-ful to treat molecular orbitals as formed by overlap of atomic orbitals.

It is possible to calculate the shapes and energies of atomic and molecular orbitals by quantum theory. The shapes of atomic orbitals depend on the orbital angular mo-mentum (the sub-shell). For each shell there is one s orbital, three p orbitals, five

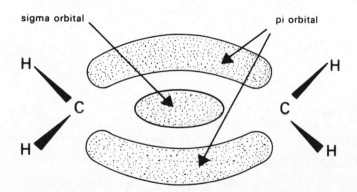

The bonding in ethene. Each carbon atom has three sp² hybrid orbitals. Overlap of these produces a sigma orbital between the carbon atoms. The lateral overlap of the two p orbitals produces a pi orbital. Note that this is a single orbital with two lobes, above and below the plane of the molecule. The double bond thus consists of a sigma bond and a pi bond.

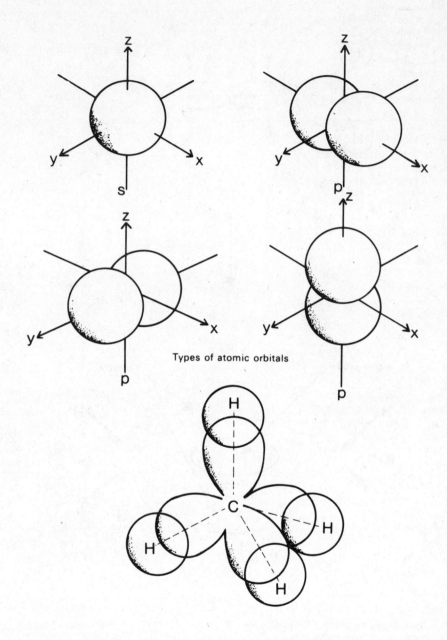

Types of atomic orbitals

Bonding in methane: four sp³ orbitals of the carbon overlap with the s orbitals of hydrogen atoms, forming four sigma bonds.

d orbitals, etc. The s orbitals are spherical, the p orbitals each have two lobes; d orbitals have more complex shapes, typically with four lobes.

Molecular orbitals are formed by overlap of atomic orbitals, and again there are different types. If the orbital is completely symmetrical about an axis between the nuclei, it is a *sigma orbital*. This can occur, for instance, by overlap of two s orbitals, as in the hydrogen atom, or two p orbitals with their lobes along the axis. However, two p orbitals overlapping at right angles to the axis form a different type of molecular orbital – a *pi orbital* – with regions above and below the axis. Pi orbitals are also formed by overlap of d orbitals. Each molecular orbital can contain a pair of electrons, forming a sigma bond or pi bond. A double bond, for example the bond in ethene, is a combination of a sigma bond and a pi bond.

Hybrid orbitals are atomic orbitals formed by combinations of s, p, and d atomic orbitals, and are useful in describing the bonding in compounds. There are various types. In carbon, for instance, the electron configuration is $1s^2 2s^2 2p^2$. Carbon, in its outer (valence) shell, has one filled s orbital, two filled p orbitals, and one 'empty' p orbital. These four orbitals may *hybridize* (sp^3 hybridization) to act as four equal orbitals arranged tetrahedrally, each with one electron. In methane, each hybrid orbital overlaps with a hydrogen s orbital to form a sigma bond. Alternatively, the s and two of the p orbitals may hybridize (sp^2 hybridization) and act as three orbitals in a plane at 120°. The remaining p orbital is at right angles to the plane, and can form pi bonds. Finally, sp hybridization may occur, giving two orbitals in a line. More complex types of hybridization, involving d orbitals, explain the geometries of inorganic complexes.

The combination of two atomic orbitals in fact produces two molecular orbitals. One – the *bonding orbital* – has a concentration of electron density between the nuclei, and thus tends to hold the atoms together. The other – the *antibonding orbital* – has slightly higher energy and tends to repel the atoms. If both atomic orbitals are filled, the two molecular orbitals are also filled and cancel each other out – there is no net bonding effect. If each atomic orbital has one electron, the pair occupies the lower energy bonding orbital, producing a net attraction.

order The sum of the indices of the concentration terms in the expression that determines the rate of a chemical reaction. For example, in the expression
$$\text{rate} = k[A]^x[B]^y$$
x is called the order with respect to A, y the order with respect to B, and $(x + y)$ the order overall. The values of x and y are not necessarily equal to the coefficients of A and B in the molecular equation. Order is an experimentally determined quantity derived without reference to any equation or mechanism. Fractional orders do occur. For example, in the reaction
$$CH_3CHO \rightarrow CH_4 + CO$$
the rate is proportional to $[CH_3CHO]^{1.5}$ i.e. it is of order 1.5.

ore A mineral source of a chemical element.

organic chemistry The chemistry of compounds of carbon. Originally the term *organic chemical* referred to chemical compounds present in living matter, but now it is any carbon compound with the exception of certain simple ones, such as the carbon oxides, carbonates, cyanides, and cyanates. These are generally studied in inorganic chemistry. The vast numbers of synthetic and natural organic compounds exist because of the ability of carbon to form chains of atoms (catenation). Other elements are involved in organic compounds; principally hydrogen and oxygen but also nitrogen, halogens, sulphur, and phosphorus.

organometallic compound An organic compound containing a carbon–metal bond. Tetraethyl lead, $(C_2H_5)_4Pb$, is an example of an organometallic compound, used as an additive in petrol.

ortho- 1. Designating the form of a diatomic molecule in which both nuclei have

the same spin direction; e.g. orthohydrogen, orthodeuterium.

2. Designating a benzene compound with
substituents in the 1,2 positions. The position next to a substituent is the ortho position on the benzene ring. This was used in
the systematic naming of benzene derivatives. For example, orthodinitrobenzene
(or o-dinitrobenzene) is 1,2-dinitroben-
zene.

3. Certain acids, regarded as formed from
an anhydride and water, were named
ortho acids to distinguish them from the
less hydrated meta acids. For example,
H_4SiO_4 (from $SiO_2 + 2H_2O$) is orthosilicic
acid; H_2SiO_3 ($SiO_2 + H_2O$) is metasilicic
acid.

See also meta-, para-.

orthohydrogen See hydrogen.

orthophosphoric acid See phos-
phoric(V) acid.

orthophosphorous acid See phosphonic acid.

orthorhombic crystal See crystal system.

osmium A transition metal that is found
associated with platinum. It oxidizes readily and is used in pen nibs, pivots, and electrical contacts.
Symbol: Os; m.p. 3045°C; b.p. 5027°C;
r.d. 22.6; p.n. 76; r.a.m. 190.2.

osmosis Systems in which a solvent is
separated from a solution by a semipermeable membrane approach equilibrium
by solvent molecules on the solvent side of
the membrane migrating through it to the
solution side; this process is called osmosis and always leads to dilution of the solution. The phenomenon is quantified by
measurement of the osmotic pressure.
The process of osmosis is of fundamental
importance in transport and control mechanisms in biological systems; for example,
plant growth and general cell function.
Osmosis is a colligative property and its
theoretical treatment is similar to that for
the lowering of vapour pressure. The

membrane can be regarded as equivalent
to the liquid-vapour interface, i.e. one that
permits free movement of solvent molecules but restricts the movement of solute
molecules. The solute molecules occupy a
certain area at the interface and therefore
inhibit solvent egress from the solution.
Just as the development of a vapour pressure in a closed system is necessary for
liquid−vapour equilibrium, the development of an osmotic pressure on the solution side is necessary for equilibrium at the
membrane.

See osmotic pressure.

osmotic pressure Symbol: π The
pressure that must be exerted on a solution to prevent the passage of solvent molecules into it when the solvent and
solution are separated by a semipermeable membrane. The osmotic pressure is
therefore the pressure required to maintain equilibrium between the passage of
solvent molecules through the membrane
in either direction and thus prevent the
process of osmosis proceeding. The osmotic pressure can be measured by plac-
ing the solution, contained in a small
perforated thimble covered by a semipermeable membrane and fitted with a length
of glass tubing, in a beaker of the pure
solvent. Solvent molecules pass through
the membrane, diluting the solution and
thereby increasing the volume on the solution side and forcing the solution to rise up
the glass tubing. The process continues
until the pressure exerted by the solvent
molecules on the membrane is balanced
by the hydrostatic pressure of the solution
in the tubing. A sample of the solution is
then removed and its concentration determined. Osmosis is a colligative property,
therefore the method can be applied to
the determination of relative molecular
masses, particularly for large molecules,
such as proteins, but it is restricted by the
difficulty of preparing good semipermeable membranes.

As the osmotic pressure is a colligative
property it is directly proportional to the
molar concentration of the solute if the
temperature remains constant; thus π is
proportional to the concentration n/V,

where n is the number of moles of solute, and V the solvent volume. The osmotic pressure is also proportional to the absolute temperature. Combining these two proportionalities gives $\pi V = nCT$, which has the same form as the gas equation, $PV = nRT$, and experimental values of C are similar to those for R, the universal gas constant. This gives considerable weight to the kinetic description of colligative properties.

Ostwald's dilution law *See* dissociation.

oxidant An oxidizing agent. In rocket fuels, the oxidant is the compound that provides the oxygen for combustion (e.g. liquid oxygen or hydrogen peroxide).

oxidation An atom, an ion, or a molecule is said to undergo oxidation or to be oxidized when it loses electrons. The process may be effected chemically, i.e. by reaction with an *oxidizing agent*, or electrically, in which case oxidation occurs at the anode. For example,
$$2Na + Cl_2 \rightarrow 2Na^+ + 2Cl^-$$
where chlorine is the oxidizing agent and sodium is oxidized, and
$$4CN^- + 2Cu^{2+} \rightarrow C_2N_2 + 2CuCN$$
where Cu^{2+} is the oxidizing agent and CN^- is oxidized.
The *oxidation state* of an atom is indicated by the number of electrons lost or effectively lost by the neutral atom, i.e. the oxidation number. The oxidation number of a negative ion is negative. The process of oxidation is the converse of reduction. *See also* redox.

oxidation-reduction *See* redox.

oxidizing acid An acid that acts as an oxidizing agent, e.g. sulphuric(VI) acid or nitric(V) acid. Because it is oxidizing, nitric(V) acid can dissolve metals below hydrogen in the electrochemical series:
$$2HNO_3 + M \rightarrow MO + 2NO_2 + H_2O$$
$$MO + 2HNO_3 \rightarrow H_2O + M(NO_3)_2$$

oxidizing agent *See* oxidation.

oxime A type of organic compound containing the C:NOH group, formed by reaction of an aldehyde or ketone with hydroxylamine (NH_2OH).

oxo process A method of manufacturing aldehydes by passing a mixture of carbon monoxide, hydrogen, and alkanes over a cobalt catalyst at high pressure (100 atmospheres and 150°C). The aldehydes can subsequently be reduced to alcohols, making the process a useful source of alcohols of high molecular weight.

oxygen A colourless odourless diatomic gas; the first member of group VI of the periodic table. It has the electronic configuration $[He]2s^22p^4$ and its chemistry involves the acquisition of electrons to form either the di-negative ion, O^{2-} or two covalent bonds. In each case the oxygen atom attains the configuration of the noble gas neon.
Oxygen is the most plentiful element in the Earth's crust accounting for over 40% by weight. It is present in the atmosphere (20%) and is a constituent of the majority of minerals and rocks (e.g. sandstones, SiO_2, carbonates, $CaCO_3$, clays-aluminosilicates) as well as the major constituent of the sea. Oxygen is an essential element for almost all living things. Elemental oxygen has two forms: the diatomic molecule O_2 and the less stable molecule trioxygen (ozone), O_3, which is formed by passing an electric discharge through oxygen gas. There are several convenient laboratory routes to oxygen:
(1) Heat on unstable oxides,
$$2HgO \rightarrow 2Hg + O_2$$
(2) Decomposition of peroxides,
$$2H_2O_2 \rightarrow 2H_2O + O_2$$
(3) Heat on '-ate' compounds such as chlorates or permanganates
$$2KClO_3 \rightarrow KCl + 3O_2$$
$$2KMnO_4 \rightarrow K_2MnO_4 + MnO_2 + O_2$$
Industrially oxygen is obtained by the fractionation of liquid air.
Oxygen reacts with elements from all groups of the periodic table except group 0 and even this unreactive group forms several compounds containing oxygen (XeO_3, XeO_4, $XeOF_2$, XeO_2^{2-}).

The nature of the oxides has long been a convenient guide to metallic and non-metallic character. Thus, metals typically combine with oxygen to form oxides, which if soluble react with water to form *bases* (via OH^- ion); the more electropositive the metal the more vigorous the reaction. Most of the group I and group II (main-group) metals behave like this forming oxides, O^{2-}, (e.g., Na_2O, CaO), peroxides, O_2^{2-}, (e.g., Na_2O_2, BaO_2), and superoxides, O_2^-, (e.g., RbO_2, CsO_2). Oxides of metallic elements are solids and X-ray studies show that discrete O^{2-} ions do exist in the solid state even though they do not exist in solution:

$$O^{2-} + H_2O \rightarrow 2OH^-$$

The non-metals typically form molecular oxides such as SO_2, ClO_2, and NO_2. When soluble in water they generally dissolve to form *acids*. The less electronegative elements among the non-metals generally form weaker acids, e.g. $B(OH)_3$ and H_2CO_3. The more electronegative elements form stronger acids. The higher the oxidation state of the central atom the stronger the acid ($HOCl < HClO_2 < HClO_4$). Non-metal oxides may be gaseous (SO_2, CO_2), liquid (N_2O_4), or solid (P_2O_5). The weakly non-metallic elements such as silicon may require bases for dissolution and in some cases the oxides have extended polymeric structures, e.g., SiO_2. In between the acidic and basic oxides are the *amphoteric oxides*, which behave acidically to strong bases and basically towards strong acids. This is associated with elements which are physically metallic but which are chemically only weakly cationic, e.g., ZnO gives Zn^{2-} with acids and the zincates, $Zn(OH)_4^{2-}$, with bases. There is also a small class of *mixed oxides* that can be regarded as salts between a 'basic' and an 'acidic' form of the same metal: for example,

$$FeO + Fe_2O_3 \rightarrow Fe(FeO_2)_2 = Fe_3O_4$$
$$2PbO + PbO_2 \rightarrow Pb_3O_4$$

The neutral oxides are essentially insoluble non-metal oxides such as N_2O, NO, CO, F_2O, although extreme conditions can lead to some degree of acidic behaviour in the case of CO, for example,

$$CO + OH^- \rightarrow HCOO^-$$

Oxygen occurs in three natural isotopic forms, ^{16}O (99.76%), ^{17}O (0.0374%), ^{18}O (0.2039%); the rarer isotopes are used in detailed studies of the behaviour of oxygen-containing groups during reactions (tracer studies).

Oxygen is very limited in its catenation but an interesting range of O–O species is formed by the following: O_2^{2-} peroxide, O_2^- superoxide, O_2 oxygen gas, O_2^+ oxygenyl cation. The O–O bond gets both progressively shorter and stronger in the series. The peroxides are formed with Na, Ca, Sr, and Ba and are formally related to hydrogen peroxide (prepared by the action of acids on ionic peroxides). The superoxides of K, Rb, and Cs are prepared by burning the metal in air (the less electropositive metals, Na, Mg, and Zn form less stable superoxides). Oxygenyl cation, O_2^+, is formed by reaction of PtF_6 (high electron affinity) with oxygen at room temperature. Other oxygenyl cation species can be made with group V fluorides, e.g. $O_2^-AsF_6^-$.

Symbol: O; m.p. $-218.4°C$; b.p. $-183°C$; d. 1.429 kg m^{-3}; p.n. 8; r.a.m. 15.99.

ozone (trioxygen, O_3) A poisonous, blue-coloured allotrope of oxygen made by passing oxygen through a silent electric discharge. Ozone is unstable and decomposes to oxygen on warming. It is present in the upper layers of the atmosphere, where it screens the Earth from harmful short-wave ultraviolet radiation. There is concern that the ozone layer is possibly being depleted by fluorocarbons and other compounds produced by industry. Ozone is a strong oxidizing agent. Its concentration can be determined by reacting it with iodide ions and titrating the iodine formed with standard sodium thiosulphate(VI). It forms ozonides with alkenes. *See* ozonolysis.

ozonide *See* ozonolysis.

ozonolysis The addition of ozone (O_3) to alkenes and the subsequent hydrolysis of the ozonide into hydrogen peroxide and a mixture of carbonyl compounds. The carbonyl compounds can be separated and identified, which in turn, identifies the groups and locates the position of the double bond in the original alkene. Ozonolysis was formerly an important analytical technique.

Ozonolysis of an alkene to form a mixture of carbonyl compounds

P

spotted samples

Apparatus for paper chromatography.

palladium A transition metal occurring in platinum ores in Canada. It is used in electrical relays and as a catalyst in hydrogenation processes.
Symbol: Pd; m.p. 1552°C; b.p. 3140°C; r.d. 12.0; p.n. 46; r.a.m. 106.4.

palmitic acid *See* hexadecanoic acid.

paper chromatography A technique widely used for the analysis of mixtures. Paper chromatography usually employs a specially produced paper as the stationary phase. A base line is marked in pencil near the bottom of the paper and a small sample of the mixture is spotted onto it using a capillary tube. The paper is then placed vertically in a suitable solvent, which rises up to the base line and beyond by capillary action. The components within the sample mixture dissolve in this mobile phase and are carried up the paper. However, the paper holds a quantity of moisture and some components will have a greater tendency to dissolve in this moisture than in the mobile phase. In addition, some components will cling to the surface of the paper. Therefore, as the solvent moves through the paper, certain components will be left behind and separated from each other. When the solvent has almost reached the top of the paper, the paper is removed and quickly dried. The paper is developed to locate the positions of colourless fractions by spraying with a suitable chemical, e.g. ninhydrin, or by exposure to ultraviolet radiation. The components are identified by comparing the distance they have travelled up the paper with standard solutions that have been run simultaneously, or by computing an R_F value. A simplified version of paper chromatography uses a piece of filter paper. The sample is spotted at the centre of the paper and solvent passed through it. Separation of the components of the mixture again takes place as the mobile phase spreads out on the paper.
Paper chromatography is an application of the partition law.

para- 1. Designating the form of diatomic molecule in which both nuclei have opposite spin directions; e.g. parahydrogen, paradeuterium.
2. Designating a benzene compound with substituents in the 1,4 positions. The position on a benzene ring directly opposite a substituent is the para position. This was used in the systematic naming of benzene compounds. For example, para-dinitrobenzene (or p-dinitrobenzene) is 1,4-dinitrobenzene.
See also meta-, ortho-.

paraffin 1. See petroleum.
2. See alkane.

paraffin wax A solid mixture of hydrocarbons, obtained from petroleum.

paraformaldehyde See methanal.

parahydrogen See hydrogen.

paraldehyde See ethanal.

paramagnetism See magnetism.

partial ionic character The electrons of a covalent bond between atoms or groups with different electronegativities will be polarized towards the more electronegative constituent; the magnitude of this effect can be measured by the ionic character of the bond. When the effect is small the bond is referred to simply as a polar bond and is adequately treated using dipole moments; as the effect grows the theoretical treatment requires other contributions to ionic character. Examples of partial ionic character (from nuclear quadrupole resonance) are H−I, 21%; H−Cl, 40%; Li−Br, 90%; compare NaCl, 100%.

partial pressure In a mixture of gases, the contribution that one component makes to the total pressure. It is the pressure that the gas would have if it alone were present in the same volume. See Dalton's law of partial pressures.

partition coefficient If a solute dissolves in two non-miscible liquids, the partition coefficient is the ratio of the concentration in one liquid to the concentration in the other liquid.

pascal Symbol: Pa The SI unit of pressure, equal to a pressure of one newton per square metre ($1 \text{ Pa} = 1 \text{ N m}^{-2}$).

Paschen series A series of lines in the infrared spectrum emitted by excited hydrogen atoms. The lines correspond to the atomic electrons falling into the third lowest energy level and emitting energy as radiation. The wavelength (λ) of the radiation in the Paschen series is given by $1/\lambda = R(1/3^2 - 1/n^2)$ where n is an integer and R is the Rydberg constant. See also spectral series.

passive Describing a metal that is unreactive because of the formation of a thin protective layer. Iron, for example, does not dissolve in concentrated nitric(V) acid because of the formation of a thin oxide layer.

Pauli exclusion principle See exclusion principle.

p-block elements The elements of the main groups IIIA (B to Tl), IVA (C to Pb), VA (N to Bi), VIA (O to Po), VIIA (F to At), and 0 (He to Rn). They are so called because they have outer electronic configurations that have occupied p levels.

PEC See photoelectrochemical cell.

pentahydrate A crystalline compound that has five molecules of water of crystallization per molecule of compound.

pentane (C_5H_{12}) A straight-chain alkane obtained by distillation of crude oil.

pentanoic acid (valeric acid, $CH_3(CH_2)_3COOH$) A colourless liquid carboxylic acid, used in making perfumes.

pentavalent (quinquevalent) Having a valence of five.

167

pentose A sugar that has five carbon atoms in its molecules. *See* sugar.

pentyl group The group
$CH_3CH_2CH_2CH_2CH_2-$.

peptide A compound formed by linkage of two or more amino-acid groups together.

perchloric acid *See* chloric(VII) acid.

perfect gas *See* gas laws, kinetic theory.

period One of the horizontal rows in the conventional periodic table. Each period represents the elements arising from progressive filling of the outer shell (i.e. the addition of one extra electron for each new element), the elements being arranged in order of ascending proton number. In a strict sense hydrogen and helium represent one period but convention refers to the elements lithium to neon (8 elements) as the first *short period* ($n = 2$), and the elements sodium to argon (8 elements) as the second *short period* ($n = 3$). With entry to the $n = 4$ level there is filling of the 4s, then back filling of the 3d, before the 4p are filled. Thus this set contains a total of 18 electrons (potassium to krypton) and is called a *long period*. The next set, rubidium to xenon, is similarly a long period. *See also* Aufbau principle.

periodic acid *See* iodic(VII) acid.

periodic law The law upon which the modern periodic table is based. Enunciated in 1869 by Mendeléev, this law stated that the properties of the elements are a periodic function of their atomic weights: if arranged in order of increasing atomic weight then elements having similar properties occur at fixed intervals. Certain exceptions or gaps in the table lead to the view that the nuclear charge is a more characteristic function, thus the modern statement of the periodic law is that the physical and chemical properties of elements are a periodic function of their proton number.

periodic table A table of the elements arranged in order of increasing proton number to show similarities in chemical behaviour between elements. Horizontal rows of elements are called *periods*. Across a period there is a general trend from metallic to non-metallic behaviour. Vertical columns of related elements are called *groups*. Down a group there is an increase in atomic size and in electropositive behaviour. *See also* group, period.

permanent hardness A type of water hardness caused by the presence of dissolved calcium, iron, and magnesium sulphates or chlorides. This form of hardness cannot be removed by boiling. *Compare* temporary hardness. *See also* water softening.

permanganate *See* manganate(VII).

Permutit (*Tradename*) A substance used to remove unwanted chemicals that have dissolved in water.

It is a zeolite consisting of a complex chemical compound, sodium aluminium silicate (Na^+ Alsilicate$^-$). When hard water is passed over this material, calcium and magnesium ions exchange with sodium ions in the Permutit. This is a good example of ion exchange. Once all the available sodium ions have been used up, the Permutit can be regenerated by washing it with a saturated solution of sodium chloride. The excess of sodium ions now exchange with the calcium and magnesium ions in the Permutit.

peroxide 1. An oxide containing the $^-O-O^-$ ion.
2. A compound containing the $-O-O-$ group.

peroxosulphuric(VI) acid (Caro's acid, H_2SO_5) A white crystalline solid formed by reacting hydrogen peroxide with sulphuric acid. It is a powerful oxidizing agent.

Perspex (*Trademark*) A widely-used acrylic resin, polymethylmethacrylate.

peta- Symbol: P A prefix denoting 10^{15}.

petrochemicals Chemical obtained from crude oil or from natural gas.

petroleum A mixture of hydrocarbons formed originally from marine animals and plants, found beneath the ground trapped between layers of rock. It is obtained by drilling (also called *crude oil*). Different oilfields produce petroleum with differing compositions. The mixture is separated into fractions by fractional distillation in a vertical column. The main fractions are:
Diesel oil (*gas oil*) in the range 220–350°C, consisting mainly of C_{13}–C_{25} hydrocarbons. It is used in diesel engines.
Kerosine (*paraffin*) in the range 160–250°C, consisting mainly of C_{11} and C_{12} hydrocarbons. It is a fuel both for domestic heating and jet engines.
Gasoline in the range 40–180°C, consisting mainly of C_5–C_{10} hydrocarbons. It is used as motor fuel (petrol) and as a raw material for making other chemicals.
Refinery gas, consisting of C_1–C_4 gaseous hydrocarbons.
In addition lubricating oils and paraffin wax are obtained from the residue. The black material left is bitumen tar.

pewter A tarnish-resistant alloy of tin and lead, usually containing 80–90% tin. Modern pewter is hardened by antimony and copper, which in the best grades replace the lead content. It is lighter than old pewter and having a bright or a soft satiny finish is often used for decorative and ceremonial purposes.

pH The logarithm to base 10 of the reciprocal of the hydrogen-ion concentration of a solution. In pure water at 25°C the concentration of hydrogen ions is 1.00×10^{-7} mol 1^{-1}, thus the pH equals 7 at neutrality. An increase in acidity increases the value of $[H^+]$, decreasing the value of the pH below 7. An increase in the concentration of hydroxide ion $[OH^-]$ proportionately decreases $[H^+]$, therefore increasing the value of the pH above 7 in basic solutions. pH values can be obtained approximately by using indicators. More precise measurements use electrode systems.

phase One of the physically separable parts of a chemical system. For example, a mixture of ice (solid phase) and water (liquid phase) consists of two phases. A system consisting of only one phase is said to be homogeneous. A system consisting of more than one phase is said to be heterogeneous.

phase diagram A graphical representation of the state in which a substance will occur at a given pressure and temperature.
The lines show the conditions under which more than one phase can coexist at equilibrium. For one-component systems (e.g. water) the point at which all three phases can coexist at equilibrium is called the triple point and is the point on the graph at which the pressure-temperature curves intersect.

phase equilibrium A state in which the proportions of the various phases in a chemical system are fixed. When two or more phases are present at fixed temperature and pressure a dynamic condition will be established in which individual particles will leave one phase and enter another; at equilibrium this will be balanced by an equal number of particles making the reverse change, leaving the total composition unchanged.

phase rule In a system at equilibrium the number of phases (P), number of components (C), and number of degrees of freedom (F) are related by the formula:
$$P + F = C + 2$$

phenol 1. (carbolic acid, C_6H_5OH) A white crystalline solid used to make a variety of other organic compounds. **2.** A type of organic compound in which at least one hydroxyl group is bound directly to one of the carbon atoms of an aromatic ring. Phenols do not show the behaviour typical of alcohols. In particular they are more acidic because of the electron-withdrawing effect of the aromatic ring. The preparation of phenol itself is by fusing the sodium salt of the sulphonic acid with sodium hydroxide:

$$C_6H_5SO_2.ONa + 2NaOH \rightarrow C_6H_5ONa$$
$$+ Na_2SO_3 + H_2O$$

The phenol is then liberated by sulphuric acid:

$$2C_6H_5ONa + H_2SO_4 \rightarrow 2C_6H_5OH +$$
$$Na_2SO_3$$

Reactions of phenol include:
(1) Replacement of the hydroxyl group with a chlorine atom using phosphorus(V) chloride.
(2) Reaction with acyl halides to form esters of carboxylic acids.
(3) Reaction with haloalkanes under alkaline conditions to give mixed alkyl-aryl ethers.
In addition phenol can undergo further substitution on the benzene ring. The hydroxyl group directs other substituents into the 2- and 4-positions.

phenolphthalein An acid-base indicator that is colourless in acid solutions and becomes red if the pH rises above the transition range of 8−9.6. It is used as the indicator in titrations for which the end point lies clearly on the basic side (pH>7), e.g. oxalic acid or potassium hydrogen tartrate against caustic soda.

phenylalanine See amino acids.

phenylethene (styrene, $C_6H_5CHCH_2$) A liquid hydrocarbon used as the starting material for the production of polystyrene. The manufacture of phenylethene is by dehydrogenation of ethyl benzene:

$$C_6H_5C_2H_5 \rightarrow C_6H_5CH:CH_2 + H_2$$

phenyl group The group C_6H_5-, derived from benzene.

phenylhydrazone See hydrazone.

phenylmethanol (benzyl alcohol, $C_6H_5CH_2OH$) An aromatic primary alcohol. Phenylmethanol is synthesized by Cannizzaro's reaction, which involves the simultaneous oxidation and reduction of benzene-carbaldehyde (benzaldehyde) by

refluxing in an aqueous solution of sodium hydroxide:

$$2C_6H_5CHO \rightarrow C_6H_5CH_2OH +$$
$$C_6H_5COOH$$

Phenylmethanol undergoes the reactions characteristic of alcohols, especially those in which the formation of a stable carbonium ion as an intermediate $(C_6H_5CH_2{}^+)$ enhances the reaction. Substitution onto the benzene ring is also possible; the $-CH_2OH$ group directs into the 2- or 4- position by the donation of electrons to the ring.

Phillips process A method for the manufacture of high-density polyethene using a catalyst of chromium(III) oxide on a promoter of silica and alumina. The reaction conditions are 150°C and 30 atm pressure. See also Ziegler process.

phlogiston theory A former theory of combustion (disproved in the eighteenth century by Lavoisier). Flammable compounds were supposed to contain a substance (phlogiston), which was released when they burned, leaving ash.

phosphates See phosphoric(V) acid.

phosphide A compound of phosphorus with a more electropositive element.

phosphine (PH_3) A colourless gas that is slightly soluble in water. It has a characteristic fishy smell. It can be made by reacting water and calcium phosphide or by the action of yellow phosphorus on a concentrated alkali. Phosphine usually ignites spontaneously in air because of contamination with diphosphine. It decomposes into its elements if heated to 450°C in the absence of oxygen and it burns in oxygen or air to yield phosphorus oxides. It reacts with solutions of metal salts to precipitate phosphides. Like its analogue ammonia it forms salts, called *phosphonium salts*. It also forms complex addition compounds with metal ions. As in ammonia, one or more of the hydrogen atoms can be replaced by alkyl groups.

phosphinic acid (hypophosphorous acid, H_3PO_2) A white crystalline solid. It is a monobasic acid forming the anion $H_2PO_2^-$ in water. The sodium salt, and hence the acid, can be prepared by heating yellow phosphorus with sodium hydroxide solution. The free acid and its salts are powerful reducing agents.

phosphonic acid (phosphorous acid, orthophosphorous acid, H_3PO_3) A colourless deliquescent solid that can be prepared by the action of water on phosphorus(III) oxide or phosphorus(III) chloride. It is a dibasic acid producing the anions $H_2PO_3^-$ and HPO_3^{2-} in water. The acid and its salts are slow reducing agents. On warming, phosphonic acid decomposes to phosphine and phosphoric(V) acid.

phosphonium ion The ion PH_4^+.

phosphor A substance that shows luminescence or phosphorescence.

phosphorescence 1. The absorption of energy by atoms followed by emission of electromagnetic radiation. Phosphorescence is a type of luminescence, and is distinguished from fluorescence by the fact that the emitted radiation continues for some time after the source of excitation has been removed. In phosphorescence the excited atoms have relatively long lifetimes before they make transitions to lower energy states. However, there is no defined time distinguishing phosphorescence from fluorescence.
2. In general usage the term is applied to the emission of 'cold light' — light produced with a high temperature. The name comes from the fact that white phosphorus glows slightly in the dark as a result of a chemical reaction with oxygen. The light comes from excited atoms produced directly in the reaction — not from the heat produced. It is thus an example of *chemiluminescence*. There are also a number of biochemical examples termed bioluminescence; for example, phosphorescence is sometimes seen in the sea from marine organisms, or on rotting wood from certain fungi (known as 'fox fire').

phosphoric(V) acid (orthophosphoric acid, H_3PO_4) A white solid that can be made by reacting phosphorus(V) oxide with water or by heating yellow phosphorus with nitric acid. The naturally occurring *phosphates* (orthophosphates, M_3PO_4) are salts of phosphoric(V) acid. An aqueous solution of phosphoric(V) acid has a sharp taste and has been used in the manufacture of soft drinks. At about 220°C phosphoric(V) acid is converted to *pyrophosphoric acid* ($H_4P_2O_7$). Pure pyrophosphoric acid in the form of colourless crystals is made by warming solid phosphoric(V) acid with phosphorus(III) chloride oxide. *Metaphosphoric acid* is obtained as a polymer, $(HPO_3)_n$, by heating phosphoric(V) acid to 320°C. Both meta- and pyrophosphoric acids change to phosphoric(V) acid in aqueous solution; the change will take place faster at higher temperatures. Phosphoric(V) acid is tribasic, forming three series of salts with the anions $H_2PO_4^-$, HPO_4^{2-}, and PO_4^{3-}. These series of salts are acidic, neutral, and alkaline respectively. The alkali metal phosphates (except lithium phosphate) are soluble in water but other metal phosphates are insoluble. Phosphates are useful in water softening and as fertilizers.

phosphorous acid *See* phosphonic acid.

phosphorus A reactive solid non-metallic element; the second element in group V of the periodic table. It has the electronic configuration $[Ne]3s^23p^3$ and is therefore formally similar to nitrogen. It is however very much more reactive than nitrogen and is never found in nature in the uncombined state. Phosphorus is widespread throughout the world; economic sources are phosphate rock ($Ca_3(PO_4)_2$) and the apatites, variously found as fluoroapatite ($3Ca_3(PO_4)_2.CaF_2$) and as chloroapatite ($3Ca_3(PO_4)_2CaCl_2$). Guano formed from the skeletal phosphate of fish in sea-bird droppings is also an important source of phosphorus. The largest tonnages of phosphorus compounds produced are used as fertilizers, with the detergents industry producing increasingly large tonnages of phosphates. Phosphorus is an essential constituent of living tissue and bones, and it plays a very important part in metabolic processes and muscle action.

The element is obtained industrially by the reduction of phosphate rock using sand (SiO_2) and coke (C) in an electric furnace. The phosphorus is distilled off as P_4 molecules and collected under water as white phosphorus.

$$2Ca_3(PO4)_2 + 6SiO_2 + 10C \rightarrow P_4 + 6CaSiO_3 + 10CO$$

There are three common allotropes of phosphorus and several other modifications of these, some of which have indefinite structures. The common forms are:
(1) white (or yellow) phosphorus (distillation of phosphorus; soluble in some organic solvents)
(2) red phosphorus (heat on white P, insoluble in organic solvents)
(3) black phosphorus (heat on white phosphorus under high pressures).

Phosphorus, being in group V is non-metallic in character and has a sufficiently high ionization potential for the chemistry of bonds to phosphorus to be almost entirely covalent. Phosphorus compounds are formally P(III) and P(V) compounds. The element forms a hydride, PH_3 (phosphine), which is analogous to ammonia, but it is not formed by direct combination. Phosphorus(V) hydrides are not known.

When phosphorus is burned in air the product is the highly hygroscopic white solid P_4O_{10} (commonly called phosphorus pentoxide, P_2O_5); this is the P(V) oxide. In a limited supply of air the P(III) oxide, P_4O_6 (called phosphorus trioxide) contaminated with unreacted phosphorus is obtained.

The oxides both react with water to form acids;

$$P(III), P_4O_6 + 6H_2O \rightarrow$$
$$4H_3PO_3 \text{ phosphorous acid}$$
$$P(V), P_4O_{10} + 2H_2O \rightarrow$$
$$4HPO_3 \text{ metaphosphoric acid}$$
$$P_4O_{10} + 6H_2O \rightarrow$$

$4H_3PO_4$ orthophosphoric acid.

There are also many poly-phosphoric acids and salts with different ring sizes and different chain lengths. Pyrophosphoric acid is obtained by melting phosphoric acid, further dehydration above 200°C leads to metaphosphoric acids,

$$2H_3PO_4 \rightarrow H_4P_2O_7 + H_2O$$
$$H_4P_2O_7 \rightarrow 2HPO_3 + H_2O$$

Polyphosphates with molecular weights from 1000 to 10 000 are industrially important in detergent formulations and the lower members are used to 'complex' metal ions (sequestration).

Phosphorus forms the P(III) halides, PX_3, with all the halogens, and P(V) halides, PX_5, with X = F, Cl, Br. The P(III) halides are formed by direct combination but because of hazards in the reaction of phosphorus and fluorine, PF_3 is prepared from PCl_3

The P(III) halides are all pyramidal (as is PH_3) and are readily hydrolysed. The P(V) halides are also readily hydrolysed by water in two stages giving first the oxyhalide, then phosphoric acid,

$$PX_5 + H_2O \rightarrow POX_3 + 2HX$$
$$POX_3 + 3H_2O \rightarrow H_3PO_4 + 3HX$$

In the gas phase the trihalides are pyramidal and the pentahalides are trigonal bipyramids. However, in carbon tetrachloride solution phosphorus pentachloride is a dimer with two chlorine bridges exhibiting the ready expansion of the coordination number to six. In the solid state, (PCl_5) is $[PCl_4]^+[PCl_6]^-$ (the first tetrahedral, the second octahedral) obtained by transfer of a chloride ion.

Phosphorus interacts directly with several metals and metalloids to give:
(1) Ionic phosphides such as Na_3P, Ca_3P_2, Sr_3P_2.
(2) Molecular phosphides such as P_4S_3 and P_4S_5.
(3) Metallic phosphides such as Fe_2P.
Symbol: P; m.p. 44.1°C (white P); b.p. 280.5°C; r.d. 2.70; p.n. 15; r.a.m. 30.9.

phosphorus(III) bromide (phosphorus tribromide, PBr_3) A colourless liquid made by reacting phosphorus with bromine. It is readily hydrolysed by water to phosphonic acid and hydrogen bromide. Phosphorus(III) bromide is important in organic chemistry, being used to replace a hydroxyl group with a bromine atom.

phosphorus(V) bromide (phosphorus pentabromide, PBr_5) A yellow crystalline solid that sublimes easily. It can be made by the reaction of bromine and phosphorus(III) bromide. Phosphorus(V) bromide is readily hydrolysed by water to phosphoric(V) acid and hydrogen bromide. Its main use is in organic chemistry to replace a hydroxyl group with a bromine atom.

phosphorus(III) chloride (phosphorus trichloride, PCl_3) A colourless liquid formed from the reaction of phosphorus with chlorine. It is rapidly hydrolysed by water to phosphonic acid and hydrogen chloride. Phosphorus(III) chloride is used in organic chemistry to replace a hydroxyl group with a chlorine atom.

phosphorus(V) chloride (phosphorus pentachloride, PCl_5) A white easily sublimed solid formed by the action of chlorine on phosphorus(III) chloride. It is hydrolysed by water to phosphoric(V) acid and hydrogen chloride. Its main use is as a chlorinating agent in organic chemistry to replace a hydroxyl group with a chlorine atom.

phosphorus(III) chloride oxide (phosphorus trichloride oxide, phosphorus oxychloride, phosphoryl chloride, $POCl_3$) A colourless liquid that can be obtained by reacting phosphorus(III) chloride with oxygen or by distilling phosphorus(III) chloride with potassium chlorate. The reactions of phosphorus(III) chloride oxide are similar to those of phosphorus(III) chloride. The chlorine atoms can be replaced by alkyl groups using Grignard reagents or by alkoxo groups using alcohols. Water hydrolysis yields phosphoric(V) acid. Phosphorus(III) chloride oxide forms complexes with many metal ions.

phosphorus(III) oxide (phosphorus trioxide, P_2O_3) A white waxy solid with a characteristic smell of garlic; it usually exists as P_4O_6 molecules. It is soluble in organic solvents, such as benzene and carbon disulphide, and is made by burning phosphorus in a limited supply of air. Phosphorus(III) oxide oxidizes in air to phosphorus(V) oxide at room temperature and inflames above 70°C. It dissolves slowly in cold water or dilute alkalis to give phosphonic acid or one of its salts. Phosphorus(III) oxide reacts violently with hot water to form phosphine and phosphoric(V) acid.

phosphorus(V) oxide (phosphorus pentoxide, P_2O_5) A white powder that is soluble in organic solvents. It usually exists as P_4O_{10} molecules. Phosphorus(V) oxide can be prepared by burning phosphorus in a plentiful supply of oxygen. It readily combines with water to form phosphoric(V) acid and is therefore used as a drying agent for gases. It is a useful dehydrating agent because it is able to remove the elements of water from certain oxyacids and other compounds containing oxygen and hydrogen; for example, it reacts with nitric acid (HNO_3) to give nitrogen pentoxide (N_2O_5) on heating.

phosphorus oxychloride See phosphorus(III) chloride oxide.

phosphorus pentabromide See phosphorus(V) bromide.

phosphorus pentachloride See phosphorus(V) chloride.

phosphorus pentoxide See phosphorus(V) oxide.

phosphorus tribromide See phosphorus(III) bromide.

phosphorus trichloride See phosphorus(III) chloride.

phosphorus trichloride oxide See phosphorus(III) chloride oxide.

phosphorus trioxide See phosphorus(III) oxide.

phosphoryl chloride See phosphorus(III) chloride oxide.

phot A unit of illumination in the c.g.s. system, equal to an illumination of one lumen per square centimetre. It is equal to 10^4 lx.

photochemical reaction A reaction brought about by light; examples include the bleaching of coloured material, the reduction of silver halides, and the photosynthesis of carbohydrates. Chemical changes only occur when the reacting atoms or molecules absorb photons of the appropriate energy.
The amount of substance that reacts is proportional to the quantity of energy absorbed. For example, in the reaction between hydrogen and chlorine, it is not the concentrations of hydrogen or chlorine that dictate the rate of reaction but the intensity of the radiation.

photochemistry The branch of chemistry dealing with reactions induced by light or ultraviolet radiation.

photoelectric effect The emission of electrons from a solid (or liquid) surface when it is irradiated with electromagnetic radiation. For most materials the photoelectric effect occurs with ultraviolet radiation or radiation of shorter wavelength; some show the effect with visible radiation. In the photoelectric effect, the number of electrons emitted depends on the intensity of the radiation and not on its frequency. The kinetic energy of the electrons that are ejected depends on the frequency of the radiation. This was explained, by Einstein, by the idea that electromagnetic radiation consists of streams of photons. The *photon energy* is $h\nu$, where h is the Planck constant and ν the frequency of the radiation. To remove an electron from the solid a certain minimum energy must be supplied, known as the *work function*, ϕ. Thus, there is a certain minimum threshold frequency ν_0 for radiation to eject electrons: $h\nu_0 = \theta$. If the frequency is higher than this threshold the electrons are ejected. The maximum kinetic energy (W) of the electrons is given by *Einstein's equation*:
$$W = h\nu - \phi$$
The photoelectric effect also occurs with gases. See photoionization.

photoelectrochemical cell (PEC) A type of voltaic cell in which one of the electrodes is a light-sensitive semiconductor (such as gallium arsenide). A current is generated by light falling on the semiconductor, causing electrons to pass between the electrode and the electrolyte and produce chemical reactions in the cell. PECs are potentially useful because they can store energy as well as generate it.

photoelectron An electron ejected from a solid, liquid, or gas by the photoelectric effect or by photoionization.

photoemission The emission of photoelectrons by the photoelectric effect or by photoionization.

photoionization The ionization of atoms or molecules by electromagnetic radiation. Photons absorbed by an atom may have sufficient photon energy to free an electron from its attraction by the nucleus. The process is
$$M + h\nu \rightarrow M^+ + e^-$$
As in the photoelectric effect, the radiation must have a certain minimum threshold frequency. The energy of the photoelectrons ejected is given by $W = h\nu - I$, where I is the ionization potential of the atom or molecule.

photolysis A chemical reaction that is produced by electromagnetic radiation (light or ultraviolet radiation). Many photolytic reactions involve the formation of free radicals.

photosynthesis The photochemical process green plants use to synthesize organic compounds from carbon dioxide and water.

physical change A change to a substance that does not alter its chemical properties. Physical changes (e.g. melting, boiling, and dissolving) are comparatively easy to reverse.

physical chemistry The branch of chemistry concerned with physical properties of compounds and how these depend on the chemical bonding.

physisorption *See* adsorption.

pico- Symbol: p A prefix denoting 10^{-12}. For example, 1 picofarad (pF) = 10^{-12} farad (F).

piezoelectricity The generation of an electric charge by certain crystals when subjected to pressure.

pig iron The crude iron produced from a blast furnace, containing carbon, silicon, and other impurities. The molten iron is run out of the furnace into channels (called 'sows'), which branch out into a number of offshoots (called 'pigs') in which the metal is allowed to cool.

pigment An insoluble, particulate colouring material.

pi bond *See* orbital.

pi orbital *See* orbital.

pipette A device used to transfer a known volume of solution from one container to another; in general, several samples of equal volume are transferred for individual analysis from one stock solution. Pipettes are of two types, bulb pipettes, which transfer a known and fixed volume, and graduated pipettes, which can transfer variable volumes. Pipettes were at one time universally mouth-operated but safety pipettes using a plunger or rubber bulb are now preferred.

pK The logarithm to the base 10 of the reciprocal of an acid's dissociation constant:

$$\log_{10}(1/K)$$

Planck constant Symbol: h A fundamental constant; the ratio of the energy (W) carried by a photon to its frequency (v). A basic relationship in the quantum theory of radiation is $W = hv$. The value of h is $6.626\ 196 \times 10^{-34}$ J s. The Planck constant appears in many relationships in which some observable measurement is quantized (i.e. can take only specific separate values rather than any of a range of values).

plane polarization A type of polarization of electromagnetic radiation in which the vibrations take place entirely in one plane.

plasma A mixture of ions and electrons, as in an electrical discharge. Sometimes, a plasma is described as a fourth state of matter.

plaster of Paris *See* calcium sulphate.

plastics A common term for synthetic polymers.

platinic chloride *See* chloroplatinic acid.

platinum A transition metal. It occurs naturally in Australia and Canada, either free or in association with other platinum metals. It is resistant to oxidation and is not attacked by acids (except aqua regia) or alkalis. Platinum is used as a catalyst for the oxidation of ammonia to manufacture nitric acid and in jewellery.
Symbol: Pt; m.p. 1770°C; b.p. 3830°C; r.d. 21.5; p.n. 78; r.a.m. 195.09.

platinum black A finely divided black form of platinum produced, as a coating, by evaporating platinum onto a surface in an inert atmosphere. Platinum-black coatings are used as pure absorbent electrode coatings in experiments on electric cells. They are also used, like carbon-black coatings, to improve the ability of a surface to absorb radiation.

platinum(II) chloride ($PtCl_2$) A grey-brown powder prepared by the partial decomposition of platinum(IV) chloride. It may also be prepared by passing chlorine over heated platinum. Platinum(II) chloride is insoluble in water but dissolves in concentrated hydrochloric acid to form a complex acid.

platinum(IV) chloride ($PtCl_4$) A reddish hygroscopic solid prepared by the action of heat on chloroplatinic acid. Crystals of the pentahydrate, $PtCl_4.5H_2O$, are formed; anhydrous platinum(IV) chloride is prepared by treating the hydrated crystals with concentrated sulphuric acid. The chloride dissolves in water to produce a strongly acidic solution. Platinum(IV) chloride forms a series of hydrates having 1, 2, 4, and 5 molecules of water; the tetrahydrate is the most stable. The strongly acidic solution produced on dissolving in water probably contains $H_2[PtCl_4(OH)_2]$.

platinum-iridium An alloy of platinum containing up to 30% iridium. Hardness and resistance to chemical attack increase as the iridium content is increased. It is used in jewellery, electrical contacts, and hypodermic needles.

platinum metals The group of transition metals ruthenium (Ru), osmium (Os), rhodium (Rh), iridium (Ir), palladium (Pd), and platinum (Pt). They are sometimes classed together as they have related properties and compounds.

Plexiglas (*Trademark*) A widely-used acrylic resin, polymethylmethacrylate.

plumbane (lead(IV) hydride, PbH_4) A colourless unstable gas that can be obtained by the action of acids on a mixture of magnesium and lead pellets.

plumbic Designating a lead(IV) compound.

plumbous Designating a lead(II) compound.

plutonium A highly toxic radioactive silvery element of the actinoid series of metals. It is a transuranic element found on

Earth only in minute quantities in uranium ores but readily obtained, as ^{239}Pu, by neutron bombardment of natural uranium. The readily fissionable ^{239}Pu is a major nuclear fuel and nuclear explosive.
Symbol: Pu; m.p. 641°C; b.p. 3232°C; r.d. 19.84; p.n. 94; stablest isotope ^{244}Pu.

poison 1. A substance that destroys catalyst activity.
2. Any substance which endangers biological activity, whether by physical or chemical means.

polar Describing a compound with molecules that have a permanent dipole moment. Hydrogen chloride and water are examples of polar compounds.

polar bond A covalent bond in which the bonding electrons are not shared equally between the two atoms. A bond between two atoms of different electronegativity is said to be polarized in the direction of the more electronegative atom, i.e. the electrons are drawn preferentially towards the atom. This leads to a small separation of charge and the development of a bond dipole moment as in, for example, hydrogen fluoride, represented as H←F or as $H^{\delta+}-F^{\delta-}$ (F is more electronegative).
The charge separation is much smaller than in ionic compounds; molecules in which bonds are strongly polar are said to display partial ionic character. The effect of the electronegative element can be transmitted beyond adjacent atoms, thus the C−C bonds in, for example, $CCl_3.CH_3$ and $CH_3.CHO$ are slightly polar. *See also* dipole moment, intermolecular forces.

polarizability The ease with which an electron cloud is deformed (polarized). In ions, an increase in size or negative charge leads to an increase in polarizability. The concept is of particular use in the treatment of covalent contributions to predominantly ionic bonds embodied in Fajans' Rules. Thus ions, such as I^-, Se^{2-}, Te^{2-}, are especially prone to covalent character, particularly in combination with ions of small size and high positive charge (i.e. high ionic potential). Molecular polarizability is a measure of the ease with which electrons in the molecule are deformed, particularly by electromagnetic radiation.

polarization 1. The restriction of the vibrations in a transverse wave so that the vibration occurs in a single plane. Electromagnetic radiation, for instance, is a transverse wave motion. It can be thought of as an oscillating electric field and an oscillating magnetic field, both at right angles to the direction of propagation and at right angles to each other. Usually, the electric vector is considered since it is the electric field that interacts with charged particles of matter and causes the effects. In 'normal' unpolarized radiation, the electric field oscillates in all possible directions perpendicular to the wave direction. On reflection or on transmission through certain substances (e.g. Polaroid) the field is confined to a single plane. The radiation is then said to be *plane-polarized*. **2.** *See* polarizability.
3. The reduction of current in a voltaic cell, caused by the build-up of products of the chemical reaction. Commonly, the cause is the build-up of a layer of bubbles (e.g. of hydrogen). This reduces the effective area of the electrode, causing an increase in the cell's internal resistance. It can also produce a back e.m.f. Often a substance such as manganese(IV) oxide (a *depolarizer*) is added to prevent hydrogen build-up.

polar molecule A molecule in which the individual polar bonds are not perfectly symmetrically arranged and are therefore not 'in balance'. Thus the charge separation in the bonds gives rise to an overall charge separation in the molecule as for example, in water. Such molecules possess a dipole moment.

polarography An analytical method in which current is measured as a function of potential. A special type of cell is used in which there is a small easily polarizable cathode (the dropping mercury electrode) and a large non-polarizable anode (reference cell). The analytical reaction takes place at the cathode and is essentially a reduction of the cations, which are discharged according to the order of their electrode potential (E) values. The data is expressed in the form of a *polarogram*, which is a plot of current against applied voltage. As the applied potential is increased a point is reached at which the ion is discharged. There is a step-wise increase in current, which levels off because of polarization effects. The potential at half the step height (called the *half-wave potential*) is used to identify the ion. Most elements can be determined by polarography. The optimum concentrations are in the range 10^{-2}–10^{-4} M; modified techniques allow determinations in the parts per million range.

polar solvent *See* solvent.

pollution Any damaging or unpleasant change in the environment that results from the physical, chemical, or biological side-effects of man's industrial or social activities. Pollution can affect the atmosphere, rivers, seas, and the soil.
Air pollution is caused by the domestic and industrial burning of carbonaceous fuels, by industrial processes, and by car exhausts. Among recent problems are industrial emissions of sulphur dioxide causing *acid rain*, and the release into the atmosphere of chlorofluorocarbons, used in refrigeration, aerosols, etc., has been linked to the depletion of ozone in the stratosphere. Carbon dioxide, produced by burning fuel and by car exhausts, is slowly building up in the atmosphere, which could result in an overall increase in the temperature of the atmosphere. Car exhausts also contain carbon monoxide and lead. The former has not yet reached dangerous levels, but vegetation near main roads contains a high proportion of lead and levels are sufficiently high in urban areas to cause concern about the effects on children. Lead-free petrol is now becoming available. Photochemical smog, caused by the action of sunlight on hydrocarbons and nitrogen oxides from car exhausts, is a problem in the USA.

Water pollutants include those that are *biodegradable*, such as sewage effluent, which cause no permanent harm if adequately treated and dispersed, as well as those which are nonbiodegradable, such as certain chlorinated hydrocarbon pesticides (e.g. DDT) and heavy metals, such as lead, copper, and zinc in some industrial effluents (causing *heavy-metal pollution*). When these accumulate in the environment they can become very concentrated in food chains. The pesticides DDT, aldrin, and dieldrin are now banned. Water supplies can become polluted by leaching of nitrates from agricultural land. The discharge of waste heat can cause thermal pollution of the environment, but this is reduced by the use of cooling towers. In the sea, oil spillage from tankers and the inadequate discharge of sewage effluent are the main problems.

Other forms of pollution are noise from aeroplanes, traffic, and industry and the disposal of radioactive waste.

polonium A radioactive metallic element belonging to group VI of the periodic table. It occurs in very minute quantities in uranium ores. Over 30 radioisotopes are known, nearly all alpha-particle emitters. A suggested use for ^{210}Po is as a thermoelectric power source in satellites.
Symbol: Po; m.p. 254°C; b.p. 962°C; r.d. 9; p.n. 84; stablest isotope ^{209}Po.

polyamide A synthetic polymer in which the monomers are linked by the group −NH−CO−. Nylon is an example of a polyamide.

polyatomic Describing a molecule (or ion or radical) that consists of several atoms (three or more). Examples are benzene (C_6H_6) and methane (CH_4).

polychloroethene (polyvinyl chloride, PVC) A synthetic polymer made from chloroethene. It is a strong material with a wide variety of uses.

polycrystalline Describing a substance composed of very many minute interlocking crystals that have solidified together.

polycyclic Describing a compound that has two or more rings in its molecules.

polyene An alkene with more than two double bonds in its molecules.

polyester A synthetic polymer made by reacting alcohols with acids, so that the monomers are linked by the group −O−CO−. Synthetic fibres such as Terylene and Dacron are polyesters.

polyethene (polyethylene, polythene) A synthetic polymer made from ethene. It is produced in two forms — a soft material of low density and a harder, higher density form, which is more rigid. It can be made by the Ziegler process.

polyethylene *See* polyethene.

polyhydric alcohol An alcohol that has several −OH groups in its molecules.

polymer A compound in which there are very large molecules made up of repeating molecular units (monomers). Polymers do not usually have a definite relative molecular mass, as there are variations in the lengths of different chains. They may be natural substances (e.g. polysaccharides or proteins) or synthetic materials (e.g. nylon or polyethene). The two major classes of synthetic polymers are thermosetting (e.g. Bakelite) and thermoplastic (e.g. polyethene). The former are infusible, and heat may only make them harder, whilst the latter soften on heating. *See also* polymerization.

polymerization The process in which one or more compounds react to form a polymer. *Homopolymerization* is polymerization of one monomer (e.g. the formation of polyethene). *Copolymerization* involves two or more monomers (as for nylon). *Addition polymerization* occurs when the monomers undergo addition reactions, with no other substance formed. *Condensation polymerization* involves the polyethene

polypropene

PVC

Addition polymers

elimination of small molecules in formation of the polymer. *See* nylon.

polymorphism The ability of certain
chemical substances to exist in more than
one physical form. For elements this is
called allotropy. The existence of two
forms is called *dimorphism*.
Crystalline structures of compounds can
vary with temperature as a result of different packing arrangements of the particles.
There is a transition temperature between
two forms and usually a marked change in
density.

polypropene (polypropylene) A synthetic polymer made from propene. It is
similar in properties to polythene, but
stronger and lighter. The propene is
polymerized by the Ziegler process.

polypropylene *See* polypropene.

polysaccharide High molecular
polymers of the monosaccharides or sugars. They contain many repeated units in
their molecular structures. They can be
broken down to smaller polysaccharides,
disaccharides, and monosaccharides by
hydrolysis or by the appropriate enzyme.
Important polysaccharides are inulin,
starch, glycogen (also known as animal
starch), and cellulose. *See also* carbohydrate, sugar.

polystyrene A synthetic polymer made
from styrene (phenylethene). Expanded
polystyrene is a rigid foam used in packing
and insulation.

polytetrafluoroethene (PTFE) A
synthetic polymer made from te-
trafluoroethene (i.e. $CF_2:CF_2$). It is able to
withstand high temperatures without de-

179

composing and also has a very low coefficient of friction, hence its use in non-stick pans, bearings, etc.

polyurethane A synthetic polymer containing the group $-NH-CO-O$ linking the monomers. Polyurethanes are made by condensation of isocyanates ($-NCO$) with alcohols.

polyvinyl chloride See polychloroethene.

potassium A soft reactive metal; the third member of the alkali metals (group I of the periodic table). The atom has the argon electronic configuration plus an outer $4s^1$ electron. The element gives a distinct lilac colour to flames but, due to the intense yellow coloration from any trace impurity of sodium, the flame must be viewed through a cobalt glass. The ionization potential is low and the chemistry of potassium is largely the chemistry of the unipositive K^+ ion. Potassium accounts for 2.4% of the lithosphere and occurs in large salt deposits such as carnallite (KCl.$MgCl_2$.$6H_2O$) and arcanite (K_2SO_4). Large amounts also occur in mineral forms that are not of much use for recovery of potassium, e.g., orthoclase, $K_2Al_2Si_6O_{10}$. The commercial usage of potassium is much smaller than that of sodium; industrially the element is obtained by electrolysis of the fused hydroxide. Potassium hydroxide is obtained by electrolysis of carnallite solutions (the magnesium is initially precipitated as $Mg(OH)_2$), and both hydrogen and chlorine are recovered as by-products. The chemistry of potassium and its compounds is very similar to that of the other group I elements;

Particular points of distinction from sodium are:

(1) Burning in air produces the superoxide KO_2.

(2) Because of the larger size of the K^+ ion and hence lower lattice energies, potassium salts are often more soluble than corresponding sodium salts.

(3) Potassium salts are usually less heavily hydrated than corresponding sodium salts.

The most abundant isotope of potassium is ^{39}K (93.1%), with ^{41}K also a stable isotope (6.8%), but there is a naturally occurring radioactive isotope ^{40}K (0.11%).
Like sodium, potassium forms a number of ionic alkyl- and aryl-derivatives of the type RK.
Symbol: K; m.p. 63.7°C; b.p. 760°C; r.d. 0.86; p.n. 19; r.a.m. 39.09.

potassium bicarbonate See potassium hydrogencarbonate.

potassium bromide (KBr) A white solid that is extremely soluble in water. It is formed by the action of bromine on hot potassium hydroxide solution or by the neutralization of the carbonate with hydrobromic acid. Potassium bromide forms colourless cubic crystals. It is used extensively in the manufacture of photographic plates, films, and papers and (formerly) as a sedative in medicine.

potassium carbonate (pearl ash, K_2CO_3) A white deliquescent solid manufactured from potassium chloride by the Leblanc process or, more often, using the Precht process. It cannot be made by the Solvay process. In the laboratory, it can be prepared by thermal decomposition of potassium hydrogencarbonate. Potassium carbonate is very soluble in water, its solutions being strongly alkaline due to salt hydrolysis. It is used in the laboratory as a drying agent and industrially in the manufacture of soft soap, hard glass, and in the dyeing industry. Potassium carbonate crystallizes out between 10 and 25°C as K_2CO_3.$3H_2O$; it dehydrates at 100°C to K_2CO_3.H_2O and at 130°C to K_2CO_3. Potassium carbonate is useful in salting out alcohol from aqueous solutions.

potassium chlorate ($KClO_3$) A white soluble solid prepared by the electrolysis of a concentrated solution of potassium chloride. Industrially, it is prepared by the fractional crystallization of a solution containing sodium chlorate and potassium chloride. When heated, it decomposes to yield oxygen and potassium chloride. Potassium chlorate is a powerful oxidizing agent and is used in explosives, matches, weedkillers, and fireworks, and as a disinfectant. It oxidizes iodide ions to iodine when in an acidic medium. When heated just above its melting point, potassium perchlorate is formed.

potassium chloride (KCl) A white ionic solid prepared by neutralizing hydrochloric acid with potassium hydroxide solution. Potassium chloride occurs naturally as the minerals sylvine and carnallite. It is more soluble than sodium chloride in hot water but less soluble in cold water. On evaporation of an aqueous solution colourless cubic crystals similar to those of sodium chloride are produced. Potassium chloride is used as a fertilizer and in the manufacture of potassium hydroxide.

potassium chromate (K_2CrO_4) A bright yellow solid prepared by adding potassium hydroxide solution to a solution of potassium dichromate. It is extremely soluble in water. Addition of an acid to an aqueous solution of potassium chromate converts the chromate ions into dichromate ions. The salt is used as an indicator in silver nitrate titrations. The crystals are isomorphous with potassium sulphate.

potassium cyanide (KCN) A white ionic solid that is very soluble in water and extremely poisonous. It is made industrially by the Castner process and is used as a source of cyanide and hydrocyanic acid. Aqueous solutions of potassium cyanide are strongly hydrolysed, the solutions being alkaline. On standing, these solutions slowly evolve hydrocyanic acid.

potassium dichromate ($K_2Cr_2O_7$) An orange-red solid prepared by adding potassium chloride solution to a concentrated solution of sodium dichromate and crystallizing out or by acidifying a solution of potassium chromate and evaporating. It is less soluble than sodium dichromate in cold water but more soluble in hot water. Potassium dichromate is used as an oxidizing agent both in volumetric analysis and in organic chemistry. The crystals are triclinic, anhydrous, and non-deliquescent.

Addition of an alkali to a solution of the dichromate yields the chromate.

potassium hydride (KH) A white ionic solid prepared by passing hydrogen over heated potassium, the metal being suspended in an inert medium. It is an excellent reducing agent.

potassium hydrogencarbonate (potassium bicarbonate, $KHCO_3$) A white solid prepared by passing carbon dioxide through a saturated solution of potassium-carbonate. It occurs naturally in the mineral calcinite. Potassium hydrogencarbonate is more soluble in water than sodium hydrogencarbonate. When heated, it undergoes thermal decomposition to give the carbonate, water, and carbon dioxide. The salt forms monoclinic crystals. Its solutions are strongly alkaline due to salt hydrolysis.

potassium hydroxide (caustic potash, KOH) A white solid manufactured by the electrolysis of potassium chloride solution in a mercury-cathode cell. It can also be prepared by heating either potassium carbonate or potassium sulphate with slaked lime. The process closely resembles the Gossage process for making sodium hydroxide. In the laboratory, it can be made by reacting potassium, potassium monoxide, or potassium superoxide with water. Potassium hydroxide resembles sodium hydroxide but is more soluble in water and alcohol. It is preferred to sodium hydroxide for the absorption of carbon dioxide and sulphur dioxide because of its greater solubility. Potassium hydroxide is used as an electrolyte in the Ni−Fe electric storage battery and in the production of soft soaps. It forms crystalline hydrates with 1, 1½, and 2 molecules of water.

potassium iodate (KIO_3) A white solid formed either by adding iodine to a hot concentrated solution of potassium hydroxide or by the electrolysis of potassium iodide solution. No hydrates are known. It is a source of iodide and iodic acid. When treated with a dilute acid and a reducing agent, the iodate ions are reduced to iodine.

potassium iodide (KI) A white ionic solid readily soluble in water. It is prepared by dissolving iodine in hot concentrated potassium hydroxide solution. Both the iodide and iodate are formed but the latter is removed by fractional crystallization. Potassium iodide has a sodium chloride cubic lattice. In solution with iodine it forms potassium tri-iodide. It is used in medicine, particularly in the treatment of goitre resulting from iodine deficiency. Dilute acidified solutions of potassium iodide can act as reducing agents, manganate(VII) ions being reduced to manganese(II) ions, copper(II) ions to copper(I) ions, and iodate ions to iodine.

potassium manganate(VII) (potassium permanganate, $KMnO_4$) A purple solid soluble in water. It is prepared by oxidizing potassium manganate(VI) with chlorine. Potassium permanganate is used in volumetric analysis as an oxidizing agent, as a bactericide, and as a disinfectant. In aqueous solution its behaviour as an oxidizing agent depends on the pH of the solution.

potassium monoxide (K_2O) An ionic solid that is white when cold and yellow when hot. It is prepared by heating potassium with potassium nitrate. Potassium monoxide dissolves violently in water to form potassium hydroxide solution. The hydrate $K_2O.3H_2O$ is known. Potassium monoxide dissolves in liquid ammonia with the formation of potassium hydroxide and potassamide.

potassium nitrate (saltpetre, nitre, KNO_3) A white solid, soluble in water, formed by fractional crystallization of sodium nitrate and potassium chloride solutions. It occurs naturally as nitre (saltpetre) in rocks in India, South Africa, and Brazil. When heated it decomposes to give the nitrite and oxygen. Unlike sodium nitrate it is non-deliquescent. Potassium nitrate is used in gunpowder, fertilizers, and in the laboratory preparation of nitric acid.

potassium nitrite (KNO_2) A creamy deliquescent solid that is readily soluble in water. It reacts with cold dilute mineral acids to produce solutions of nitrous acid. Potassium nitrite is used in organic chemistry in the process of diazotization.

potassium permanganate See potassium manganate(VII).

potassium sulphate (K_2SO_4) A white solid prepared by the neutralization of either potassium hydroxide or potassium carbonate with dilute sulphuric acid. It occurs naturally in Stassfurt deposits as schönite. Potassium sulphate is soluble in water, forming a neutral solution. It is used as a fertilizer, particularly for tobacco plants, and in the chemical industry in the preparation of alums. The anhydrous form crystallizes in the rhombic form.

potassium sulphide (K_2S) A yellowish-brown solid prepared by saturating an aqueous solution of potassium hydroxide with hydrogen sulphide and then adding an equal volume of potassium hydroxide. Industrially it is produced by heating potassium sulphate and carbon at high temperature. The sulphide crystallizes out from aqueous solution as $K_2S.5H_2O$. Its aqueous solutions undergo hydrolysis, the solution being strongly alkaline.

potassium superoxide (K_2O) A yellow paramagnetic solid prepared by burning potassium in excess oxygen. When treated with cold water or dilute mineral acids, hydrogen peroxide is produced. If heated strongly, it yields oxygen and potassium monoxide. Potassium superoxide is a powerful oxidizing agent.

potentiometric titration A titration in which an electrode is used in the reaction mixture. The end point can be found by monitoring the potential of this during the titration.

poundal Symbol: pdl The unit of force in the f.p.s. system. It is equal to 0.138 255 N.

praseodymium A soft ductile malleable silvery element of the lanthanoid series of metals. It occurs in association with other lanthanoids. Praseodymium is used in several alloys, as a catalyst, and in compound form in carbon-arc searchlights, etc., and in the glass and enamel industries.
Symbol: Pr; m.p. 931°C; b.p. 3212°C; r.d. 6.77; p.n. 59; r.a.m. 140.91.

precipitate A suspension of small particles of a solid in a liquid formed by a chemical reaction.

pressure Symbol: p The pressure on a surface due to forces from another surface or from a fluid is the force acting at 90° to unit area of the surface:
$$\text{pressure} = \text{force}/\text{area}$$
The unit is the pascal (Pa).

primary cell A voltaic cell in which the chemical reaction that produces the e.m.f. is not reversible. *Compare* accumulator.

primary standard A substance that can be used directly for the preparation of standard solutions without reference to some other concentration standard. Primary standards should be easy to purify, dry, capable of preservation in a pure state, unaffected by air or CO_2, of a high molecular weight (to reduce the significance of weighing errors), stoichiometric, and readily soluble. Any likely impurities should be easily identifiable.

producer gas (air gas) A mixture of carbon monoxide (25–30%), nitrogen (50–55%), and hydrogen (10–15%), prepared by passing air with a little steam through a thick layer of white-hot coke in a furnace or 'producer'. The air gas is used while still hot to prevent heat loss and finds uses in industrial heating, for example the firing of retorts and in glass furnaces. *Compare* water gas.

proline *See* amino acids.

promethium A radioactive element of the lanthanoid series of metals. It is not found naturally on Earth but can be produced artificially by the fission of uranium. Possible uses involve its radioactive properties.
Symbol: Pm; m.p. 1080°C; b.p. 2460°C (approx.); r.d. 7.22; p.n. 61; stablest isotope ^{145}Pm (half-life 18 years).

promoter (activator) A substance that improves the efficiency of a catalyst. It does not itself catalyse the reaction but assists the catalytic activity. For example, alumina or molybdenum promotes the catalytic activity of finely divided iron in the Haber process.
The manner in which a promoter functions is not fully understood; no one theory covers all the examples.
See coenzymes.

proof A measure of the ethanol content of intoxicating drinks. In the UK, proof spirit contains 42.28% of ethanol by weight (57.1% by volume at 16°C). In the US it contains 50% by volume. The degree of proof gives the number of parts of proof spirit per 100 parts of the total. 100° of proof is 57.1% by volume, 80° of proof is 0.8 × 57.1%, etc.

propanal (propionaldehyde, C_2H_5CHO) A colourless liquid aldehyde.

propane (C_3H_8) A gaseous alkane obtained either from the gaseous fraction of crude oil or by the cracking of heavier fractions. The principal use of propane is as a fuel for heating and cooking, since it can be liquefied under pressure, stored in cylinders, and transported easily. Propane is the third member of the homologous series of alkanes.

propane-1,2,3-triol (glycerol, glycerine, $CH_2(OH)CH(OH)CH_2(OH)$) A colourless viscous liquid obtained as a by-product from the manufacture of soap by the reaction of animal fats with sodium hydroxide. It is used as a solvent and plasticizer. *See also* glyceride.

propanoic acid (propionic acid, C_2H_5COOH) A colourless liquid carboxylic acid.

propanol Either of two alcohols: propan-1-ol ($CH_3CH_2CH_2OH$) and propan-2-ol ($CH_3CH_2(OH)CH_3$). Both are colourless volatile flammable liquids.

propanone (acetone, CH_3COCH_3) A colourless liquid ketone, used as a solvent and in the manufacture of methyl 2-methyl-propanoate (from which Perspex is produced). Propanone is manufactured from propene, either by the air-oxidation of propan-2-ol or, more recently, as a by-product from the cumene process.

propene (propylene, C_3H_6) A gaseous alkene. Propene is not normally present in the gaseous crude-oil fraction but can be obtained from heavier fractions by catalytic cracking. This is the principal industrial source. Propene is the organic starting material for the production of propan-2-ol, required for the manufacture of propanone (acetone), and the starting material for the production of polypropene (polyprene).

propenonitrile (acrylonitrile, $H_2C:CH(CN)$) An organic compound from which acrylic-type polymers are produced.

propionaldehyde *See* propanal.

propionic acid *See* propanoic acid.

propylene *See* propene.

propyl group The group $CH_3CH_2CH_2-$.

protactinium A toxic radioactive element of the actinoid series of metals. It occurs in minute quantities in uranium ores as a radioactive decay-product of actinium. Symbol: Pa; m.p. 1565°C; r.d. 15.4 (calc.); p.n. 91; stablest isotope ^{231}Pa (half-life 32 480 years).

protein A naturally occurring compound found in all living matter, consisting of

Peptide link

amino acids linked into long chains. The primary structure is the particular sequence of amino acids present. The protein also has a secondary structure, with the chains coiled in a helix or held together in pleated sheets. The tertiary structure is the way in which the chain or sheet is arranged in space. Secondary structures are held by hydrogen bonds between N−H and O=C groups. Tertiary structures are held by hydrogen bonds or cross linkages of the type −S−S− (cystine links).
See also amino acid, peptide.

proton An elementary particle with a positive charge (+1.602 192 C) and rest mass $1.672 614 \times 10^{-27}$ kg. Protons are nucleons, found in all nuclides.

proton number (atomic number) Symbol: Z The number of protons in the nucleus of an atom. The proton number determines the chemical properties of the element because the electron structure, which determines chemical bonding, depends on the electrostatic attraction to the positively charged nucleus.

pseudoaromatic *See* aromatic compound.

pseudo-first order Describing a reaction that appears to exhibit first-order kinetics under special conditions, even though the 'true' order is greater than one. For example, in the hydrolysis of an ester in the presence of a *large* volume of water, the concentration of water remains approximately constant. The rate of reaction is thus found experimentally to be proportional to the concentration of the ester only (even though it also depends on the amount of water present). Such a reaction is described as 'bimolecular of the first order'.

pseudohalogens A small group of simple inorganic compounds with symmetrical molecules that resemble the halogens in some reactions and compounds. For example, $(CN)_2$ (cyanogen) has compounds analogous to halogen compounds (e.g. HCN, KCN, CH_3CN, etc.). Thiocyanogen, $(SCN)_2$, is another example.

PTFE *See* polytetrafluoroethene.

PVC *See* polychloroethene.

pyranose A sugar that has a six-membered ring form (five carbon atoms and one oxygen atom). *See also* sugar.

Pyrex (*Trademark*) A particularly strong, heat resistant, and chemically inert borosilicate glass commonly used for laboratory glassware.

pyridine (C_5H_5N) An organic liquid of formula C_5H_5N. The molecules have a hexagonal planar ring and are isoelectronic with benzene. Pyridine is an example of an aromatic heterocyclic compound, with the electrons in the carbon–carbon pi bonds and the lone pair of the nitrogen delocalized over the ring of atoms. The

Pyridine

compound is extracted from coal tar and used as a solvent and raw material for organic synthesis.

pyrites A mineral sulphide of a metal; e.g. iron pyrites FeS_2.

pyrolysis The decomposition of chemical compounds by subjecting them to very high temperature.

pyrometer An instrument used in the chemical industry to measure high temperature, e.g. in reactor vessels.

pyrophoric 1. Describing a compound that ignites spontaneously in air.
2. Describing a metal or alloy that gives sparks when struck. (Lighter flints are made of pyrophoric alloy.)

pyrosulphuric acid *See* oleum.

Q

quadrivalent Having a valence of four.

qualitative analysis Analysis carried out with the purpose of identifying the components of a sample. Classical methods involved simple preliminary tests followed by a carefully devised scheme of systematic tests and procedures. Modern methods include the use of such techniques as infrared spectroscopy and emission spectrography. *Compare* quantitative analysis.

quantitative analysis Analysis carried out with the purpose of determining the concentration of one or more components of a sample. Classical wet methods include volumetric and gravimetric analysis. A wide range of more modern instrumental techniques are also used, including polarography and various types of spectroscopy. *Compare* qualitative analysis.

quantized Describing a physical quantity that can only take certain discrete values, and not a continuous range of values. Thus, in an atom or molecule the electrons around the nucleus can have certain energies, E_1, E_2, etc., and cannot have intermediate values. Similarly, in atoms and molecules, the electrons have quantized values of spin angular momentum and orbital angular momentum.

quantum (plural *quanta*) A definite amount of energy released or absorbed in a process. Energy often behaves as if it were 'quantized' in this way. The quantum of electromagnetic radiation is the photon.

quantum electrodynamics The use of quantum mechanics to describe how particles and electromagnetic radiation interact.

quantum mechanics *See* quantum theory.

quantum number An integer or half integer that specifies the value of a quantized physical quantity (energy, angular momentum, etc.). *See* atom, Bohr theory, spin.

quantum states States of an atom, electron, particle, etc., specified by a unique set of quantum numbers. For example, the hydrogen atom in its ground state has an electron in the K shell specified by the four quantum numbers: $n = 1$, $l = 0$, $m = 0$, $m_s = \frac{1}{2}$. In the helium atom there are two electrons:

$$n = 1, l = 0, m = 0, m_s = \frac{1}{2}$$
$$n = 1, l = 0, m = 0, m_s = -\frac{1}{2}$$

quantum theory A mathematical theory originally introduced by Max Planck (1900) to explain the radiation emitted from hot bodies. Quantum theory is based on the idea that energy (or certain other physical quantities) can be changed only in certain discrete amounts for a given system. Other early applications were the explanations of the photoelectric effect and the Bohr theory of the atom.
Quantum mechanics is a system of mechanics that developed from quantum theory and is used to explain the behaviour of atoms, molecules, etc. In one form it is based on de Broglie's idea that particles can have wavelike properties – this branch of quantum mechanics is called *wave mechanics*. Wave mechanics is the basis of modern theory. *See* orbital.

quartz A natural crystalline form of silica (SiO_2).

quaternary ammonium compound A compound formed from an amine by addition of a proton to produce a positive ion. Quaternary compounds are salts, the simplest example being ammonium compounds formed from ammonia and an acid, for example:

$$NH_3 + HCl \rightarrow NH_4{}^+Cl^-$$

Other amines can also add protons to give analogous compounds. For instance, methylamine (CH_3NH_2) forms the compound

$$[CH_3NH_3]^+X^-$$

where X^- is an acid radical.

The formation of quarternary compounds occurs because the lone pair on the nitrogen atom can form a coordinate bond with a proton. This can also occur with heterogeneous nitrogen compounds, such as adenine, cytosine, thymine, and guanine.

Such compounds are known as *nitrogenous bases*.

quenching A method used to alter the mechanical properties of metals. Hot metal is lowered quickly into a bath of oil, water, or brine and the rapid cooling results in a fine grain structure. This treatment increases the hardness of a metal but often makes it brittle.

quicklime *See* calcium oxide.

R

racemate *See* optical activity.

racemic mixture *See* optical activity.

racemization The conversion of an optical isomer into an equal mixture of isomers, which is not optically active.

rad A unit of absorbed dose of ionizing radiation, defined as being equivalent to an absorption of 10^{-2} joule of energy in one kilogram of material.

radian Symbol: rad The SI unit of plane angle; 2π radian is one complete revolution (360°).

radiation In general the emission of energy from a source, either as waves (light, sound, etc.) or as moving particles (beta rays or alpha rays).

radical A group of atoms in a molecule. *See also* free radical.

radioactive Describing an element or nuclide that exhibits natural radioactivity.

radioactive dating (radiometric dating) Any method of measuring the age of materials that depends on radioactivity. *See* carbon dating.

radioactive decay series The series of definite isotopes to which a given radioactive element is transformed as it decays.

radioactivity The disintegration of certain unstable nuclides with emission of radiation. The emission of alpha particles, beta particles, and gamma waves are the three most important forms of radiation which occur during decay.

radiocarbon dating *See* carbon dating.

radiochemistry The chemistry of radioactive isotopes of elements. Radio-

chemistry involves such topics as the preparation of radioactive compounds, the separation of isotopes by chemical reactions, the use of radioactive labels in studies of mechanisms, and experiments on the chemical reactions and compounds of transuranic elements.

radioisotope A radioactive isotope of an element. Tritium, for instance, is a radioisotope of hydrogen. Radioisotopes are extensively used in research as souces of radiation and as tracers in studies of chemical reactions. Thus, if an atom in a compound is replaced by a radioactive nuclide of the element (a *label*) it is possible to follow the course of the chemical reaction. Radioisotopes are also used in medicine or diagnosis and treatment.

radiolysis A chemical reaction produced by high-energy radiation (X-rays, gamma rays, or particles).

radiometric dating *See* radioactive dating.

radio waves A form of electromagnetic radiation with wavelengths greater than a few millimetres. *See also* electromagnetic radiation.

radium A white radioactive luminescent metallic element of the alkaline-earth group. It has several short-lived radioisotopes and one long-lived isotope, radium-226 (half-life 1602 years). Radium is found in uranium ores, such as the oxides pitchblende and carnotite. it is used in luminous paints and neutron sources, and (progressively less) in radiotherapy.
Symbol: Ra; m.p. 700°C; b.p. 1140°C; r.d. 5(?); p.n. 88; r.a.m. 226.02.

radon A colourless monatomic radioactive element of the rare-gas group, now known to form unstable compounds. It has 19 short-lived radioisotopes; the most stable, radon-222, is a decay product of radium-226 and itself disintegrates into an isotope of polonium with a half-life of 3.82 days. ^{222}Rn is used in radiotherapy. Symbol: Rn; m.p. $-71\,°C$; b.p. $-61.8\,°C$; d. 9.73 kg m^{-3}; p.n. 86.

Raney nickel A catalytic form of nickel produced by treating a nickel-aluminium alloy with caustic soda. The aluminium dissolves (as aluminate) and Raney nickel is left as a spongy mass, which is pyrophoric when dry. It is used especially for catalysing hydrogenation reactions.

Raoult's law A relationship between the pressure exerted by the vapour of a solution and the presence of a solute. It states that the partial vapour pressure of a solvent above a solution (p) is proportional to the mole fraction of the solvent in the solution (X) and that the proportionality constant is the vapour pressure of pure solvent, (p_0), at the given temperature: i.e. $p = p_0X$. Solutions that obey Raoult's law are said to be *ideal*. There are some binary solutions for which Raoult's law holds over all values of X for either component. Such solutions are said to be *perfect* and this behaviour occurs when the intermolecular attraction between molecules within one component is almost identical to the attraction of molecules of one component for molecules of the other (e.g. chlorobenzene and bromobenzene). Because of solvation forces this behaviour is rare and in general Raoult's law holds only for dilute solutions.

For solutions that are ideal but not perfect the solute behaviour is similar in that the partial pressure of the solute, p_s, is proportional to the mole fraction of the solute, X_s, but in this case the proportionality constant, p', is not the vapour pressure of the pure solute but must be determined experimentally for each system. This solute equivalent of Raoult's law has the form $p_s = p'X_s$, and is called Henry's law. Because of intermolecular attractions p' is usually less than p_0.

rare earths See lanthanoids.

rare gases See noble gases.

Rashig process A method for the manufacture of chlorobenzene, and thence phenol, from benzene. Benzene vapour, hydrogen chloride, and air are passed over a copper(II) chloride catalyst (230°C):
$$2C_6H_6 + 2HCl + O_2 \rightarrow 2C_6H_5Cl + 2H_2O$$
The conversion to phenol is by a silicon catalyst (425°C):
$$C_6H_5Cl + H_2O \rightarrow HCl + C_6H_5OH$$

rate constant (velocity constant, specific reaction rate) Symbol: k The constant of proportionality in the rate expression for a chemical reaction. For example, in a reaction $A + B \rightarrow C$, the rate may be proportional to the concentration of A multiplied by that of B; i.e.
$$\text{rate} = k[A][B]$$
where k is the rate constant for this particular reaction. The constant is independent of the concentrations of the reactants but depends on temperature; consequently the temperature at which k is recorded must be stated. The units of k vary depending on the number of terms in the rate expression, but are easily determined remembering that rate has the units s^{-1}.

rate determining step (limiting step) The slowest step in a multistep reaction. Many chemical reactions are made up of a number of steps in which the one with the lowest velocity is the one that determines the rate of the overall process.
The overall rate of a reaction cannot exceed the rate of the slowest step. For example, the first step in the reaction between acidified potassium iodide solution and hydrogen peroxide is the rate determining step:
$$H_2O_2 + I^- \rightarrow H_2O + OI^- \text{ (slow)}$$
$$H^+ + OI^- \rightarrow HOI \text{ (fast)}$$
$$HOI + H^+ + I^- \rightarrow I_2 + H_2O \text{ (fast)}$$

rate of reaction A measure of the amount of reactant consumed in a chemical reaction in unit time. It is thus a measure of the number of effective collisions between reactant molecules. The rate at which a reaction proceeds can be measured by the rate the reactants disappear

rayon

or by the rate at which the produces are formed. The principal factors affecting the rate of reaction are temperature, pressure, concentration of reactants, light, and the action of a catalyst. The units usually used to measure the rate of a reaction are $mol\,dm^{-3}\,s^{-1}$. *See also* law of mass action.

rayon An artificial fibre formed from wood pulp (cellulose). There are two types. *Viscose rayon* is made by dissolving the cellulose in carbon disulphide and sodium hydroxide. The solution is forced through a fine nozzle into an acid bath, which regenerates the fibres. *Acetate rayon* is made by dissolving cellulose acetate in an organic solvent, and forcing the solution through a nozzle. The solvent is evaporated, and the cellulose acetate thus obtained as fibres.

rationalized units A system of units in which the equations have a logical form related to the shape of the system. SI units form a rationalized system of units. In it formulae concerned with circular symmetry contain a factor of 2π; those concerned with radial symmetry contain a factor of 4π.

reactant A compound taking part in a chemical reaction.

reagent A compound that reacts with another (the substrate). The term is usually used for common laboratory chemicals – sodium hydroxide, hydrochloric acid, etc. – used for experiment and analysis.

rearrangement A reaction in which the groups of a compound rearrange themselves to form a different compound.

recrystallization The repeated crystallization of a compound to ensure purity of the sample.

red lead *See* dilead(II) lead(IV) oxide.

redox Relating to the process of oxidation and reduction, which are intimately connected in that during oxidation by chemical agents the oxidizing agent itself becomes reduced, and vice versa. Thus an oxidation process is always accompanied by a reduction process. In electrochemical processes this is equally true, oxidation taking place at the anode and reduction at the cathode. These systems are often called *redox systems*, particularly when the interest centres on both compounds. Oxidizing and reducing power is indicated quantitatively by the *redox potential* or standard electrode potential, $E^{\ominus}$. Redox potentials are normally expressed as reduction potentials. They are obtained by electrochemical measurements and the values are referred to the H^+/H_2 couple for which $E^{\ominus}$ is set equal to zero. Thus increasingly negative potentials indicate increasing ease of oxidation or difficulty of reduction. Thus in a redox reaction the half reaction with the most positive value of $E^{\ominus}$ is the reduction half and the half reaction with the least value of $E^{\ominus}$ (or most highly negative) becomes the oxidation half. *See also* electrode potential.

reducing agent *See* reduction.

reduction The gain of electrons by such species as atoms, molecules, or ions. It often involves the loss of oxygen from a compound, or addition of hydrogen. Reduction can be effected chemically, i.e. by the use of *reducing agents*, or electrically, in which case the reduction process occurs at the cathode. For example,

$$2Fe^{3+} + Cu \rightarrow 2Fe^{2+} + Cu^{2+}$$

where Cu is the reducing agent and Fe^{3+} is reduced, and

$$2H_2O + SO_2 + 2Cu^{2+} \rightarrow$$
$$4H^+ + SO_4{}^{2-} + 2Cu^+$$

where SO_2 is the reducing agent and Cu^{2+} is reduced.

Stannous chloride, sulphur dioxide, the bisulphite ion, and hydroxylamine hydrochloride are common reducing agents. *See also* redox.

reduction potential *See* electrode potential.

refining The process of removing impurities from a substance or of extracting a substance from a mixture.

refluxing The process of boiling a liquid in a vessel connected to a condenser, so that the condensed liquid runs back into the vessel. By using a reflux condenser, the liquid can be maintained at its boiling point for long periods of time, without loss. The technique is a standard method of carrying out reactions in organic chemistry.

reforming The cyclization of straight-chain hydrocarbons from crude oil by heating under pressure with a catalyst, usually platinum on alumina. For example, the manufacture of methylbenzene from heptane:
$$C_7H_{16} \rightarrow C_6H_{11}CH_3$$
This first step is the production of methyl cyclohexane, which then loses six hydrogen atoms to give methylbenzene:
$$C_6H_{11}CH_3 \rightarrow C_6H_5CH_3 + 3H_2$$
See also steam reforming.

refractory Describing a compound (e.g. an inorganic oxide) that has a very high melting point.

Regnault's method A method used for the determination of the density of gases. A bulb of known volume is evacuated and weighed then the gas is admitted at a known pressure (from a vacuum line) and the bulb weighed again. The temperature is also noted and the data corrected to STP. The method is readily applicable to the obtaining of approximate molecular weights of gaseous samples.

relative atomic mass Symbol: A_r The ratio of the average mass per atom of the naturally occurring element to 1/12 of the mass of an atom of nuclide ^{12}C. It was formerly called *atomic weight*.

relative density Symbol: d The ratio of the density of a given substance to the density of some reference substance. The relative densities of liquids are usually measured with reference to the density of water at 4°C. Relative densities are also specified for gases; usually with respect to air at STP. The temperature of the substance is stated or is understood to be 20°C. Relative density was formerly called *specific gravity*.

relative molecular mass Symbol: M_r The ratio of the average mass per molecule of the naturally occurring form of an element or compound to 1/12 of the mass of an atom of nuclide ^{12}C. This was formerly called *molecular weight*. It does not have to be used only for compounds that have discrete molecules; for ionic compounds (e.g. NaCl) and giant-molecular structures (e.g. BN) the formula unit is used.

relaxation The process by which an excited species loses energy and falls to a lower energy level (such as the ground state).

rem (radiation equivalent *man*) A unit for measuring the effects of radiation dose on the human body. One rem is equivalent to an average adult male absorbing one rad of radiation. The biological effects depend on the type of radiation as well as the energy deposited per kilogram. The statutory limit for occupational exposure to radiation is 5 rem per year.

resolution (of racemates) The separation of a racemate into the two optical isomers. This cannot be done by normal methods, such as crystallization or distillation, because the isomers have identical physical properties. The main methods are:
(1) Mechanical separation. Certain optically active compounds form crystals with distinct left- and right-handed shapes. The crystals can be sorted by hand.
(2) Chemical separation. The mixture is reacted with an optical isomer. The products are then not optical isomers of each other, and can be separated by physical means. For instance, a mixture of D- and L-forms of an acid, acting with a pure L-base, produces two salts that can be sepa-

rated by fractional crystallization and then reconverted into the acids.

(3) Biochemical separation. Certain organic compounds can be separated by using bacteria that feed on one form only, leaving the other.

resonance The behaviour of many compounds cannot be adequately explained by a single structure using simple single and double bonds. The bonding electrons of the compound have a different distribution in the molecules. The actual bonding in the molecule can be regarded as a hybrid of two or more conventional forms of the molecule, called *resonance forms* or *canonical forms*. The result is a resonance hybrid. For example, the carbonyl group in a ketone has negative charge on the oxygen atom. It can be described as a resonance hybrid, somewhere between $=C=O$, in which a pair of electrons is shared between the C and the O, and $=C^+-O^-$, in which the electrons are localized on the O atom. Note that the two canonical forms do not contribute equally in the hybrid. The bonding of benzene can be represented by a resonance hybrid of two Kekulé structures and, to a lesser extent, three Dewar structures.

retort A piece of laboratory apparatus consisting of a glass bulb with a long narrow neck. In industrial chemistry, various vessels in which distillations or reactions take place are called retorts.

reverbatory furnace A furnace for smelting metals. It has a curved roof so that heat is reflected downwards, the fuel being in one part of the furnace and the ore in the other.

reversible change A change in the pressure, volume, or other properties of a system, in which the system remains at equilibrium throughout the change. Such processes could be reversed; i.e. returned to the original starting position through the same series of stages. They are never realized in practice. An isothermal reversible compression of a gas, for example, would have to be carried out infinitely

slowly and involve no friction, etc. Ideal energy transfer would have to take place between the gas and the surroundings to maintain a constant temperature.

In practice, all real processes are *irreversible changes* in which there is not an equilibrium throughout the change. In an irreversible change, the system can still be returned to its original state, but not through the same series of stages. For a closed system, there is an entropy increase involved in an irreversible change.

reversible reaction A chemical reaction that can proceed in either direction. For example, the reaction:
$$N_2 + 3H_2 = 2NH_3$$
is reversible. In general, there will be an equilibrium mixture of reactants and products.

R_F value A measure of the relative distance travelled by a sample in a chromatography experiment. It is obtained by dividing the distance travelled by the solvent front into the distance travelled by the sample. Under standard conditions, the R_F value is characteristic of a particular substance. *See* paper chromatography, thin-layer chromatography.

rhenium A very rare transition metal that usually occurs naturally with molybdenum. It is chemically similar to manganese and is used in alloys and catalysts. Symbol: Re; m.p. 3180°C; b.p. 5630°C; r.d. 20.0; p.n. 75; r.a.m. 186.21.

rhodium A transition metal, extracted in Canada. It is difficult to work and highly resistant to corrosion. Rhodium is used in protective finishes, alloys, and in mirrors formed by electrolytic deposition. Symbol: Rh; m.p. 1970°C; b.p. 3720°C; r.d. 12.4; p.n. 45; r.a.m. 102.91.

rhombic crystal *See* crystal system.

ribose ($C_5H_{10}O_5$) A monosaccharide; a component of RNA.

ring A closed loop of atoms in a molecule, as in benzene or cyclohexane. A *fused ring* is one joined to another ring in such a way that they share two atoms. Naphthalene is an example of a fused-ring compound.

ring closure A reaction in which one part of an open chain in a molecule reacts with another part, so that a ring of atoms is formed.

rock salt A transparent naturally-occurring mineral form of sodium chloride.

roentgen Symbol: R A unit of radiation, used for X-rays and γ-rays, defined in terms of the ionizing effect on air. One roentgen induces 2.58×10^{-4} coulomb of charge in one kilogram of dry air.

Rose's metal A fusible alloy containing 50% bismuth, 25–28% lead, and tins. Its low melting point (about 100°C) leads to its use in fire-protection devices.

rotary dryers Devices commonly used in the chemical industry for the drying, mixing, and sintering of solids. They consist essentially of a rotating inclined cylinder, which is longer in length than in diameter. Gases flow through the cylinder in either a countercurrent or cocurrent direction to regulate the flow of solids, which are fed into the end of the cylinder. Rotary dryers can be applied to both batch and continuous processes.

rubber A natural or synthetic polymeric elastic material. Natural rubber is a polymer of methylbuta-1,3-diene (isoprene).

Various synthetic rubbers are made by polymerization; for example chloroprene rubber (from 2-chlorobuta-1,3-diene) and silicone rubbers. *See also* vulcanization.

rubidium A soft silvery highly reactive element of the alkali-metal group. Naturally occurring rubidium comprises two isotopes, one of which, ^{87}Rb, is radioactive (half-life $5. \times 10^{10}$ years). It is found in small amounts in several complex silicate minerals, including lepidolite. Rubidium is used in vacuum tubes, photocells, and in making special glass.
Symbol: Rb; m.p. 38.89°C; b.p. 688°C; r.d. 1.53 (solid), 1.47 (liquid); p.n. 37; r.a.m. 85.47.

rusting The corrosion of iron in air to form an oxide. Both oxygen and moisture must be present for rusting to occur. The process is electrolytic, involving electrochemical reactions on the wet iron:
$$Fe(s) \rightarrow Fe^{2+}(aq) + 2e^-$$
$$2H_2O + O_2(aq) + 4e^- \rightarrow 4OH^-(aq)$$
Iron(II) hydroxide precipitates and is oxidized to the red hydrated iron(III) oxide $Fe_2O_3.H_2O$.

ruthenium A transition metal that occurs naturally with platinum. It forms alloys with platinum that are used in electrical contacts. Ruthenium is also used in jewellery alloyed with palladium.
Symbol: Ru; m.p. 2310°C; b.p. 3900°C; r.d. 12.4; p.n. 44; r.a.m. 101.07.

rutherfordium *See* element 104.

saccharide *See* sugar.

Sachse reaction A process for the manufacture of ethyne from natural gas (methane). Part of the methane is burned in two stages to raise the furnace temperature to about 1500°C. Under these conditions, the rest of the methane is converted to ethyne and hydrogen:
$$2CH_4 \rightarrow C_2H_2 + 3H_2$$
The process is important since it provides a source of ethyne from readily available natural gas, thus avoiding the expensive 'carbide process'.

sacrificial protection A method of protection against electrolytic corrosion (especially rusting). In protecting steel pipelines, for instance, zinc or magnesium rods are buried in the ground at points along the line and connected to the pipeline. Rusting of iron is an electrochemical process in which Fe^{2+} ions dissolve in the water in contact with the surface. A more electropositive element, such as zinc, protects against this because Zn^{2+} ions dissolve in preference. In other words, the zinc rod sacrifices itself for the steel pipeline. The same effect occurs with galvanized iron. If the zinc coating is scratched, the iron exposed does not rust until all the zinc has dissolved away. (Note that tin, being less electropositive than iron, has the opposite effect.)

sal ammoniac *See* ammonium chloride.

saline Containing a salt, especially an alkali–metal halide such as sodium chloride.

salt A compound with an acidic and a basic radical, or a compound formed by total or partial replacement of the hydrogen in an acid by a metal. In general terms a salt is a material that has identifiable cationic and anionic components.

salt bridge An electrical contact between two half cells, used to prevent mixing. A glass U-tube filled with potassium chloride in agar is commonly used.

salt hydrate A metallic salt in which the metal ions are surrounded by a fixed number of water molecules. The water molecules are bound to the metal ions by ion–dipole interactions. In the solid state the water molecules form part of the crystal structure. It is thus a clathrate.
Salt hydrates that contain several molecules of water for each metal ion can lose them progressively (effloresce) if the pressure of water vapour is kept below the dissociation pressure of the system; eventually an anhydrous salt is formed. Alternatively, if the vapour pressure is continuously raised the system will add molecules of water (deliquesce) until a saturated solution forms.

saltpetre *See* potassium nitrate.

samarium A silvery element of the lanthanoid series of metals. It occurs in association with other lanthanoids. Samarium is used in the metallurgical, glass, and nuclear industries.
Symbol: Sm; m.p. 1077°C; b.p. 1790°C; r.d. 7.5; p.n. 62; r.a.m. 150.35.

Sandmeyer reaction A method for the synthesis of chloro- and bromoarenes by heating the diazonium salt with the copper(I) halide, for example:
$$C_6H_5N_2{}^+ + CuBr \rightarrow C_6H_5Br$$

sandwich compound A type of complex formed between transition-metal ions and aromatic compounds, in which the metal ion is 'sandwiched' between the rings. The bonding is between the d orbitals of the metal and the pi electrons on the ring. Four-, five-, six-, seven-, and eight-membered rings are known to complex with V, Cr, Mn, Co, Ni, and Fe. Ferrocene $(Fe(C_5H_5)_2)$ is the best-known example.

saponification The process of hydrolysing an ester with a hydroxide. The carboxylic acid forms a salt. For instance, with sodium hydroxide:
$$RCOOR' + NaOH \rightarrow NaOOCR + R'OH$$
The saponification of fats, which are esters of 1,2,3-trihydroxypropane with long-chain carboxylic acids, forms soaps.

saturated compound An organic compound that does not contain any double or triple bonds in its structure. A saturated compound will undergo substitution reactions but not addition reactions since each atom in the structure will already have formed its maximum possible number of single bonds.
Compare unsaturated compound.

saturated solution A solution that contains the maximum equilibrium amount of solute at a given temperature. A solution is saturated if it is in equilibrium with its solute. If a saturated solution of a solid is cooled slowly, the solid may stay temporarily in solution; i.e. the solution may contain *more* than the equilibrium amount of solute. Such solutions are said to be *supersaturated*.

saturated vapour A vapour that is in equilibrium with the solid or liquid. A saturated vapour is at the maximum pressure (the saturated vapour pressure) at a given temperature. If the temperature of a saturated vapour is lowered, the vapour condenses. Under certain circumstances, the substance may stay temporarily in the vapour phase; i.e. the vapour contains *more* than the equilibrium concentration of the substance. The vapour is then said to be *supersaturated*.

s-block elements The elements of the first two main groups; i.e. IA (H, Li, Na, K, Rb, Cs, Fr) and IIA (Be, Mg, Ca, Sr, Ba, Ra). They are so called because their outer shells have the electronic configurations ns^1 or ns^2. The s-block excludes those with inner $(n-1)$d levels occupied (i.e. excluding transition elements, which also have s^2 and occasionally s^1 configurations).

scandium A light-weight silvery element belonging to the first transition series. It is found in minute amounts in over 800 minerals, often associated with lanthanoids. Scandium is used in high-intensity lights and in electronic devices.
Symbol: Sc; m.p. 1541°C; b.p. 2831°C; r.d. 2.99; p.n. 21; r.a.m. 44.96.

Schiff's base A type of compound formed by reacting an aldehyde or ketone (e.g. RCOR) with an aryl amine (e.g. $ArNH_2$). The product, an *N*-arylimide, which is usually crystalline, has the formula R_2C:NAr.

Schiff's reagent An aqueous solution of magenta dye decolorized by reduction with sulphur(IV) oxide. It is a test for aldehydes and ketones. Aliphatic aldehydes restore the colour quickly; aliphatic ketones and aromatic aldehydes slowly; aromatic ketones give no reaction.

Schotten-Baumann reaction A method for the preparation of *N*-phenylbenzamide (benzanilide) in which a mixture of phenylamine (aniline) and sodium hydroxide is stirred while benzoyl chloride is added slowly. The *N*-phenylbenzamide precipitates and can be recrystallized from hot ethanol:
$$C_6H_5NH_2 + C_6H_5COCl \rightarrow$$
$$C_6H_5NH.COC_6H_5 + HCl$$

Schottky defect *See* defect.

scrubber The part of a chemical plant that removes impurities from a gas by passing it through a liquid.

second Symbol: s The SI base unit of time. It is defined as the duration of 9 192 631 770 cycles of a particular wavelength of radiation corresponding to a transition between two hyperfine levels in the ground state of the caesium-133 atom.

second-order reaction A reaction in which the rate of reaction is proportional to the product of the concentrations of two of the reactants or to the square of the concentration of one of the reactants; i.e.

$$\text{rate} = k[A][B]$$

or

$$\text{rate} = k[A]^2$$

For example, the hydrolysis by dilute alkali of an ester is a second-order reaction:

$$\text{rate} = k[\text{ester}][\text{alkali}]$$

The rate constant for a second-order reaction has the units $\text{mol}^{-1}\,\text{dm}^3\,\text{s}^{-1}$. Unlike a first-order reaction, the time for a definite fraction of the reactants to be consumed is dependent on the original concentrations.

sedimentation The settling of a suspension, either under gravity or in a centrifuge. The speed of sedimentation can be used to estimate the average size of the particles. This technique is used with an ultracentrifuge to find the relative molecular masses of macro-molecules.

seed A small crystal added to a gas or liquid to assist solidification or precipitation from a solution. The seed, usually a crystal of the substance to be formed, enables particles to pack into predetermined positions so that a larger crystal can form.

selenium A metalloid element existing in several allotropic forms and belonging to group VI of the periodic table. It occurs in minute quantities in sulphide ores and industrial sludges. The common grey allotrope is very light-sensitive and is used in photocells, solar cells, some glasses, and in xerography. The red allotrope is unstable and reverts to the grey form under normal conditions.
Symbol: Se; m.p. 217°C (grey); b.p. 684.9°C (grey); r.d. 4.79 (grey); p.n. 34; r.a.m. 78.96.

semicarbazone A type of organic compound containing the $C:N.NH.CO.NH_2$ grouping, formed by reaction of an aldehyde or ketone with semicarbazide ($H_2N.NH.CO.NH_2$). The compounds are crystalline solids with sharp melting points, which can be used to characterize the original aldehyde or ketone.

semiconductor A crystalline material which only conducts under certain conditions. The conductivity, unlike in metals, increases with temperature, since the highest occupied and lowest unoccupied energy levels are very close. A semiconductor can be formed by introducing small amounts of an impurity to an ultrapure material. If these impurities can withdraw electrons from the occupied energy level, leaving 'holes' which permit conduction, the resultant semiconduction is known as *p-type*. Alternatively, these impurities may donate electrons which enter the unoccupied energy level, thus forming an *n-type* semiconductor.
Semiconductors are used extensively by the electronics industry in such devices as integrated circuits.

semipermeable membrane A membrane that, when separating a solution from a pure solvent, permits the solvent molecules to pass through it but does not allow the transfer of solute molecules. Synthetic semipermeable membranes are generally supported on a porous material, such as unglazed porcelain or fine wire screens, and are commonly formed of cellulose or related materials. They are used in osmotic studies, gas separations, and medical applications.
Equilibrium is reached at a semipermeable membrane if the chemical potentials on both sides become identical; migration of solvent molecules towards the solution is an attempt by the system to reach equilibrium. The pressure required to halt this migration is the osmotic pressure.

semipolar bond A coordinate bond.

septivalent (hexavalent) Having a valence of seven.

sequestration The formation of a complex with an ion in solution, so that the ion does not have its normal activity. Sequestering agents are often chelating agents, such as edta.

serine *See* amino acids.

sesqui- Prefix indicating a 2/3 ratio. A sesquioxide, for example, would have the formula M_2O_3.

shell A group of electrons that share the same principal quantum number. Early work on X-ray emission studies used the terms K, L, M, and these are still sometimes used for the first three shells: $n = 1$ K-shell; $n = 2$ L-shell; $n = 3$ M-shell.

sheradizing A technique used to obtain corrosion-resistant articles or iron or steel by heating with zinc dust in a sealed rotating drum for several hours at about 375°C. At this temperature the two metals amalgamate forming internal layers of zinc–iron and zinc–steel alloys and an external layer of pure zinc. Sheradizing is principally used for small parts, such as springs, washers, nuts, and bolts.

short period *See* period.

SI *See* SI units.

side chain *See* chain.

side reaction A chemical reaction that takes place to a limited extent at the same time as a main reaction. Thus the main product of a reaction may contain small amounts of other compounds.

siemens (mho) Symbol: S The SI unit of electrical conductance, equal to a conductance of one ohm^{-1}.

sigma bond *See* orbital.

silane Any of a small number of hydrides of silicon: SiH_4, Si_2H_6, Si_3H_8, etc. Silanes can be made by the action of acids on magnesium silicide (Mg_2Si). They are unstable compounds and ignite spontaneously in air. Only the first few members of the series are known and there is not the range of 'hydrosilicon' compounds to compare with the hydrocarbons.

silica *See* silicon(IV) oxide.

silica gel A gel made by coagulating sodium silicate sol. The gel is dried by heating and used as a catalyst support and as a drying agent. The silica gel used in desiccators and in chemical-balance cases to remove moisture is usually coloured with a cobalt salt to indicate whether it is still active (blue = dry; pink = moist).

silicates A large number of compounds containing metal ions and complex silicon–oxygen compounds. The negative ions in silicates are of the form SiO_4^{4-}, $Si_2O_7^{6-}$, etc., and many are polymeric, containing SiO_4 units linked in long chains, sheets, or three-dimensional arrays. *Aluminosilicates* and *borosilicates* are similar materials containing aluminium or boron atoms in the structure. Many silicates occur naturally in rocks and minerals.

silicide A compound of silicon with a more electropositive element.

silicon A hard brittle grey metalloid element; the second element in group IV of the periodic table. It has the electronic configuration of neon with four additional outer electrons; i.e., $[Ne]3s^23p^2$.
Silicon accounts for 27.7% of the mass of the Earth's crust and occurs in a wide variety of silicates with other metals, clays, micas, and sand, which is largely SiO_2. The element is obtained on a small scale by the reduction of silicon(IV) oxide (SiO_2) by carbon or calcium carbide. For semiconductor applications very pure silicon is produced by direct reaction of silicon with an HCl/Cl_2 mixture to give silicon tetrachloride ($SiCl_4$), which can be purified by distillation. This is then decomposed on a hot wire in an atmosphere of hydrogen. For ultra-pure samples zone refining is used. Unlike carbon, silicon does not form allotropes but has only the diamond type of structure.
Silicon does not react with hydrogen except under extreme conditions in the presence of chlorine, in which case compounds such as trichlorosilane ($HSiCl_3$) are obtained. The hydride SiH_4 itself may be prepared in the laboratory by hydrolysis of magnesium silicide (a

number of other hydrides up to Si_6H_{14} are obtained), or more conveniently by reduction of silicon tetrachloride with lithium tetrahydridoaluminate:

$$SiCl_4 + LiAlH_4 \rightarrow SiH_4 + LiCl + AlCl_3$$

The silanes are mildly explosive and the chloroalkyl silanes extremely so.

Silicon combines with oxygen when heated in air to form silicon(IV) oxide. The mineral silicates constitute a vast collection of materials with a wide range of structures including chains, rings, lattices, twisted chains, sheets, and varyingly cross-linked structures all based on different assemblies of $[SiO_4]^{2-}$ tetrahedra. Being a metalloid, silicon is somewhat amphoteric and consequently silica is weakly acidic, dissolving in fused alkalis to form the appropriate silicate.

Silicon also forms an immense number of organic derivatives. In many cases where single bonding only is involved these compounds are rather similar to organic analogues with the general distinction that nucleophilic attachment occurs at silicon (rather than at hydrogen) and bond angles at oxygen and nitrogen are often larger than the organic materials. Optical activity and inversion of activity is observed. The major differences are due to the type of 'double bond' available to carbon ($p\pi-p\pi$), and to silicon ($p \rightarrow d\pi$ or back bonding). Thus silicon does not form conventional double bonds and silicon bonds to oxygen are of the ether type (as in siloxanes) rather than the ketone type. The Si–OH bond is also very much less stable than the alcohol link, R–OH, ready condensation to the siloxane occurring in the silicon case.

$$2R_3SiOH \rightarrow R_3Si-O-SiR_3 + H_2O$$

Symbol: Si; m.p. 1410°C; b.p. 2355°C; r.d. 2.32–2.34; p.n. 14; r.a.m. 28.

silicon carbide (carborundum, SiC) A black very hard crystalline solid made by heating silicon(IV) oxide (sand) with carbon (coke) in an electric furnace. It is used as an abrasive.

silicon dioxide See silicon(IV) oxide.

silicones Polymeric synthetic silicon compounds containing chains of alternating silicon and oxygen atoms, with organic groups bound to the silicon atoms. Silicones are used as lubricants and water repellants and in waxes and varnishes. Silicone rubbers are superior to natural rubbers in their resistance to both high and low temperatures and chemicals. See also siloxanes.

silicon(IV) oxide (silicon dioxide, silica, SiO_2) A hard crystalline compound occurring naturally in three crystalline forms (quartz, tridymite, and crystobalite). Sand is mostly silicon(IV) oxide. Fused silica is a glassy substance used in laboratory apparatus. Silica is used in the manufacture of glass.

siloxanes Compounds containing Si–O–Si groups with organic groups bound to the silicon atoms. The silicones are polymers of siloxanes.

silver A transition metal that occurs native and as the sulphide (Ag_2S) and chloride (AgCl). It is extracted as a by-product in refining copper and lead ores. Silver darkens in air due to the formation of silver sulphide. It is used in coinage alloys, tableware, and jewellery. Silver compounds are used in photography.
Symbol: Ag; m.p. 961°C; b.p. 2210°C; r.d. 10.5; p.n. 47; r.a.m. 107.87.

silver(I) bromide (AgBr) A pale yellow precipitate obtained by adding a soluble bromide solution to a solution of silver(I) nitrate. The halide dissolves in concentrated ammonia solution but not in dilute ammonia solution. It is used extensively in photography for making the sensitive emulsions for plates and films. Silver(I) bromide crystallizes in a similar manner to sodium(I) chloride. Unlike silver(I) chloride and silver(I) iodide, silver(I) bromide does not absorb ammonia gas.

silver(I) chloride (AgCl) A compound prepared as a curdy white precipitate by the addition of a soluble chloride solution to a solution of silver(I) nitrate. It occurs in nature as the mineral 'horn silver'. Silver(I) chloride is insoluble in water but dissolves in concentrated hydrochloric acid and concentrated ammonia solution. It is rather unreactive but is affected by sunlight, being used extensively in photography. Silver(I) chloride crystallizes in the sodium chloride lattice pattern. It absorbs ammonia gas forming ammines, e.g. $AgCl.2NH_3$ and $AgCl.3NH_3$.

silver(I) iodide (AgI) A pale yellow precipitate prepared by the addition of a soluble iodide solution to a solution of silver(I) nitrate. Silver(I) iodide occurs in nature as iodoargyrite in the form of hexagonal crystals. It is virtually insoluble in ammonia solution and is used in the photographic industry. Silver(I) iodide will absorb ammonia gas. It is trimorphic and is found to contract when heated and expand on cooling.

silver-mirror test A test for the aldehyde group. A few drops of the sample are warmed with Tollen's reagent. An aldehyde reduces Ag^+ to silver metal, causing a brilliant silver mirror to coat the inside wall of the tube. *See also* Tollen's reagent.

silver(I) nitrate ($AgNO_3$) A white crystalline solid prepared by dissolving silver in dilute nitric acid and crystallizing the solution. Silver(I) nitrate is dimorphous, crystallizing in the orthorhombic and hexagonal systems. It undergoes thermal decomposition to yield silver, nitrogen(IV) oxide, and oxygen. Probably the most important silver salt, it is used in the medical industry for the treatment of warts and in the photographic industry. In the laboratory, it is used both as an analytical and volumetric reagent. Silver(I) nitrate in solution and in the solid state can oxidize organic compounds, e.g. aldehydes can be oxidized to carboxylic acids; the silver ions are reduced to silver.

silver(I) oxide (argentous oxide, Ag_2O) A brown amorphous solid formed by the addition of sodium or potassium hydroxide solution to a solution of silver nitrate. Silver(I) oxide turns moist red litmus blue and decomposes (at 160°C) to give silver and oxygen. Silver(I) oxide dissolves in concentrated ammonia solution, forming the complex ion $[Ag(NH_3)_2]^+$. In organic chemistry it is used when moist to convert alkyl halides to the corresponding alcohol and when dry to convert alkyl halides into ethers.

silver(II) oxide (argentic oxide, AgO) A black diamagnetic substance prepared by the oxidation of either silver or silver(I) oxide using ozone. Alternatively it can be prepared as a precipitate by the addition of a solution of potassium persulphate to one of silver(I) nitrate. Silver(II) oxide has a structure similar to that of zinc sulphide.

single bond A covalent bond between two elements that involves one pair of electrons only. It is represented by a single line, for example $H-Br$, and is usually a sigma bond, although it can be a pi bond. *Compare* multiple bond. *See also* orbital.

sintering A process in which certain powdered substances (e.g. metals, ceramics) coagulate into a single mass when heated to a temperature below the substance's melting point. Sintered glass is a porous material used for laboratory filtration.

SI units (*Système International d'Unités*) The internationally adopted system of units used for scientific purposes. It has seven base units (the metre, kilogram, second, kelvin, ampere, mole, and candela) and two supplementary units (the radian and steradian). Derived units are formed by multiplication and/or division of base units; a number have special names. Standard prefixes are used for multiples and submultiples of SI units. The SI system is a coherent rationalized system of units.

slag Glasslike compounds of comparatively low melting point formed during the extraction of metals when the impurities in an ore react with a flux. The fact that the slag is a liquid of comparatively low density means that it can be separated from the

Base and Supplementary SI Units

physical quantity	name of SI unit	symbol for unit
length	metre	m
mass	kilogram(me)	kg
time	second	s
electric current	ampere	A
thermodynamic temperature	kelvin	K
luminous intensity	candela	cd
amount of substance	mole	mol
*plane angle	radian	rad
*solid angle	steradian	sr
*supplementary units		

Derived SI Units with Special Names

physical quantity	name of SI unit	symbol for SI unit
frequency	hertz	Hz
energy	joule	J
force	newton	N
power	watt	W
pressure	pascal	Pa
electric charge	coulomb	C
electric potential difference	volt	V
electric resistance	ohm	Ω
electric conductance	siemens	S
electric capacitance	farad	F
magnetic flux	weber	Wb
inductance	henry	H
magnetic flux density	tesla	T
luminous flux	lumen	lm
illuminance (illumination)	lux	lx
absorbed dose	gray	Gy

Decimal Multiples and Submultiples to be used with SI Units

submultiple	prefix	symbol	multiple	prefix	symbol
10^{-1}	deci-	d	10^{1}	deca-	da
10^{-2}	centi-	c	10^{2}	hecto-	h
10^{-3}	milli-	m	10^{3}	kilo-	k
10^{-6}	micro-	μ	10^{6}	mega-	M
10^{-9}	nano-	n	10^{9}	giga-	G
10^{-12}	pico-	p	10^{12}	tera-	T
10^{-15}	femto-	f	10^{15}	peta-	P
10^{-18}	atto-	a	10^{18}	exa-	E

liquid metal on which it floats. In a blast furnace, the flux used is limestone ($CaCO_3$) and the main impurity is silica (SiO_2). The slag formed is mainly calcium silicate:

$$CaCO_3 + SiO_2 \rightarrow CaSiO_3$$

It is used as railway ballast and in fertilizers and concrete.

slaked lime See calcium hydroxide.

slip planes The planes of weakness in a crystal, e.g. the boundaries between the octahedral layers in graphite.

slurry A thin paste of suspended solid particles in a liquid.

smelting An industrial process for extracting metals from their ores at high temperatures. Generally the ore is reduced with carbon (for zinc and tin) or carbon monoxide (for iron). Copper and lead are obtained by reduction of the oxide with the sulphide; for example:

$$2Cu_2O + Cu_2S \rightarrow 6Cu + SO_2$$

A flux is also used to combine with impurities and form a slag on top of the molten metal.

soap One of a number of sodium or potassium compounds of fatty acids that are commonly used to improve the cleansing properties of water. Soap was the earliest known detergent. It is made by first reacting vegetable oils and animal fats with a strong solution of sodium or potassium hydroxide to produce organic acids, such as octadecanoic acid. The soap is then precipitated out as the salt, e.g. sodium octadecanoate, by adding excess sodium chloride. In water, soap molecules break up to produce ions, which are responsible for the cleansing properties. See also detergents.

soda ash See sodium carbonate.

soda lime A grey solid produced by adding sodium hydroxide solution to calcium oxide, to give a mixture of $Ca(OH)_2$ and NaOH. It is used in the laboratory as a drying agent and to absorb carbon dioxide.

sodamide ($NaNH_2$) A white ionic solid formed by passing dry ammonia over sodium at $300-400°C$. The compound reacts with water to give sodium hydroxide and ammonia. It reacts with red-hot carbon to form sodium cyanide and with nitrogen(I) oxide to form sodium azide. Sodamide is used in the Castner process and in the explosives industry.

sodium A soft reactive metal; the second member of the alkali metals (group I of the periodic table). It has the electronic configuration of a neon structure plus an additional outer 3s electron. Electronic excitation in flames or the familiar sodium lamps gives a distinctive yellow colour arising from intense emission at the so called 'sodium-D' line pair. The ionization potential is rather low and the sodium atom readily loses its electron (i.e. the metal is strongly reducing). The chemistry of sodium is largely the chemistry of the monovalent Na^+ ion.

Sodium occurs widely as NaCl in seawater and as deposits of halite in dried up lakes etc., (2.6% of the lithosphere). The element is obtained commercially by electrolysis of NaCl melts in which the melting point is reduced by the addition of calcium chloride; sodium is produced at the iron cathode (the Downs cell). The metal is extremely reactive. It reacts vigorously with the halogens, and also with water to give hydrogen and sodium hydroxide. The chemistry of sodium is very similar to that of the other members of group I.

Nearly all sodium compounds are soluble in water. Sodium hydroxide is produced commercially by the electrolysis of brine using diaphragm cells (largely in the US) or mercury-cathode cells (in the UK); chlorine being a coproduct.

Solutions of sodium metal in liquid ammonia are blue and have high electrical conductivities; the main current carrier of such solutions is the solvated electron. Such solutions are used in both organic and inorganic chemistry as efficient reducing agents. Sodium also forms a number of alkyl and aryl derivatives by reaction with the appropriate mercury compound, e.g.:

$(CH_3)_2Hg + 2Na \rightarrow 2CH_3Na + Na/Hg$

These materials are used as catalysts for polymerization reactions. The metal has a body-centred structure.

Symbol: Na; m.p. 97.8°C; b.p. 883°C; r.d. 0.97; p.n. 11; r.a.m. 22.98.

sodium acetate *See* sodium ethanoate.

sodium aluminate ($NaAlO_2$) A white solid produced by adding excess aluminium to a hot concentrated solution of sodium hydroxide. Once the reaction has been initiated, sufficient heat is liberated to keep the reaction going; hydrogen is also produced. In solution the aluminate ions have the structure $Al(OH)_4^-$ and $NaAl(OH)_4$ is known as sodium(I) tetrahydroxoaluminate(III). Addition of sodium hydroxide to an aluminium salt solution produces a white gelatinous precipitate of aluminium hydroxide, which dissolves in excess alkali to give a solution of sodium(I) tetrahydroxoaluminate(III).

sodium azide (NaN_3) A white solid prepared by passing nitrogen(I) oxide over heated sodamide. On heating, sodium azide decomposes to give sodium and nitrogen. It is used as a reagent in organic chemistry and in the preparation of lead azide (for use in detonators).

sodium benzenecarboxylate (sodium benzoate, C_6H_5COONa) A white crystalline powder made by neutralizing benzoic acid with sodium hydroxide solution and then evaporating the solution. It is used as an urinary antiseptic, in the dyeing industry, and as a food preservative.

sodium benzoate *See* sodium benzenecarboxylate.

sodium bicarbonate *See* sodium hydrogencarbonate.

sodium bisulphate *See* sodium hydrogensulphate.

sodium bisulphite *See* sodium hydrogensulphite.

sodium bromide (NaBr) A white solid formed by the action of bromine on hot sodium hydroxide solution. Alternatively, it can be made by the action of dilute hydrobromic acid on sodium carbonate or hydroxide. Sodium bromide is soluble in water and forms crystals similar in shape to sodium chloride. It is used in medicine and photography. Sodium bromide forms a hydrate, $NaBr.2H_2O$.

sodium carbonate (soda ash, Na_2CO_3) A white amorphous powder, which aggregates on exposure to air owing to the formation of hydrates. Industrially it is prepared by the ammonia–soda (Solvay) process. On crystallizing from aqueous solution large translucent crystals of the decahydrate ($Na_2CO_3.10H_2O$) are formed (*washing soda*). These effloresce forming the monohydrate, $Na_2CO_3.H_2O$. Large quantities of sodium carbonate are used in the manufacture of sodium hydroxide. Washing soda is used as a domestic cleanser. The salt produces an alkaline solution in water by hydrolysis:

$Na_2CO_3 + 2H_2O \rightarrow 2NaOH + H_2CO_3$

It is used in volumetric analysis to standardize strong acids.

sodium chlorate(V) ($NaClO_3$) A white solid formed by the action of chlorine on hot concentrated sodium hydroxide solution, or by the electrolysis of a concentrated sodium chloride solution. It is soluble in water. Sodium chlorate is a powerful oxidizing agent. If heated it yields oxygen and sodium chloride. The chlorate is used in explosives, in matches, as a weed killer, and in the textile industry.

sodium chloride (common salt, salt, NaCl) A white solid prepared by the neutralization of hydrochloric acid with aqueous sodium hydroxide. It occurs in sea water and natural brines. Natural solid deposits of *rock salt* are also found in certain places. The compound dissolves in water with the absorption of heat, its solubility changing very little with temperature. It is used to season and preserve food. Industrially, it is used as a raw material in the manufacture of sodium carbonate (Solvay process), sodium hydroxide (electrolysis), and soap. Sodium chloride is almost insoluble in alcohol.

○ chloride ion

● sodium ion

Sodium-chloride structure

sodium-chloride structure A form of crystal structure that consists of a face-centred cubic arrangement of sodium ions with chloride ions situated at the middle of each edge and in the centre of the cube. Electrostatic attraction holds the oppositely charged ions together.

The lattice can also be thought of as two interpenetrating face-centred cubes, one composed of sodium ions and the other of chloride ions. Other compounds (e.g. sodium bromide and potassium chloride) having their ions arranged in the same positions, are also described as having the sodium-chloride structure.

sodium cyanide (NaCN) A white solid formed either by the action of carbon on sodamide at high temperature or by passing ammonia over sodium at 300–400°C to form sodamide and reacting it with carbon. The industrial salt is prepared by the Castner process and purified by recrystallization from liquid ammonia. It is extremely poisonous and used as a source of cyanide and hydrocyanic acid. In the cyanide process it is used in the extraction of silver and gold. Its aqueous solutions are alkaline due to salt hydrolysis.

sodium dichromate ($Na_2Cr_2O_7$) An orange deliquescent solid that is very soluble in water. It is made from finely powdered chromite, which is heated with calcium oxide and sodium carbonate in a furnace. If an alkali is added to a solution of the dichromate, the chromate is produced. At high temperatures, sodium dichromate decomposes to give the chromate, chromic oxide, and oxygen. It is used as an oxidizing agent, particularly in organic chemistry and in volumetric analysis to estimate iron(II) ions and iodide ions.

sodium dihydrogen phosphate(V) (sodium dihydrogen orthophosphate, NaH_2PO_4) A white solid prepared by titrating phosphoric acid with sodium hydroxide solution using methyl orange as the indicator. On evaporation white crystals of the monohydrate are formed. It is used in some baking powders. The monohydrate crystallizes in the orthorhombic system.

sodium dioxide See sodium superoxide.

sodium ethanoate (sodium acetate, CH_3COONa) A white solid prepared by the neutralization of ethanoic acid with either sodium carbonate or sodium hydroxide. Sodium ethanoate reacts with sulphuric acid to form sodium hydrogensulphate and ethanoic acid; with sodium hydroxide it gives rise to sodium carbonate and methane. Sodium ethanoate is used in the dyeing industry.

sodium fluoride (NaF) A white solid formed by the action of hydrofluoric acid on sodium carbonate or sodium hydroxide (the reaction between metallic sodium and fluorine is too violent). Sodium fluoride is soluble in water and reacts with concentrated sulphuric acid to yield hydrogen fluoride. It is used as a constituent of ceramic enamels, as an antiseptic, and as an agent to prevent fermentation.

sodium formate *See* sodium methanoate.

sodium hexafluoroaluminate (cryolite, Na_3AlF_6) A compound found in large quantities in South Greenland. It is white or colourless, but may be reddish or brown due to impurities. Sodium hexafluoroaluminate is used as a flux in the manufacture of aluminium. It crystallizes in the monoclinic system but in forms that closely resemble cubes and isometric octahedrals.

sodium hydride (NaH) A white crystalline solid prepared by passing a pure stream of dry hydrogen over sodium at 350°C; the sodium is usually suspended in an inert medium. Electrolysis of fused sodium hydride yields hydrogen at the anode. Sodium hydride reacts with water to form sodium hydroxide solution and hydrogen. It is used as a powerful reducing agent to convert water to hydrogen, concentrated sulphuric acid to hydrogen sulphide, and iron(III) oxide to iron. Sodium hydride bursts into flames spontaneously on contact with the halogens at room temperature. It dissolves in liquid ammonia to give sodamide, $NaNH_2$.

sodium hydrogencarbonate (sodium bicarbonate, $NaHCO_3$) A white solid formed either by passing an excess of carbon dioxide through sodium carbonate or hydroxide solution, or by precipitation when cold concentrated solutions of sodium chloride and ammonium hydrogencarbonate are mixed. Sodium hydrogencarbonate decomposes on heating to give sodium carbonate, carbon dioxide, and water. With dilute acids, it yields carbon dioxide. It is used as a constituent of baking powder, in effervescent beverages, and in fire extinguishers. Its aqueous solutions are alkaline due to salt hydrolysis. Sodium hydrogencarbonate forms monoclinic crystals.

sodium hydrogensulphate (sodium bisulphate, $NaHSO_4$) A white solid formed either by the partial neutralization of sodium hydroxide solution with sulphuric acid, by the action of concentrated sulphuric acid on sodium nitrate, or by heating equimolar quantities of sodium chloride and concentrated sulphuric acid. In aqueous solution sodium hydrogensulphate is strongly acidic. It crystallizes as the monohydrate ($NaHSO_4.H_2O$), which dehydrates on warming. If heated strongly the pyrosulphate ($Na_2S_2O_7$) is formed, which decomposes to give the sulphate and sulphur trioxide. Sodium hydrogensulphate is used as a cheap source of sulphuric acid and in the dyeing industry.

sodium hydrogensulphite (sodium bisulphite, $NaHSO_3$) A white powder prepared by saturating a solution of sodium carbonate with sulphur dioxide. It is isolated from the aqueous solution by precipitation with alcohol. If heated it undergoes thermal decomposition to give sodium sulphate, sulphur dioxide, and sulphur. Sodium hydrogensulphite is used to sterilize wine casks and in medicine as an antiseptic.

sodium hydroxide (caustic soda, NaOH) A white deliquescent slightly translucent solid. In solution it is a strong alkali and electrolyte. In the laboratory it can be prepared by reacting sodium, sodium monoxide, or sodium peroxide with water. Industrially it can be prepared by the electrolysis of sodium chloride using a mercury cathode (Castner-Kellner) or 'diaphragm' cell. It can also be manufactured by the Gossage process. Sodium hydroxide dissolves readily in water with the evolution of heat. Its solutions have a soapy feel and are very corrosive. It is used to absorb acidic gases, such as carbon dioxide and sulphur dioxide. Industrially it is used in soap and paper manufacture and in the purification of bauxite.

sodium iodide (NaI) A white solid formed by reacting sodium carbonate or sodium hydroxide with hydroiodic acid; the solution is then evaporated. Sodium iodide forms colourless crystals having a cubic shape. It is used as a source of iodine and in medicine and photography. Acidified solutions of sodium iodide exhibit reducing properties owing to the formation of hydroiodic acid.

sodium methanoate (sodium formate, HCOONa) A white solid produced by reacting carbon monoxide with solid sodium hydroxide at 200°C and 10 atmospheres pressure. Alternatively it can be prepared by the neutralization of methanoic acid with sodium hydroxide solution.

sodium monoxide (Na_2O) A white solid formed by burning sodium in a deficiency of oxygen or alternatively by reducing sodium peroxide or sodium hydroxide with the requisite amount of sodium. It reacts violently with water to form sodium hydroxide and with acids to form solutions of their salts. Sodium monoxide forms cubic crystals. It dissolves in liquid ammonia to give sodamide and sodium hydroxide.

sodium nitrate (Chile saltpetre, $NaNO_3$) A white solid formed by the neutralization of nitric acid with either sodium carbonate or sodium hydroxide. It occurs naturally in large quantities in South America. Impure industrial sodium nitrate is called *caliche*. Sodium nitrate is very soluble in water and on crystallization forms colourless deliquescent crystals. On heating it decomposes to give sodium nitrite and oxygen. When heated with concentrated sulphuric acid, nitric acid is produced. Sodium nitrate is used as a fertilizer and as a source of nitrates and nitric acid. The salt crystallizes in the rhombohedral system and is isomorphous with the iodate.

sodium nitrite ($NaNO_2$) A yellowish-white solid formed by the thermal decomposition of sodium nitrate. It is readily soluble in water. The anhydrous salt forms orthorhombic crystals. When treated with cold dilute hydrochloric acid, sodium nitrite forms nitrous acid. It is used in organic chemistry in the process of diazotization and industrially as a corrosion inhibitor.

sodium orthophosphate *See* trisodium phosphate(V).

sodium peroxide (Na_2O_2) A yellowish-white ionic solid formed by the direct combination of burning sodium with excess oxygen. It reacts with water to form sodium hydroxide and hydrogen peroxide; the latter decomposes rapidly in the alkaline solution to give oxygen. Sodium peroxide is used as a bleaching agent and an oxidizing agent for such materials as wool and wood pulp. Sodium peroxide can convert nitrogen(II) oxide to sodium nitrate and iodine to sodium iodate.

sodium sulphate (Na_2SO_4) A white soluble solid formed by heating a mixture of sodium chloride and concentrated sulphuric acid. Industrially it is prepared by the first stage of the Leblanc process. It forms two hydrates, the decahydrate ($Na_2SO_4.10H_2O$) and the heptahydrate ($Na_2SO_4.7H_2O$). Sodium decahydrate, known as *Glauber's salt*, is used in the manufacture of glass and as a purgative in medicine. It effloresces on exposure to air to form the anhydrous salt, which is used as a drying agent in organic chemistry. At room temperature sodium sulphate crystallizes in the orthorhombic system; around 250°C there is a transition to the hexagonal system. The decahydrate crystallizes in the monoclinic form.

sodium sulphide (Na_2S) A yellow-red solid formed by the reduction of sodium sulphate using carbon (or carbon monoxide or hydrogen) at a high temperature. It is a corrosive material and deliquesces to release hydrogen sulphide. Sodium sulphide forms hydrates containing 4.5, 5, and $9H_2O$. In aqueous solution it is readily hydrolysed, the solution being strongly alkaline.

sodium sulphite (Na_2SO_3) A white solid formed by reacting the exact amount of sulphur dioxide with either sodium carbonate or sodium hydroxide. It is readily soluble in water and crystallizes as colourless crystals of the heptahydrate ($Na_2SO_3.7H_2O$). When treated with dilute mineral acids, sulphur dioxide is evolved. At high temperatures, sodium sulphite undergoes thermal decomposition to give sodium sulphate and sodium sulphide.

sodium superoxide (sodium dioxide, NaO_2) A pale yellow solid obtained by heating sodium peroxide in oxygen at 490°C. It reacts with water to give hydrogen peroxide, sodium hydroxide solution, and oxygen. Commercial sodium peroxide contains 10% sodium superoxide.

sodium thiosulphate ($Na_2S_2O_3$) A white solid prepared either by boiling sodium sulphite with flowers of sulphur or by passing sulphur dioxide into a suspension of sulphur in boiling sodium hydroxide. Sodium thiosulphate is readily soluble in water and crystallizes as large colourless crystals of the pentahydrate ($Na_2S_2O_3 \cdot 5H_2O$). It reacts with dilute acids to give sulphur and sulphur dioxide. It is used in photography as 'hypo' and industrially as an antichlor. In volumetric analysis solutions of sodium thiosulphate are usually prepared from the pentahydrate. On heating it disproportionates to give sodium sulphate and sodium sulphide.

soft soap A liquid soap made by saponification with potassium hydroxide (rather than sodium hydroxide).

soft water *See* hardness (of water).

sol A colloid consisting of solid particles distributed in a liquid medium. A wide variety of sols are known; the colours often depend markedly on the particle size. The term *aerosol* is used for solid or liquid phases dispersed in a gaseous medium. *See also* colloid.

solder An alloy used in joining metals. The molten solder alloys with (wets) the surfaces to be joined, without melting them, and solidifies on cooling to form a hard joint. The surfaces must be clean and free of oxide and a *flux* is often used to achieve this. *Soft solders* consist of lead with up to 60% tin and melt in the range 183–250°C. Soft-soldering is used, for

example, in plumbing and making electrical contacts. *Brazing solders* are copper–zinc alloys that have higher melting points and produce stronger joints than soft solders; silver can be added to produce *silver solders*.

solid The state of matter in which the particles occupy fixed positions, giving the substance a definite shape. The particles are held in these positions by bonds. Three kinds of attraction fix the positions of the particles: ionic, covalent, and intermolecular. Since these bonds act over short distances the particles in solids are packed closely together. The strengths of these three types of bonds are different and so, therefore, are the mechanical properties of different solids.

solid solution A solid composed of two or more substances mixed together at the molecular level. Atoms, ions, or molecules of one component in the crystal are at lattice positions normally occupied by the other component. Certain alloys are solid solutions of one metal in another. Isomorphic salts can also sometimes form solid solutions, as in the case of crystalline alums.

solubility The amount of one substance that could dissolve in another to form a saturated solution under specified conditions of temperature and pressure. Solubilities are stated as moles of solute per 100 grams of solvent, or as mass of solute per unit volume of solvent.

solubility product Symbol: K_s If an ionic solid is in contact with its saturated solution, there is a dynamic equilibrium between solid and solution:
$$AB(s) = A^+(aq) + B^-(aq)$$
The equilibrium constant for this is given by
$$[A^+][B^-]/[AB]$$
The concentration of undissolved solid [AB] is also constant, so
$$K_s = [A^+][B^-]$$
K_s is the solubility product of the salt (at a given temperature). For a salt A_2B_3, for instance:

$K_s = [A^+]^2[B^-]^3$, etc.
Solubility products are meaningful only for sparingly soluble salts. If the product of ions exceeds the solubility product, precipitation occurs.

solute A material that is dissolved in a solvent to form a solution.

solution A liquid system of two or more species that are intimately dispersed within each other at a molecular level. The system is therefore totally homogeneous. The major component is called the solvent (generally liquid in the pure state) and the minor component is called the solute (gas, liquid, or solid).
The process occurs because of a direct intermolecular interaction of the solvent with the ions or molecules of the solute. This interaction is called solvation. Part of the energy of this interaction appears as a change in temperature on dissolution. *See also* solid solution, solubility.

solvation The attraction of a solute species (e.g. an ion) for molecules of solvent. In water, for example, a positive ion will be surrounded by water molecules, which tend to associate around the ion because of attraction between the positive charge of the ion, and the negative part of the polar water molecule. The energy of this solvation (hydration in the case of water) is the 'force' needed to overcome the attraction between positive and negative ions when an ionic solid dissolves. The attraction of the dissolved ion for solvent molecules may extend for several layers. In the case of transition metal elements, ions may also form complexes by coordination to the nearest layer of molecules.

Solvay process (ammonia soda process) An industrial process for making sodium carbonate. The raw materials are calcium carbonate and sodium chloride (with ammonia). The calcium carbonate is heated:
$$CaCO_3 \rightarrow CaO + CO_2$$
Carbon dioxide is bubbled into a solution of sodium chloride saturated with ammonia, precipitating sodium hydrogen carbonate and leaving ammonium chloride in solution. The sodium hydrogen carbonate is then heated:
$$2NaHCO_3 \rightarrow Na_2CO_3 + H_2O + CO_2$$
The ammonia is regenerated by heating the ammonium chloride with the calcium oxide:
$$2NH_4Cl + CaO \rightarrow CaCl_2 + 2NH_3 + H_2O$$

solvent A liquid capable of dissolving other materials (solids, liquids, or gases) to form a solution. The solvent is generally the major component of the solution.
Solvents can be divided into classes, the most important being:
Polar. A solvent in which the molecules possess a moderate to high dipole moment and in which polar and ionic compounds are easily soluble. Polar solvents are usually poor solvents for non-polar compounds. For example, water is a good solvent for many ionic species, such as sodium chloride or potassium nitrate, and polar molecules, such as the sugars, but does not dissolve paraffin wax.
Non-polar. A solvent in which the molecules do not possess a permanent dipole moment and consequently will solvate non-polar species in preference to polar species. For example, benzene and carbon tetrachloride are good solvents for iodine and paraffin wax, but do not dissolve sodium chloride.
Amphiprotic. A solvent which undergoes self-ionization and can act both as a proton donor and as an acceptor. Water is a good example and ionizes according to:
$$2H_2O = H_3O^+ + OH^-$$
Aprotic. A solvent which can neither accept nor yield protons. An aprotic solvent is therefore the opposite to an amphiprotic solvent.

solvolysis A reaction between a compound and the solvent in which it is dissolved. *See also* hydrolysis.

sorption Absorption of gases by solids.

specific Denoting a physical quantity per unit mass. For example, volume (V) per unit mass (m) is called specific volume:

$$V = V_m$$

In certain physical quantities the term does not have this meaning: for example, *specific gravity* is more properly called relative density.

specific gravity *See* relative density.

specific rotatory power Symbol: α_m The rotation of plane-polarized light in degrees produced by a 10 cm length of solution containing 1 g of a given substance per millilitre of stated solvent. The specific rotatory power is a measure of the optical activity of substances in solution. It is measured at 20°C using the D-line of sodium.

spectral line A particular wavelength of light emitted or absorbed by an atom, ion, or molecule. *See* line spectrum.

spectral series A group of related lines in the absorption or emission spectrum of a substance. The lines in a spectral series occur when the transitions all occur between one particular energy level and a set of different levels. *See also* Bohr theory.

spectrograph An instrument for producing a photographic record of a spectrum.

spectrographic analysis A method of analysis in which the sample is excited electrically (by an arc or spark) and emits radiation characteristic of its component atoms. This radiation is passed through a slit, dispersed by a prism or a grating, and recorded as a spectrum, either photographically or photoelectrically. The photographic method was widely used for qualitative and semiquantitative work but photoelectric detection also allows wide quantitative application.

spectrometer 1. An instrument for examining the different wavelengths present in electromagnetic radiation. Typically, spectrometers have a source of radiation, which is collimated by a system of lenses and/or slits. The radiation is dispersed by a prism or grating, and recorded photographically or by a photocell. There are many types for producing and investigating spectra over the whole range of the electromagnetic spectrum. Often spectrometers are called *spectroscopes*.
2. Any of various other instruments for analysing the energies, masses, etc., of particles. *See* mass spectrometer.

spectrophotometer A form of spectrometer able to measure the intensity of radiation at different wavelengths in a spectrum, usually in the visible, infrared, or ultraviolet regions.

spectroscope An instrument for examining the different wavelengths present in electromagnetic radiation. *See also* spectrometer.

spectroscopy 1. The production and analysis of spectra. There are many spectroscopic techniques designed for investigating the electromagnetic radiation emitted or absorbed by substances. Spectroscopy, in various forms, is used for analysis of mixtures, for identifying and determining the structures of chemical compounds, and for investigating energy levels in atoms, ions, and molecules. In the visible and longer wavelength ultraviolet, transitions correspond to electronic energy levels in atoms and molecules. The shorter wavelength ultraviolet corresponds to transitions in ions. In the X-ray region, transitions in the inner shells of atoms or ions are involved. The infrared region corresponds to vibrational changes in molecules, with rotational changes at longer wavelengths.
2. Any of various techniques for analysing the energy spectra of beams of particles or for determining mass spectra.

spectrum (plural *spectra*) **1.** A range of electromagnetic radiation emitted or absorbed by a substance under particular circumstances. In an *emission spectrum*, light or other radiation emitted by the body is analysed to determine the particular wavelengths produced. The emission of radiation may be induced by a variety of methods; for example, by high temperature, bombardment by electrons, absorption of higher-frequency radiation, etc. In an *absorption spectrum* a continuous flow of radiation is passed through the sample. The radiation is then analysed to determine which wavelengths are absorbed. *See also* band spectrum, continuous spectrum, line spectrum. **2.** In general, any distribution of a property. For instance, a beam of particles may have a spectrum of energies. A beam of ions may have a mass spectrum (the distribution of masses of ions). *See* mass spectrometer.

speltar Commercial zinc containing about 3% impurities, mainly lead.

spin A property of certain elementary particles whereby the particle acts as if it were spinning on an axis; i.e. it has an angular momentum. Such particles also have a magnetic moment. In a magnetic field the spins line up at an angle to the field direction and precess around this direction. Certain definite orientations to the field direction occur such that $m_s h/2\pi$ is the component of angular momentum along this direction. m_s is the spin quantum number, and for an electron it has values $+1/2$ and $-1/2$. h is the Planck constant.

square-planar *See* complex.

stabilization energy The difference in energy between the delocalized structure and the conventional structure for a compound. For example, the stabilization energy of benzene is 150 kJ per mole, which represents the difference in energy between a Kekulé structure and the delocalized structure: the delocalized form being of lower energy is therefore more stable.
The stabilization energy can be determined by comparing the experimental value for the heat of hydrogenation of benzene with that calculated for Kekulé benzene from bond-energy data.

stabilizer A substance added to prevent chemical change (i.e. a negative catalyst).

staggered conformation *See* conformation.

stainless steel *See* steel.

stalactites and stalagmites Pillars of calcium carbonate found hanging from the ceiling (stalactites) and standing on the floor (stalagmites) in limestone caverns. They form when water containing dissolved carbon dioxide forms weak carbonic acid, which is able to react with limestone rock (calcium carbonate) to give a solution of calcium hydrogen carbonate. The solution accumulates in drops on the roof of caves and as it evaporates, the reaction is reversed causing calcium carbonate to be deposited. Over a very long period, the deposits build up to give a stalactite hanging from the cavern roof. Water that drips onto the floor from the tip of the stalactite also evaporates to leave a deposit of calcium carbonate. This gradually grows to form a stalagmite. Stalagmites and stalactites can eventually meet to form a single rock pillar.

standard cell A voltaic cell whose e.m.f. is used as a standard. *See* Clark cell, Weston cadmium cell.

standard electrode A half cell used for measuring electrode potentials. The hydrogen electrode is the basic standard but, in practice, calomel electrodes are usually used.

standard pressure An internationally agreed value; a barometric height of 760 mmHg at 0°C; 101 325 Pa (approximately 100 kPa).
This is sometimes called the *atmosphere* (used as a unit of pressure). The *bar*, used mainly when discussing the weather, is 100 kPa exactly.

standard solution A solution that contains a known weight of the reagent in a definite volume of solution. A standard flask or volumetric flask is used for this purpose. The solutions may be prepared by direct weighing for primary standards. If the reagent is not available in a pure form or is deliquescent the solution must be standardized by titration against another known standard solution. *See* primary standard.

standard state The standard conditions used as a reference system in thermodynamics: pressure is 101 325 Pa; temperature is 25°C (298.15 K); concentration is 1 mol. The substance under investigation must also be pure and in its usual state, given the above conditions.

standard temperature An internationally agreed value for which many measurements are quoted. It is the melting temperature of water, 0°C (273.15 K). *See also* STP.

stannane (tin(IV) hydride, SnH_4) A colourless poisonous gas prepared by the action of lithium aluminium hydride on tin(II) chloride. It is unstable, decomposing immediately at 150°C. Stannane acts as a reducing agent.

stannic compounds Compounds of tin(IV).

stannous compounds Compounds of tin(II).

starch A polysaccharide that occurs exclusively in plants. Starches are extracted commercially from maize, wheat, barley, rice, potatoes, and sorghum. The starches are storage reservoirs for plants; they can be broken down by enzymes to simple sugars and then metabolized to supply energy needs. Starch is a dietary component of animals.
Starch is not a single molecule but a mixture of amylose (water-soluble, blue colour with iodine) and amylopectin (not water-soluble, violet colour with iodine). The composition is amylose 10–20%, amylopectin 80–90%.

states of matter The three physical conditions in which substances occur: solid, liquid, and gas. The addition or removal of energy (usually in the form of heat) enables one state to be converted into another.
The major distinctions between the states of matter depend on the kinetic energies of their particles and the distances between them. In solids, the particles have low kinetic energy and are closely packed; in gases they have high kinetic energy and are very loosely packed; kinetic energy and separation of particles in liquids are intermediate.
Solids have fixed shapes and volumes, i.e. they do not flow, like liquids and gases, and they are difficult to compress. In solids the atoms or molecules occupy fixed positions in space. In most cases there is a regular pattern of atoms — the solid is crystalline.
Liquids have fixed volumes (i.e. low compressibility) but flow to take up the shape of the container. The atoms or molecules move about at random, but they are quite close to one another and the motion is hindered.
Gases have no fixed shape or volume. They expand spontaneously to fill the container and are easily compressed. The molecules have almost free random motion.
A plasma is sometimes considered to be a fourth state of matter.

steam distillation A method of isolating or purifying substances by exploiting Dalton's law of partial pressures to lower the boiling point of the mixture. When two immiscible liquids are distilled, the boiling point will be lower than that of the more volatile component and consequently will be below 100°C if one component is water. The method is particularly useful for recovering materials from tarry mixtures.

steam reforming The conversion of a methane-steam mixture at 900°C with a nickel catalyst into a mixture of carbon monoxide and hydrogen. The mixture of gases (synthesis gas) provides a starting material in a number of processes, e.g. the manufacture of methanol.

stearate A salt or ester of stearic acid.

Steam distillation

stearic acid *See* octadecanoic acid.

steel An alloy of iron with small amounts of carbon and, often, other metals. *Carbon steels* contain 0.05–1.5% carbon — the more carbon, the harder the steel. *Alloy steels* also contain small amounts of other elements, such as chromium, manganese, and vanadium. Their properties depend on the composition. *Stainless steels*, for instance, are resistant to corrosion. The main constituent is chromium (10–25%) with up to 0.7% carbon.

step An elementary stage in a chemical reaction, in which energy may be transferred from one molecule to another, bonds may be broken or formed, or electrons may be transferred. For example, in the reaction between hypochlorite ions and iodide ions in aqueous solution there are three steps:
Step 1

$$OCl^-(aq) + H_2O(l) \rightarrow HOCl(aq) + OH^-(aq)$$

Step 2

$$I^-(aq) + HOCl(aq) \rightarrow HOI(aq) + Cl^-(aq)$$

Step 3

$$OH^-(aq) + HOI(aq) \rightarrow H_2O(l) + OI^-(aq)$$

steradian Symbol: sr The SI unit of solid angle. The surface of a sphere, for example, subtends a solid angle of 4π at its centre. The solid angle of a cone is the area intercepted by the cone on the surface of a sphere of unit radius.

stereochemistry The branch of chemistry concerned with the shapes of molecules and the way these affect the chemical properties.

stereoisomerism *See* isomerism.

steric effect An effect in which the shape of a molecule influences its reactions. A particular example occurs in molecules containing large groups, which hinder the approach of a reactant (*steric hindrance*).

steric hindrance *See* steric effect.

still An apparatus for distillation.

stoichiometry The proportions in which elements form compounds. A *stoichiometric compound* is one in which the atoms have combined in small whole numbers.

storage battery *See* accumulator.

STP (NTP) Standard temperature and pressure. Conditions used internationally when measuring quantities that vary with both pressure and temperature (such as the density of a gas). The values are 101 325 Pa (approximately 100 kPa) and 0°C (273.15 K). *See also* standard pressure, standard temperature.

straight chain *See* chain.

Strecker synthesis A method of synthesizing amino acids. Hydrogen cyanide forms an addition product (cyanohydrin) with aldehydes:
$$RCHO + HCN \rightarrow RCH(CN)OH$$
With ammonia further substitution occurs:
$$RCH(CN)OH + NH_3 \rightarrow$$
$$RCH(CN)(NH_2) + H_2O$$
Acid hydrolysis of the cyanide group then produces the amino acid
$$RCH(CN)(NH_2) \rightarrow$$
$$RCH(COOH)(NH_2)$$
The method is a general synthesis for α-amino acids (with the $-NH_2$ and $-COOH$ groups on the same carbon atom).

strontia *See* strontium oxide.

strontium A soft low-melting reactive metal; the fourth member of group II of the periodic table and classed as a typical alkaline-earth element. The electronic configuration is that of krypton with two additional outer 5s electrons. Strontium and barium are both of low abundance in the Earth's crust, strontium occurring as strontianite ($SrCO_3$) and celestine ($SrSO_4$).
The element is produced industrially by roasting the carbonate to give the oxide (800°C) and then reducing with aluminium,
$$3SrO + 2Al \rightarrow Al_2O_3 + 2Sr$$
Strontium has a low ionization potential, is large, and is therefore very electropositive. The chemistry of strontium metal is therefore characterized by high reactivity of the metal. The properties of strontium fall into sequence with other alkaline earths; viz it reacts directly with oxygen, nitrogen, sulphur, the halogens, and hydrogen to form respectively the oxide SrO, nitride

Sr_3N_2, sulphide SrS, halides SrX_2, and hydride SrH_2, all of which are largely ionic in character. The oxide SrO and the metal react readily with water to form the hydroxide $Sr(OH)_2$ which is basic and midway between $Ca(OH)_2$ and $Ba(OH)_2$ in solubility. The carbonate and sulphate are both insoluble.
As the metal is very electropositive the salts are never much hydrolysed in solution and the ions are largely solvated as $[Sr(H_2O)_6]^{2+}$
Symbol: Sr; m.p. 770°C; b.p. 1380°C; r.d. 2.6; p.n. 38; r.a.m. 87.62.

strontium bicarbonate *See* strontium hydrogen carbonate.

strontium carbonate ($SrCO_3$) A white insoluble solid that occurs naturally as the mineral strontianite. It can be prepared by passing carbon dioxide over strontium oxide or hydroxide or by passing the gas through a solution of a strontium salt. Strontium carbonate is then formed as a precipitate. It is used as a slagging agent in certain metal furnaces and to produce a red flame in fireworks.

strontium chloride ($SrCl_2$) A white solid obtained directly from strontium and chlorine or by passing chlorine over heated strontium oxide. It is used in fireworks to give a red flame. The hexahydrate ($SrCl_2.6H_2O$) is prepared by the neutralization of hydrochloric acid by strontium hydroxide or carbonate.

strontium hydrogen carbonate (strontium bicarbonate, $Sr(HCO_3)_2$) A compound present in solutions formed by the action of carbon dioxide on a suspension in cold water of strontium carbonate, to which it reverts on heating:
$$SrCO_3 + CO_2 + H_2O = Sr(HCO_3)_2$$

strontium hydroxide ($Sr(OH)_2$) A solid that normally occurs as the octahydrate ($Sr(OH)_2.8H_2O$), which is prepared by crystallizing an aqueous solution of strontium oxide. Strontium hydroxide is readily soluble in water and aqueous solutions are strongly basic. It is used in the purification of sugar.

strontium oxide (strontia, SrO) A greyish-white powder prepared by the decomposition of strontium carbonate, hydroxide, or nitrate. Strontium oxide is soluble in water, forming an alkaline solution due to the attraction between the oxide ions and protons from the water:
$$O^{2-} + H_2O \rightarrow 2OH^-$$

strontium sulphate ($SrSO_4$) A white sparingly soluble salt that occurs naturally as the mineral celestine. It can be prepared by dissolving strontium oxide, hydroxide, or carbonate in sulphuric acid. It is used as a substitute for barium sulphate in paints.

structural formula The formula of a compound showing the numbers and types of atoms present, together with the way in which these are arranged in the molecule. Often this can be done by grouping the atoms, as in the structural formula of ethanoic acid $CH_3.CO.OH$. *Compare* empirical formula, molecular formula.

structural isomerism *See* isomerism.

styrene *See* phenylethene.

sublimate A solid formed by sublimation.

sublimation The conversion of a solid into a vapour without the solid first melting. For instance (at standard pressure) iodine, solid carbon dioxide, and ammonium chloride sublime.
At certain conditions of external pressure and temperature an equilibrium can be established between the solid phase and vapour phase.

sub-shell A subdivision of an electron shell. It is a division of the orbitals that make up a shell into sets of orbitals, which are degenerate (i.e. have the same energy) in the free atom. For example, in the third shell or M-shell there are the 3s, 3p, and 3d sub-shells of which the latter two are 3- and 5-degenerate respectively.

substituent An atom for group substituted for another in a compound. Often the term is used for groups that have replaced hydrogen in organic compounds; for example, in benzene derivatives.

substitution reaction A reaction in which an atom or group of atoms in an organic molecule is replaced by another atom or group. The substitution of a hydrogen atom in an alkane by a chlorine atom is an example.
Substitution reactions fall into three major classes depending upon the nature of the attacking substituent.
Nucleophilic substitution: the attacking substituent is a nucleophile. Such reactions are very common with alcohols and halogen compounds, in which the electron-deficient carbon atom attracts the nucleophile and the leaving group readily exists alone. Examples are the hydrolysis of a haloalkane and the chlorination of an alcohol.
$$C_2H_5Cl + OH^- \rightarrow C_2H_5OH + Cl^-$$
$$C_2H_5OH + HCl \rightarrow C_2H_5Cl + H_2O$$
Electrophilic substitution: the attacking substituent is an electrophile. Such reactions are common in aromatic compounds, in which the electron-rich ring attracts the electrophile. The nitration of benzene in which the electrophile is NO_2^+ is an example:
$$C_6H_6 + NO_2^+ \rightarrow C_6H_5NO_2 + H^+$$
Free-radical substitution: a free radical is the attacking substituent. Such reactions can be used with compounds that are inert to either nucleophiles or electrophiles, for instance the halogenation of an alkane:
$$CH_4 + Cl_2 \rightarrow CH_3Cl + HCl$$
The term 'substitution' is very general and several reactions that can be considered as substitutions are more normally given special names (e.g. esterification, hydrolysis, and nitration). *See also* electrophilic substitution, nucleophilic substitution.

substrate A material that is acted on by a catalyst.

succinic acid

glucose - a pyranose ring

fructose - in a furanose ring form

ribose - a pyranose ring

Examples of monosaccharides

succinic acid *See* butanedioic acid.

sucrose (cane sugar, $C_{12}H_{22}O_{11}$) A sugar that occurs in many plants. It is extracted commercially from sugar cane and sugar beet. Sucrose is a disaccharide formed from a glucose unit and a fructose unit. It is hydrolysed to a mixture of fructose and glucose by the enzyme invertase. Since this mixture has a different optical rotation (laevorotatory) than the original sucrose, the mixture is called *invert sugar*.

sugar (saccharide) One of a class of sweet-tasting simple carbohydrates. Sugars have molecules consisting of a chain of carbon atoms with −OH groups attached, and either an aldehyde or ketone group.

They can exist in a chain form or in a ring formed by reaction of the ketone or aldehyde group with an OH group. Monosaccharides are simple sugars that cannot be hydrolysed to sugars with fewer carbon atoms. Two or more monosaccharide units can be linked in *disaccharides*, *trisaccharides*, etc.

Monosaccharides are also classified according to the number of carbon atoms: a *pentose* has five carbon atoms and a *hexose* six. Monosaccharides with aldehyde groups are *aldoses*; those with ketone groups are *ketoses*. Thus, an *aldohexose* is a hexose with an aldehyde group; a *ketopentose* is a pentose with a ketone group, etc.

The ring forms of monosaccharides are derived by reaction of the aldehyde or keα-D-glucose

β-D-glucose

Glucose

Sucrose - a disaccharide

tone group with one of the carbons at the other end of the chain. It is possible to have a six-membered (*pyranose*) ring or a five-membered (*furanose*) ring.
See also carbohydrate, polysaccharide.

sulpha drugs See sulphonamide.

sulphate A salt or ester of sulphuric(VI) acid.

sulphide 1. A compound of sulphur with a more electropositive element. Simple sulphides contain the ion S^{2-} *Polysulphides* can also be formed containing ions with chains of sulphur atoms (S_x^{2-}).
2. See thio ether.

sulphite A salt or ester of sulphurous acid.

sulphonamide A type of organic compound with the general formula $R.SO_2.NH_2$. Sulphonamides, which are amides of sulphonic acids, are active against bacteria, and some are used in pharmaceuticals ('sulpha drugs').

sulphonation A reaction introducing the $-SO_2OH$ (sulphonic acid) group into an organic compound. Sulphonation of aromatic compounds is usually accomplished by refluxing with concentrated sulphuric acid for several hours. The attacking species is SO_3 (sulphur(VI) oxide) and the reaction is an example of electrophilic substitution.

Sulphonic Acid

sulphonic acid A type of organic compound containing the $-SO_2.OH$ group. The simplest example is benzenesulphonic acid ($C_6H_5SO_2OH$). Sulphonic acids are strong acids. Electrophilic substitution can introduce other groups onto the benzene ring; the $-SO_2.OH$ group directs substituents into the 3-position.

215

sulphur A low melting non-metallic solid, yellow coloured in its common forms; the second member of group VI of the periodic table. It has the electronic configuration $[Ne]3s^23p^4$.

Sulphur occurs in the elemental form in Sicily and some southern states of the USA, and in large quantities in combined forms such as sulphide ores (FeS_2) and sulphate rocks ($CaSO_4$). It forms about 0. 5% of the Earth's crust. Elemental sulphur is extracted commercially by the Frasch process in which super-heated steam is forced down the outer of three concentric tubes leading into the deposit. This causes the sulphur to melt; then air is blown down the central tube causing molten sulphur to be forced out of the well to be collected and cooled at the well head. The large amount of sulphur obtained from smelting of sulphide ores or roasting of sulphates is not obtained as elemental sulphur but is used directly as SO_2/SO_3 for conversion to sulphuric acid (contact process).

Sulphur exhibits allotropy and its structure in all phases is quite complex. The common crystalline modification, *rhombic sulphur*, is in equilibrium with a *triclinic* modification above 96°C. Both have structures based on S_8-rings but the crystals are quite different. If molten sulphur is poured into water a dark red 'plastic' form is obtained in a semielastic form. The structure appears to be a helical chain of S atoms. Selenium and tellurium both have a grey 'metal-like' modification but sulphur does *not* have this form.

Sulphur reacts directly with hydrogen but the position of the equilibrium at normal temperatures precludes its use as a preparative method for hydrogen sulphide. The element burns readily in oxygen with a characteristic blue flame to form sulphur dioxide (SO_2) and traces of sulphur trioxide (SO_3). Sulphur also forms a large number of oxy-acid species, some containing peroxy-groups and some with two or more S-atoms. It forms four distinct fluorides, S_2F_2, SF_4, SF_6, S_2F_{10}; three chlorides, S_2Cl_2, SCl_2, SCl_4; a bromide S_2Br_2, but *no* iodide. Apart from SF_6, which is surprisingly stable and inert, the halides are generally susceptible to hydrolysis to give commonly SO_2, sometimes H_2S, and the hydrogen halide. The halides SF_4, S_2Cl_2, and SCl_2 are frequently used as selective fluorinating agents and chlorinating agents in organic chemistry. The chlorides are also industrially important for hardening rubbers. Sulphur hexafluoride may be prepared by direct reaction of the elements but SF_4 is prepared by fluorinating SCl_2 with NaF. In a limited supply of chlorine, sulphur combines to give sulphur monochloride, S_2Cl_2; in excess chlorine at higher temperatures the dichloride is formed,

$$2S + Cl_2 \rightarrow S_2Cl_2$$

$$S_2Cl_2 + Cl_2 \rightarrow 2SCl_2$$

In addition to these binary halides, sulphur forms two groups of oxyhalides; the thionyl halides, SOX_2 (F, Cl, and Br, but not I), and the sulphuryl halides SO_2X_2 (F and Cl, but not Br and I).

Most metals react with sulphur to form sulphides of which there are a large number of structural types. With electropositive elements of groups I and II the sulphides are largely *ionic* (S^{2-}) in the solid phase. Hydrolysis occurs in solution thus:

$$S^{2-} + H_2O \rightarrow SH^- + OH^-$$

Most transition metals form sulphides, which although formally treated as FeS, CoS, NiS, etc., are largely covalent and frequently non-stoichiometric.

In addition to the complexities of the binary compounds with metals, sulphur forms a range of binary compounds with elements such as Si, As, and P, which may be polymeric and generally of rather complex structure, e.g. $(SiS_4)_n$, $(SbS_3)_n$, N_4S_4, As_4S_4, P_4S_3.

Sulphur also forms a wide range of organic sulphur compounds, most of which have the typically revolting smell of H_2S.
Symbol: S; m.p. 112.8°C; b.p. 444.6°C; r.d. 2.07; p.n. 16; r.a.m. 32.06.

sulphur dichloride dioxide (sulphuryl chloride, SO_2Cl_2) A colourless fuming liquid formed by the reaction of chlorine with sulphur(IV) oxide in sunlight. It is used as a chlorinating agent.

sulphur dichloride oxide (thionyl chloride, $SOCl_2$) A colourless fuming liquid formed by passing sulphur(IV) oxide over phosphorus(V) chloride and distilling the mixture obtained. Sulphur dichloride oxide is used in organic chemistry to introduce chlorine atoms (for example, into ethanol to form monochloroethane), the organic product being easily isolated as the other products are gases.

sulphuretted hydrogen See hydrogen sulphide.

sulphuric(IV) acid See sulphurous acid.

sulphuric(VI) acid (H_2SO_4) A colourless oily liquid manufactured by the contact process. The concentrated acid is diluted by adding it slowly to water, with careful stirring. Concentrated sulphuric acid acts as an oxidizing agent, giving sulphur(IV) oxide as the main product, and also as a dehydrating agent. The diluted acid acts as a strong dibasic acid, neutralizing bases and reacting with active metals and carbonates to form sulphates. Sulphuric acid is used in the laboratory to dry gases (except ammonia), to prepare nitric acid and ethene, and to absorb alkenes. In industry, it is used to manufacture fertilizers (e.g. ammonium sulphate), rayon, and detergents, to clean metals, and in car batteries. See contact process.

sulphur monochloride See disulphur dichloride.

sulphurous acid (sulphuric(IV) acid, H_2SO_3) A weak acid found only in solution, made by passing sulphur(IV) oxide into water. The solution is unstable and smells of sulphur(IV) oxide. It is a reducing agent, converting iron(III) ions to iron(II) ions, chlorine to chloride ions, and orange dichromate(VI) ions to green chromium(III) ions.

sulphur(VI) oxide (sulphur trioxide, SO_3) A fuming volatile white solid prepared by passing sulphur(IV) oxide and oxygen over hot vanadium(V) oxide (act-

ing as a catalyst) and cooling the product in ice. Sulphur(VI) oxide reacts vigorously with water to form sulphuric acid. See also contact process.

sulphur trioxide See sulphur(VI) oxide.

sulphuryl chloride See sulphur dichloride dioxide.

superconduction The property of zero resistance exhibited by certain metals and metallic compounds at low temperatures. Metals conduct at temperatures close to absolute zero but a number of compounds are now known that conduct at higher (liquid-nitrogen) temperatures. Synthetic organic conductors have also been produced which operate at much higher temperatures.

supercooling The cooling of a liquid at a given pressure to a temperature below its melting temperature at that pressure without solidifying it. The liquid particles lose energy but do not spontaneously fall into the regular geometrical pattern of the solid. A supercooled liquid is in a metastable state and will usually solidify if a small crystal of the solid is introduced to act as a 'seed' for the formation of crystals. As soon as this happens, the temperature returns to the melting temperature until the substance has completely solidified.

superfluidity A property of liquid helium at very low temperatures. At 2.186 K liquid helium makes a transition to a superfluid state, which has a high thermal conductivity and flows without friction. See also helium.

superheating The raising of a liquid's temperature above its boiling temperature, by increasing the pressure.

supernatant Denoting a clear liquid that lies above a sediment or a precipitate.

superoxide An inorganic compound containing the O_2^- ion.

217

superphosphate A mixture — mainly calcium hydrogen phosphate and calcium sulphate — used as a fertilizer. It is made from calcium phosphate and sulphuric acid.

supersaturated solution *See* saturated solution.

supersaturated vapour *See* saturated vapour.

supplementary units The dimensionless units — the radian and the steradian — used along with base units to form derived units. *See also* SI units.

surfactant A substance that lowers surface tension and has properties of wetting, foaming, detergency, dispersion, and emulsification, e.g. a soap or other detergent.

suspension A system in which small particles of a solid or liquid are dispersed in a liquid or gas.

syn-isomer *See* isomerism.

synthesis The preparation of chemical compounds from simpler compounds.

synthesis gas A mixture of carbon monoxide and hydrogen produced by steam reforming of natural gas.
$$CH_4 + H_2O \rightarrow CO + 3H_2$$
Synthesis gas is a useful starting material for the manufacture of a number of organic compounds.

Système International d'Unités *See* SI units.

T

tantalum A transition element. It is strong, highly resistant to corrosion, and is easily worked. Tantalum is used in turbine blades and in surgical and dental work. Symbol: Ta; m.p. 2996°C; b.p. 5425°C; r.d. 16.6; p.n. 73; r.a.m. 180.95.

tartaric acid A crystalline hydroxy carboxylic acid with the formula:

HOOC(CHOH)$_2$COOH

Its systematic name is 2,3-dihydroxybutanedioic acid. *See also* optical activity.

tartrate A salt or ester of tartaric acid.

tautomerism Isomerism in which each isomer can convert into the other, so that the two isomers are in equilibrium. The isomers are called *tautomers*. Tautomerism often results from the migration of a hydrogen atom. *See* keto-enol tautomerism.

technetium A transition metal produced artificially by bombarding molybdenum with neutrons and also during the fission of uranium. It is radioactive. Symbol: Tc; m.p. 2130°C; b.p. 4620°C; r.d. 11.5; p.n. 43; r.a.m. 97.

Teflon (*Tradename*) The synthetic polymer polytetrafluoroethane.

tellurium A brittle silvery metalloid element belonging to group VI of the periodic table. It is found native and in combination with metals. Tellurium is used mainly as an additive to improve the qualities of stainless steel and various metals. Symbol: Te; m.p. 449.5°C; b.p. 989.8°C; r.d. 6.24; p.n. 52; r.a.m. 127.6.

temperature scale A practical scale for measuring temperature. A temperature scale is determined by fixed temperatures (*fixed points*), which are reproducible systems assigned an agreed temperature. On the Celsius scale the two fixed points are the temperature of pure melting ice (the *ice temperature*) and the temperature of pure boiling water (the *steam temperature*). The difference between the fixed points is the *fundamental interval* of the scale, which is subdivided into temperature units. The International Temperature Scale has 11 fixed points that cover the range 13.81 kelvin to 1337.58 kelvin.

temporary hardness A type of water hardness caused by the presence of dissolved calcium, iron, and magnesium hydrogen carbonates. This form of hardness can be removed by boiling the water. Temporary hardness arises because rainwater combines with carbon dioxide from the atmosphere to form weak carbonic acid. This reacts with carbonate rocks, such as limestone, to produce calcium hydrogen carbonate, which then goes into solution. When this solution is heated, the complete reaction sequence is reversed so that calcium carbonate is precipitated. If this is not removed, it will accumulate in boilers and hot water pipes and reduce their efficiency. *Compare* permanent hardness. *See also* water softening.

tera- Symbol: T A prefix denoting 10^{12}. For example, 1 terawatt (TW) = 10^{12} watts (W).

terbium A soft ductile malleable silvery element of the lanthanoid series of metals. It occurs in association with other lanthanoids. One of its few uses is as a dopant in solid-state devices. Symbol: Tb; m.p. 1356°C; b.p. 3123°C; r.d. 8.23; p.n. 65; r.a.m. 158.93.

ternary compound A chemical compound formed from three elements; e.g. Na_2SO_4 or $LiAlH_4$.

terpene Any of a class of natural unsaturated hydrocarbons with formulae $(C_5H_8)_n$, found in plants.

tervalent

tervalent (trivalent) Having a valence of three.

tesla Symbol: T The SI unit of magnetic flux density, equal to a flux density of one weber of magnetic flux per square metre. $1\ T = 1\ Wb\ m^{-2}$.

tetrachloromethane Carbon tetrachloride. Note that 'tetrachloromethane' is the systematic name, and is preferred in organic chemistry.

tetraethyl lead See lead tetraethyl.

tetragonal crystal See crystal system.

tetrahedral compound A compound such as methane, in which an atom has four bonds directed towards the corners of a regular tetrahedron. The angles between the bonds in such a compound are about 109°.

tetrahydrate A crystalline hydrated compound containing four molecules of water of crystallization per molecule of compound. See also complex.

tetrahydrofuran (C_4H_8O). A colourless liquid widely used as a solvent.

thallium A soft malleable greyish metallic element belonging to group III of the periodic table. It is found in lead and cadmium ores, including pyrites (FeS_2). Thallium is highly toxic and was used previously as a rodent and insect poison. Various compounds are now used in photocells, infrared detectors, and low-melting glasses.
Symbol: Tl; m.p. 303.5°C; b.p. 1457°C; r.d. 11.85; p.n. 81; r.a.m. 204.37.

thermite (thermit) A mixture of aluminium powder and (usually) iron(III) oxide. When ignited the oxide is reduced; the reaction is strongly exothermic and molten iron is formed:
$$2Al + Fe_2O_3 \rightarrow Al_2O_3 + 2Fe$$
Thermite is used in incendiary bombs and for welding steel. See also Goldschmidt process.

thermochemistry The branch of chemistry concerned with heats of reaction, solvation, etc.

thermodynamics The study of heat and other forms of energy and the various related changes in physical quantities such as temperature, pressure, density, etc.
The first law of thermodynamics states that the total energy in a closed system is conserved (constant). In all processes energy is simply converted from one form to another, or transferred from one system to another.
A mathematical statement of the first law is:
$$\delta Q = \delta U + \delta W$$
Here, δQ is the heat transferred to the system, δU the change in internal energy (resulting in a rise or fall of temperature), and δW is the external work done by the system.
The second law of thermodynamics can be stated in a number of ways, all of which are equivalent. One is that heat cannot pass from a cooler to a hotter body without some other process occurring. Another is the statement that heat cannot be totally converted into mechanical work, i.e. a heat engine cannot be 100% efficient. The third law of thermodynamics states that the entropy of a substance tends to zero as its thermodynamic temperature approaches zero.
Often a zeroth law of thermodynamics is given: that if two bodies are each in thermal equilibrium with a third body, then they are in thermal equilibrium with each other. This is considered to be more fundamental than the other laws because they assume it. See also Carnot cycle, entropy.

thermodynamic temperature Symbol: T A temperature measured in kelvins. See also absolute temperature.

thermoplastic polymer See polymer.

thermosetting polymer See polymer.

220

thin-layer chromatography A technique widely used for the analysis of mixtures. Thin-layer chromatography employs a solid stationary phase, such as alumina or silica gel, spread evenly as a thin layer on a glass plate. A base line is carefully scratched near the bottom of the plate using a needle, and a small sample of the mixture is spotted onto the base line using a capillary tube. The plate is then stood upright in solvent, which rises up to the base line and beyond by capillary action. The components of the spot of the sample will dissolve in the solvent and tend to be carried up the plate. However, some of the components will cling more readily to the solid phase than others and will not move up the plate so rapidly. In this way, different fractions of the mixture eventually become separated. When the solvent has almost reached the top, the plate is removed and quickly dried. The plate is developed to locate the positions of colourless fractions by spraying with a suitable chemical or by exposure to ultraviolet radiation. The components are identified by comparing the distance they have travelled up the plate with standard solutions that have been run simultaneously, or by computing an R_F value.

thio alcohol (mercaptan) A compound of the type RSH, similar to an alcohol with the oxygen atom replaced by sulphur.

thio ether An organic sulphide; a compound of the type RSR'.

thiosulphate A salt containing the ion $S_2O_3^{2-}$

thionyl chloride See sulphur dichloride oxide.

thixotropy The change of viscosity with time. Fluids that suffer a change in viscosity on standing are said to be thixotropic. Examples include mayonnaise and paint.

thorium A toxic radioactive element of the actinoid series that is a soft ductile silvery metal. It has several long-lived radioisotopes found in a variety of minerals including monazite. Thorium is used in magnesium alloys, in the electronics industry, and is a potential nuclear fuel. Symbol: Th; m.p. 1750°C; b.p. 4790°C (approx.); r.d. 11.72; p.n. 90; r.a.m. 232.04.

threonine See amino acids.

thulium A soft malleable ductile silvery element of the lanthanoid series of metals. It occurs in association with other lanthanoids. Symbol: Tm; m.p. 1545°C; b.p. 1947°C; r.d. 9.31; p.n. 69; r.a.m. 168.93.

tin A white lustrous metal of low melting point; the fourth member of group IV of the periodic table. Tin itself is the first distinctly metallic element of the group even though it retains some amphoteric properties. Its electronic structure has outer s^2p^2 electrons ($[Kr]4d^{10}5s^25p^2$). The element is of low abundance in the Earth's crust (0.004%) but is widely distributed, largely as cassiterite (SnO_2). The metal has been known since early bronze age civilizations when the ores used were relatively rich but currently worked ores are as low as 1–2% and considerable concentration must be carried out before roasting. The metal itself is obtained by reduction using carbon,
$$SnO_2 + C \rightarrow Sn + CO_2$$
Tin is an expensive metal and several processes are used for recovering tin from scrap tin-plate. These may involve chlorination (dry) to the volatile $SnCl_4$, or by electrolytic methods using an alkaline electrolyte,
$$Sn + 4OH^- \rightarrow Sn(OH)_4^{2-} + 2e^-$$
$$\text{(anode)}$$
$$Sn(OH)_4^{2-} \rightarrow Sn^{2+} + 4OH^- \text{ (cathode)}$$
$$Sn^{2+} + 2e \rightarrow Sn \text{ (cathode)}$$
Tin does not react directly with hydrogen but an unstable hydride, SnH_4, can be prepared by reduction of $SnCl_4$. The low stability is due to the rather poor overlap of the diffuse orbitals of the tin atom with the small H-orbitals. Tin forms both tin(II) oxide and tin(IV) oxide. Both are amphoteric, dissolving in acids to give tin(II) and tin(IV) salts, and in bases to form stannites and stannates,

tin(II) chloride

$$SnO + 4OH^- \rightarrow [SnO_3]^{4-} + 2H_2O$$
stannite (relatively unstable)
$$SnO_2 + 4OH^- \rightarrow [SnO_4]^{4-} + 2H_2O$$
stannate

The halides, SnX_2, may be prepared by dissolving tin metal in the hydrogen halide or by the action of heat on SnO plus the hydrogen halide. Tin(IV) halides may be prepared by direct reaction of halogen with the metal. Although tin(II) halides are ionized in solution their melting points are all low suggesting considerable covalency in all but the fluoride. The tin(IV) halides are volatile and essentially covalent with slight polarization of the bonds.

Tin(II) compounds are readily oxidized to tin(IV) compounds and are therefore good reducing agents for general laboratory use.

Tin has three crystalline modifications or allotropes, α-tin or 'grey tin' (diamond structure), β-tin or 'white tin', and γ-tin; the latter two are metallic with close packed structures. Tin also has several isotopes.

It is used in a large number of alloys including Babbit metal, bell metal, Britannia metal, bronze, gun metal, and pewter as well as several special solders.

Symbol: Sn; m.p. 232°C; b.p. 2362°C; p.n. 50; r.a.m. 118.7.

tin(II) chloride ($SnCl_2$) A transparent solid with a fatty lustre. It is made by dissolving tin in hydrochloric acid. Tin(II) chloride is a reducing agent, it combines with ammonia, and forms hydrates with water. It is used as a mordant.

tin(IV) chloride ($SnCl_4$) A colourless fuming liquid. Tin(IV) chloride is soluble in organic solvents but is hydrolysed by water. It dissolves sulphur, phosphorus, bromine, and iodine, and it dissolves in concentrated hydrochloric acid to give the anion $SnCl_6^{2-}$.

tin(IV) hydride *See* stannane.

tin(II) oxide (SnO) A dark green or black solid. It can also be obtained in an unstable red form, which turns black on exposure to air. Tin(II) oxide can be made by precipitating the hydrated oxide from a solution containing tin(II) ions and dehydrating the product at 100°C.

tin(IV) oxide (SnO_2) A colourless crystalline solid, which is usually discoloured owing to the presence of impurities. It exists as hexagonal or rhombic crystals or in an amorphous form. Tin(IV) oxide is insoluble in water. It is sold as 'putty powder' for polishing glass and metal.

tin(III) sulphide (SnS) A grey solid that can be prepared from tin and sulphur. Above 265°C tin(II) sulphide slowly turns into tin(IV) sulphide and tin:

$$2SnS(s) \rightarrow SnS_2(s) + Sn(s)$$

tin(IV) sulphide (SnS_2) A yellowish solid that can be precipitated by reacting hydrogen sulphide with a solution of a soluble tin(IV) salt. A crystalline form sometimes known as *mosaic gold* is obtained by heating a mixture of tin filings, sulphur, and ammonium chloride. Tin(IV) sulphide is used as a pigment.

titanium A transition metal that occurs in various ores as titanium(IV) oxide and also in combination with iron and oxygen. It is extracted by conversion of titanium(IV) oxide to the chloride, which is reduced to the metal by heating with sodium. Titanium is reactive at high temperatures. It is used in the aerospace industry as it is strong, resistant to corrosion, and has a low density. It forms compounds with oxidation states $+4$, $+3$, and $+2$, the $+4$ state being the most stable.

Symbol: Ti; m.p. 1670°C; b.p. 3280°C; r.d. 4.5; p.n. 22; r.a.m. 47.90.

titanium(IV) chloride (titanium tetrachloride, $TiCl_4$) A volatile colourless liquid formed by heating titanium(IV) oxide with carbon in a stream of dry chlorine at 700°C. The product is purified by fractional distillation. Titanium(IV) chloride fumes in moist air forming the oxychlorides of titanium. It undergoes hydrolysis in water but this can be prevented by the presence of excess hydrochloric acid. Crystallization of such a solution gives crystals of the di- and pentahydrates. Titanium(IV) chloride is used in the preparation of pure titanium and as an intermediate in the production of titanium compounds from minerals.

titanium(IV) oxide (titanium dioxide, TiO_2) A white ionic solid that occurs naturally in three crystalline forms: rutile (tetragonal), brookite (orthorhombic), and anatase (tetragonal). It reacts slowly with acids and is attacked by the halogens at high temperatures. On heating it decomposes to give titanium(III) oxide and oxygen. Titanium(IV) oxide is amphoteric; it is used as a white pigment. With hot concentrated sulphuric acid, it forms titanyl sulphate, $TiOSO_4$. When fused with alkalis, it forms titanates.

titrant See titration.

titration A procedure in volumetric analysis in which a solution of known concentration (called the *titrant*) is added to a solution of unknown concentration from a burette until the equivalence point or end point of the titration is reached. See volumetric analysis.

TNT See trinitrotoluene.

Tollen's reagent A solution of the complex ion $Ag(NH_3)_2^+$ produced by precipitation of silver oxide from silver nitrate with a few drops of sodium hydroxide solution, and subsequent dissolution of the silver oxide in aqueous ammonia. Tollen's reagent is used in the 'silver-mirror test' for aldehydes, where the Ag^+ ion is reduced to silver metal. It is also a test for alkynes with a triple bond in the 1-position. A yellow precipitate of silver carbide is formed in this case.

$$RCCH + Ag^+ \rightarrow RCC^-Ag^+ + H^+$$
See also silver-mirror test.

toluene See methylbenzene.

tonne Symbol: t A unit of mass equal to 10^3 kilograms (i.e. one megagram).

torr A unit of pressure equal to a pressure of 101 325/760 pascals (133.322 Pa). One torr is equal to one mmHg.

tracer An isotope of an element used to investigate chemical reactions or physical processes (e.g. diffusion). See isotope.

trans- Designating an isomer with groups that are on opposite sides of a bond or structure. See isomerism.

transactinide elements Those elements following ^{103}Lw. To date, element 104 (variously named rutherfordium, kurchatovium, and unnilquadium), element 105 (hahnium), and, possibly, element 106 have been synthesized. All are unstable and have very short half-lives.
There is speculation that islands of stable nuclei with very high mass numbers exist since theoretical calculations predict unusual stability for elements 114 and 184.

transition elements A class of elements occurring in the periodic table in three series: from scandium to zinc; from yttrium to cadmium; and from lanthanum to mercury. The transition elements are all metals. They owe their properties to the presence of d electrons in the atoms. In the first transition series, calcium has the configuration $3s^23p^64s^2$. The next element scandium has $3s^23p^63d^14s^2$, and the series is formed by filling the d levels up to zinc at $3d^{10}4s^2$. The other two transition series occur by filling of the 4d and 5d levels. Often the elements scandium, yttrium, and lanthanum are considered with the lanthanoids rather than the transition elements, and zinc, cadmium, and mercury are also treated as a separate group. Transition elements have certain characteristic properties resulting from the existence of d levels:
(1) They have variable valences – i.e. they can form compounds with the metal in different oxidation states (Fe^{2+} and Fe^{3+}, etc.).
(2) They form a vast number of inorganic complexes, in which coordinate bonds are formed to the metal atom or ion.
(3) Compounds of transition elements are often coloured.
(4) Because of the unpaired electrons in d levels of many complexes, transition-metal compounds are often coloured.

See also metals, d-block, lanthanoids, zinc group.

transition state (activated complex) Symbol: ‡ A short-lived high-energy molecule, radical, or ion formed during a reaction between molecules possessing the necessary activation energy. The transition state decomposes at a definite rate to yield either the reactants again or the final products. The transition state can be considered to be at the top of the energy profile. For the reaction,

$$X + YZ = X \ldots Y \ldots Z^{\ddagger} \rightarrow XY + Z$$

the sequence of events is as follows. X approaches YZ and when it is close enough the electrons are rearranged producing a weakening of the bond between Y and Z. A partial bond is now formed between X and Y producing the transition state. Depending on the experimental conditions, the transition state either breaks down to form the products or reverts back to the reactants.

transition temperature A temperature at which some definite physical change occurs in a substance. Examples of such transitions are change of state, change of crystal structure, and change of magnetic behaviour.

transmutation A change of one element into another by radioactive decay or by bombardment of the nuclei with particles.

transport number Symbol: *t* In an electrolyte, the transport number of an ion is the fraction of the total charge carried by that type of ion in conduction.

transuranic elements Those elements of the actinoid series that have higher atomic numbers than uranium. The transuranic elements are formed by adding neutrons to the lower actinoids by high-energy bombardment and they generally have such short half-lives that macroscopic isolation is impossible. Neptunium and plutonium are formed in trace quantities by natural neutron bombardment from spontaneous fission of ^{235}U.

triatomic Denoting a molecule, radical, or ion consisting of three atoms. For example, O_3 and H_2O are triatomic molecules.

tribromomethane (bromoform, $CHBr_3$) A colourless liquid compound. *See* haloform.

trichloroacetic acid *See* chloroethanoic acid.

trichloroethanal (chloral, CCl_3CHO) A colourless liquid aldehyde made by chlorinating ethanal. It was used to make the insecticide DDT. It can be hydrolysed to *2,2,2-trichloroethanediol* (chloral hydrate, $CCl_3CH(OH)_2$). Most compounds with two −OH groups on the same carbon atom are unstable. However, in this case the effect of the three chlorine atoms stabilizes the compound. It is used as a sedative.

trichloroethane (chloroform, $CHCl_3$) A colourless volatile liquid formerly used as an anaesthetic. Now its main use is as a solvent and raw material for making other chlorinated compounds. Trichloroethane is made by reacting ethanal, ethanol, or propanone with chlorinated lime.

2,2,2-trichloroethanediol *See* trichloroethanal.

trichloroethanoic acid *See* chloroethanoic acid.

triclinic crystal *See* crystal system.

trigonal bipyramid *See* complex.

triglyceride A glyceride in which esters are formed with all three −OH groups. *See* glyceride.

trigonal crystal *See* crystal system.

trihydrate A crystalline hydrated compound that contains three molecules of water of crystallization per molecule of compound.

trihydric alcohol *See* triol.

triiodomethane (iodoform, CHI_3) A yellow crystalline compound made by warming ethanal with an alkaline solution of an iodide:

$$CH_3CHO + 3I^- + 4OH^- \rightarrow$$
$$CHI_3 + HCOO^- + 3H_2O$$

The reaction also occurs with all ketones of general formula CH_3COR (R is an alkyl group) and with secondary alcohols $CH_3CH(OH)R$. Iodoform is used as a test for such reactions (the *iodoform reaction*).

triiron tetroxide (ferrosoferric oxide, magnetic iron oxide, Fe_3O_4) A black solid prepared by passing either steam or carbon dioxide over red-hot iron. It may also be prepared by passing steam over heated iron(II) sulphide. Triiron tetroxide occurs in nature as the mineral magnetite. It is insoluble in water but will dissolve in acids to give a mixture of iron(II) and iron(III) salts in the ratio 1:2. Generally it is chemically unreactive; it is, however, a fairly good conductor of electricity.

trimer A molecule (or compound) formed by addition of three identical molecules. *See* ethanol, methanol.

trimethylaluminium (aluminium trimethyl, $(CH_3)_3Al$) A colourless liquid produced by the sodium reduction of dimethyl aluminium chloride. It ignites spontaneously on contact with air and reacts violently with water, acids, halogens, alcohols, and amines. Aluminium alkyls are used in the Ziegler process for the manufacture of high-density polyethene.

trimolecular Describing a reaction or step that involves three molecules interacting simultaneously with the formation of a product. For example, the final step in reaction between hydrogen peroxide and acidified potassium iodide is trimolecular:

$$HOI + H^+ + I^- \rightarrow I_2 + H_2O$$

It is uncommon for reactions to take place involving trimolecular steps. The oxidation of nitrogen(II) oxide to nitrogen(IV) oxide,

$$2NO + O_2 \rightarrow 2NO_2$$

is often classified as a trimolecular reaction but many believe it to involve two bimolecular reactions.

trinitrotoluene (TNT, $CH_3C_6H_2$-$(NO_2)_3$) A yellow crystalline solid. It is a highly unstable substance, used as an explosive. The compound is made by nitrating methylbenzene and the nitro groups are in the 2, 4, and 6 positions.

triol (trihydric alcohol) An alcohol that has three hydroxyl groups (-OH) per molecule of compound.

trioxygen *See* ozone.

triple bond A covalent bond formed between two atoms in which three pairs of electrons contribute to the bond. One pair forms a sigma bond (equivalent to a single bond) and two pairs give rise to two pi bonds. It is conventionally represented as three lines, thus H-C≡C-H. *See* multiple bond.

triple point The only point at which the gas, solid, and liquid phases of a substance can coexist in equilibrium. The triple point of water (273.16 K at 101 325 Pa) is used to define the kelvin.

trisodium phosphate(V) (sodium orthophosphate, Na_3PO_4) A white solid prepared by adding sodium hydroxide to disodium hydrogenphosphate. On evaporation white hexagonal crystals of the dodecahydrate ($Na_3PO_4.12H_2O$) are obtained. These crystals do not effloresce or deliquesce. They dissolve readily in water to produce alkaline solutions owing to salt hydrolysis. Trisodium phosphate is used as a water softener.

tritiated compound A compound in which one or more ^1H atoms have been replaced by tritium (^3H) atoms.

tritium Symbol: T A radioactive isotope of hydrogen of mass number 3. The nucleus contains 1 proton and 2 neutrons. Tritium decays with emission of low-energy beta radiation to give ^{3}He. The half-life is 12.3 years. It is useful as a tracer in studies of chemical reactions. Compounds in which ^{3}H atoms replace the usual ^{1}H atoms are said to be *tritiated*.

trivalent (tervalent) Having a valence of three.

tropylium ion The positive ion $C_7H_7^+$, having a symmetrical seven-membered ring of carbon atoms. The tropylium ion ring shows non-benzenoid aromatic properties.

tryptophan *See* amino acids.

tungsten A transition metal occurring naturally in wolframite $((Fe,Mn)WO_4)$ and scheelite $(CaWO_4)$. It was formerly called *wolfram*. It is used as the filaments in electric lamps.
Symbol: W; m.p. 3410°C; b.p. 5930°C; r.d. 19.3; p.n. 74; r.a.m. 183.85.

tungsten carbide (W_2C and WC) Either of two carbides produced by heating powdered tungsten with carbon. The carbides are extremely strong and are used in industry to make cutting tools or as an abrasive. WC has a very high melting point (2770°C) and will conduct electricity. W_2C also has a very high melting point (2780°C) but is a less efficient conductor of electricity. It is very resistant to chemical attack and behaves in a manner very similar to that of tungsten. It is strongly attacked by chlorine to give tungsten hexachloride, WCl_6.

tyrosine *See* amino acids.

ultracentrifuge A high-speed centrifuge used for separating out very small particles. The sedimentation rate depends on the particle size, and the ultracentrifuge can be used to measure the mass of colloidal particles and large molecules (e.g. proteins).

ultrahigh vacuum *See* vacuum.

ultraviolet (UV) A form of electromagnetic radiation, shorter in wavelength than visible light. Ultraviolet wavelengths range between about 1 nm and 400 nm. Ordinary glasses are not transparent to these waves; quartz is a much more effective material for making lenses and prisms for use with ultraviolet. Like light, ultraviolet radiation is produced by electronic transitions between the outer energy levels of atoms. However, having a higher frequency, ultraviolet photons carry more energy than those of light and can induce photolysis of compounds. *See also* electromagnetic radiation.

unimolecular Describing a reaction (or step) in which only one molecule is involved. For example, radioactive decay is a unimolecular reaction:

$$RA \rightarrow Rn + \alpha$$

Only one atom is involved in each disintegration.

In a unimolecular chemical reaction, the molecule acquires the necessary energy to become activated and then decomposes. The majority of reactions involve only uni- or bimolecular steps. The following reactions are all unimolecular:

$$N_2O_4 \rightarrow 2NO_2$$
$$PCl_5 \rightarrow PCl_3 + Cl_2$$
$$CH_3CH_2Cl \rightarrow C_2H_4 + HCl$$

unit A reference value of a quantity used to express other values of the same quantity. *See also* SI units.

unit cell The smallest group of atoms, ions, or molecules that, when repeated at regular intervals in three dimensions, will produce the lattice of a crystal system. There are seven basic types of unit cells, which result in the seven crystal systems.

unit processes (chemical conversions) The recognized steps used in chemical processes, e.g. alkylation, distillation, hydrogenation, pyrolysis, and nitration. Industrial processing and the economics, design, and use of the equipment are based on these unit processes rather than consideration of each reaction separately.

univalent (monovalent) Having a valence of one.

universal indicator (multiple-range indicator) A mixture of indicator dyestuffs that shows a gradual change in colour over a wide pH range. A typical formulation contains methyl orange, methyl red, bromothymol blue, and phenolphthalein and changes through a red, orange, yellow, green, blue, and violet sequence between pH 3 and pH 10. Several commercial preparations are available as both solutions and test papers.

unnilquadium (Unq) An alternative name for element 104 (un- = 1, -nil- = 0, -quad- = 4). It has proved unpopular and is rarely encountered.

unsaturated compound An organic compound that contains at least one double or triple bond between two of its carbon atoms. The multiple bond is relatively weak; consequently unsaturated compounds readily undergo addition reactions to form single bonds. *Compare* saturated compound.

unsaturated solution *See* saturated solution.

unsaturated vapour *See* saturated vapour.

uranium A toxic radioactive silvery element of the actinoid series of metals. Its three naturally occurring radioisotopes, ^{238}U (99.283% in abundance), ^{235}U (0.711%), and ^{234}U (0.005%), are found in numerous minerals including the uranium oxides pitchblende, uraninite, and carnotite. The readily fissionable ^{235}U is a major nuclear fuel and nuclear explosive, while ^{238}U is a source of fissionable ^{239}Pu.
Symbol: U; m.p. 1132°C; b.p. 3818°C; r.d. 19.1; p.n. 92; r.a.m. 238.03.

uranium hexafluoride (UF_6) A crystalline volatile compound, used in separating uranium isotopes by differences in the rates of gas diffusion.

urea (carbamide, $CO(NH_2)_2$) A white crystalline compound made from ammonia and carbon dioxide. It is used in the manufacture of urea–formaldehyde resins. Urea is the end product of metabolism in many animals and is present in urine.

V

vacancy *See* defect.

vacuum A space containing gas below atmospheric pressure. A perfect vacuum contains no matter at all, but for practical purposes *soft (low) vacuum* is usually defined as down to about 10^{-2} pascal, and *hard (high) vacuum* as below this. *Ultrahigh vacuum* is lower than 10^{-7} pascal.

vacuum distillation The distillation of liquids under a reduced pressure, so that the boiling point is lowered. Vacuum distillation is a common laboratory technique for purifying or separating compounds that would decompose at their 'normal' boiling point.

valence (valency) The combining power of an element or radical, equal to the number of hydrogen atoms that will combine with or displace one atom of the element. For simple covalent molecules the valence is obtained directly, for example C in CH_4 is tetravalent; N in NH_3 is trivalent. For ions the valence is regarded as equivalent to the magnitude of the charge; for example Ca^{2+} is divalent, CO_3^{2-} is a divalent radical. The noble gases are zerovalent because they do not form compounds under normal conditions. As the valence for many elements is constant, the valence of some elements can be deduced without reference to compounds formed with hydrogen. Thus, as the valence of chlorine in HCl is 1, the valence of aluminium in $AlCl_3$ is 3; as oxygen is divalent (H_2O) silicon in SiO_2 is tetravalent. The product of the valence and the number of atoms of each element in a compound must be equal. For example, in Al_2O_3 for the two aluminium atoms (valence 3) the product is 6 and for the three oxygen atoms (valence 2) the product is also 6.
The valence of an element is generally equal to either the number of valence electrons or eight minus the number of valence electrons. Transition metal ions display variable valence.

valence electron An outer electron in an atom that can participate in forming chemical bonds.

valency *See* valence.

valeric acid *See* pentanoic acid.

valine *See* amino acids.

vanadium A transition element occurring in complex ores in small quantities. It is used in alloy steels. Vanadium forms compounds with oxidation states $+5$, $+4$, $+3$, and $+2$. It forms coloured ions. Symbol: V; m.p. 1900°C; b.p. 3380°C; r.d. 6.1; p.n. 23; r.a.m. 50.94.

vanadium(V) oxide (vanadium pentoxide, V_2O_5) An oxide of vanadium extensively used as a catalyst in oxidation processes, as in the contact process.

van der Waals equation An equation of state for real gases. For n moles of gas the equation is
$$(p + n^2a/V^2)(V - nb) = nRT$$
where p is the pressure, V the volume, and T the thermodynamic temperature. a and b are constants for a given substance and R is the gas constant. The equation gives a better description of the behaviour of real gases than the perfect gas equation ($pV = nRT$).
The equation contains two corrections: b is a correction for the non-negligible size of the molecules; a/V^2 corrects for the fact that there are attractive forces between the molecules, thus slightly reducing the pressure from that of an ideal gas. *See also* gas laws, kinetic theory.

van der Waals force An intermolecular force of attraction, considerably weaker than chemical bonds and arising from weak electrostatic interactions between molecules (the energies are often less than 1 J mol^{-1}).

The van der Waals interaction contains contributions from three effects; permanent dipole-dipole interactions found for any polar molecule; dipole-induced dipole interactions, where one dipole causes a slight charge separation in bonds that have a high polarizability; and dispersion forces, which result from temporary polarity arising from asymmetrical distribution of electrons around the nucleus. Even atoms of the noble gases exhibit dispersion forces.

van't Hoff factor Symbol: i The ratio of the number of particles present in a solution to the number of undissociated molecules added. It is used in studies of colligative properties, which depend on the number of entities present. For example, if n moles of a compound are dissolved and dissociation into ions occurs, then the number of particles present will be in. Osmotic pressure (Π), for instance, will be given by the equation
$$\Pi V = inRT$$

van't Hoff isochore The equation:
$$d(\log_e K)/dT = \Delta H/RT^2$$
showing how the equilibrium constant, K, of a reaction varies with thermodynamic temperature, T. ΔH is the enthalpy of reaction and R is the gas constant.

vaporization The process by which a liquid or solid is converted into a gas or vapour by heat. Unlike boiling, which occurs at a fixed temperature, vaporization can occur at any temperature. Its rate increases as the temperature rises.

vapour A gas formed by the vaporization of a solid or liquid. Some particles near the surface of a liquid acquire sufficient energy in collisions with other particles to escape from the liquid and enter the vapour; some particles in the vapour lose energy in collisions and re-enter the liquid. At a given temperature an equilibrium is established, which determines the vapour pressure of the liquid at that temperature.

vapour density The ratio of the mass of a certain volume of a vapour to the mass of an equal volume of hydrogen (measured at the same temperature and pressure). Determination of vapour densities is one method of finding the relative molecular mass of a compound (equal to twice the vapour density). Victor Meyer's method, Dumas' method, or Hofmann's method can be used.

vapour pressure The pressure exerted by a vapour. The *saturated vapour pressure* is the pressure of a vapour in equilibrium with its liquid or solid. It depends on the nature of the liquid or solid and the temperature.

vat dyes A class of insoluble dyes applied by first reducing them to derivatives that are soluble in dilute alkali. In this condition they have a great attraction for certain fibres, such as cotton. The solution is applied to the material and the insoluble dye is regenerated in the fibres by atmospheric oxidation. Indigo and indanthrene are examples of vat dyes.

verdigris The green patina which occurs on copper and bronze exposed to the atmosphere. It is usually a mixture of basic copper carbonates and chlorides.

vermilion *See* mercury(II) sulpide.

vicinal positions Positions in a molecule at adjacent atoms. For example, in 1,2-dichloroethane the chlorine atoms are in vicinal positions, and this compound can thus be named *vic*-dichloroethane.

Victor Meyer's method A method for determining vapour densities in which a given weight of sample is vaporized and the volume of air displaced by it is measured. In practice, a bulb in a heating bath is connected via a fairly long tube to a water-bath gas-collection arrangement. The system is brought to equilibrium and the sample is then added (without opening the apparatus to the atmosphere). As the air displaced by gas is collected over water a correction for the vapour pressure of water is necessary and the method may fail if the vapour is soluble in water. *See also* Dumas' method, Hofmann's method.

vinyl chloride *See* chloroethene.

vinyl group The group $CH_2{:}CH{-}$.

viscosity Symbol: η The resistance to flow of a fluid.

vitreous Resembling glass, or having the structure of a glass.

volatile Easily converted into a vapour.

volt Symbol: V The SI unit of electrical potential, potential difference, and e.m.f., defined as the potential difference between two points in a circuit between which a constant current of one ampere flows when the power dissipated is one watt. $1\ V = 1\ J\ C^{-1}$.

voltaic cell *See* cell.

volumetric analysis One of the classical wet methods of quantitative analysis. It involves measuring the volume of a solution of accurately known concentration that is required to react with a solution of the substance being determined. The solution of known concentration (standard solution) is added in small portions from a burette. The process is called a titration and the equivalence point is called the *end point*. End points are observed with the aid of indicators or by instrumental methods, such as conduction or light absorption. Volumetric analysis can also be applied to gases. The gas is typically held over mercury in a graduated tube, and volume changes are measured on reaction or after absorption of components of a mixture.

vulcanization A process of improving the quality of rubber (hardness and resistance to temperature changes) by heating it with sulphur (about 150°C). Accelerators are used to speed up the reaction. Certain sulphur compounds can also be used for vulcanization.

W

Wacker process An industrial process for making ethanal (and other carbonyl compounds). To produce ethanal, ethene and air are bubbled through an acid solution of palladium(II) chloride and copper(II) chloride (20–60°C and moderate pressure):

$$C_2H_4 + Pd^{2+} + O_2 \rightarrow CH_3CHO + Pd + 2H^+$$

The reaction involves an intermediate complex between palladium(II) ions and ethene. The purpose of the copper(II) chloride is to oxidize the palladium back to Pd^{2+} ions:

$$Pd + 2Cu^{2+} \rightarrow Pd^{2+} + 2Cu^+$$

The copper(I) ions spontaneously oxidize to copper(II) ions in air. The process provides a cheap source of ethanal (and, by oxidation, ethanoic acid) from the readily available ethene.

Walden inversion A reaction in which an optically active compound reacts to give an optically active product in which the configuration has been inverted. This happens in the S_N2 mechanism. *See* nucleophilic substitution.

washing soda *See* sodium carbonate.

water (H_2O) A colourless liquid that freezes at 0°C and, at atmospheric pressure, boils at 100°C. In the gaseous state water consists of single H_2O molecules. Due to the presence of two lone pairs the atoms do not lie in a straight line, the angle between the central oxygen atom and the two hydrogen atoms being 105°; the distance between each hydrogen atom and the oxygen atom is 0.099 nm. When ice forms, hydrogen bonds some 0.177 nm long develop between the hydrogen atom and oxygen atoms in adjacent molecules, giving ice its tetrahedral crystalline structure with a density of 916.8 kg m^{-3} at STP. Different ice structures develop under higher pressures. When ice melts to form liquid water, the tetrahedral structure breaks down, but some hydrogen bonds continue to exist; liquid water consists of groups of associated water molecules, $(H_2O)_n$, mixed with some monomers and some dimers. This mixture of molecular species has a higher density than the open-structured crystals. The maximum density of water, 999.97 kg m^{-3}, occurs at 3.98°C. This accounts for the ability of ice to float on water and for the fact that water pipes burst on freezing.

Although water is predominantly a covalent compound, a very small amount of ionic dissociation occurs ($H_2O = H^+ + OH^-$). In every litre of water at STP there is approximately 10^{-7} mole of each ionic species. It is for this reason that, on the pH scale, a neutral solution has a value of 7. As a polar liquid, water is the most powerful solvent known. This is partly a result of its high dielectric constant and partly its ability to hydrate ions. This latter property also accounts for the incorporation of water molecules into some ionic crystals as water of crystallization.

Water is decomposed by reactive metals (e.g. sodium) when cold and by less active metals (e.g. iron) when steam is passed over the hot metal. It is also decomposed by electrolysis.

water gas A mixture of carbon monoxide and hydrogen produced when steam is passed over red-hot coke or made to combine with hydrocarbons, e.g.

$$C(s) + H_2O(g) \rightarrow CO(g) + H_2(g)$$
$$CH_4(g) + H_2O(g) \rightarrow CO(g) + 3H_2(g)$$

The production of water gas using methane is an important step in the preparation of hydrogen for ammonia synthesis. *Compare* producer gas.

water of crystallization Water present in definite proportions in crystalline compounds. Compounds containing water of crystallization are called *hydrates*. Examples are copper(II) sulphate pentahydrate ($CuSO_4.5H_2O$) and sodium carbonate decahydrate ($Na_2CO_3.10H_2O$). The water can be removed by heating.

When hydrated crystals are heated the water molecules may be lost in stages. For example, copper(II) sulphate pentahydrate changes to the monohydrate ($CuSO_4.H_2O$) at $100°C$, and to the anhydrous salt ($CuSO_4$) at $250°C$. The water molecules in crystalline hydrates may be held by hydrogen bonds (as in $CuSO_4.H_2O$) or, alternatively, may be coordinated to the metal ion as a complex aquo ion.

water softening The removal from water of dissolved calcium, magnesium, and iron compounds, thus reducing the hardness of the water. The compounds are potentially damaging because they can accumulate in pipes and boilers and they react with, and therefore waste, soap. Temporary hardness can be removed by boiling the water. Permanent hardness can be removed in a number of ways: by distillation; by the addition of sodium carbonate, which causes dissolved calcium, for example, to precipitate out as calcium carbonate; and by the use of ion-exchange products such as Permutit and Calgon. *See also* hardness (of water).

watt Symbol: W The SI unit of power, defined as a power of one joule per second. $1 W = 1 J s^{-1}$.

wavelength Symbol: λ The distance between the ends of one complete cycle of a wave. Wavelength is related to the speed (c) and frequency (v) thus:
$$c = v\lambda$$

wave mechanics *See* quantum theory.

wave number Symbol: σ The reciprocal of the wavelength of a wave. It is the number of wave cycles in unit distance, and is often used in spectroscopy. The unit is the metre^{-1} (m^{-1}). The circular wave number (symbol: k) is given by:
$$k = 2\pi\sigma$$

wax A soft amorphous water-repellant organic material. The term is particularly used for fatty acid esters of monohydric alcohols.

weak acid (or base) An acid or base that is not fully dissociated in solution.

weber Symbol: Wb The SI unit of magnetic flux, equal to the magnetic flux that, linking a circuit of one turn, produces an e.m.f. of one volt when reduced to zero at a uniform rate in one second. $1 Wb = 1 V s$.

Weston cadmium cell A standard cell that produces a constant e.m.f. of 1.0186 volts at $20°C$. It consists of an H-shaped glass vessel containing a negative cadmium-mercury amalgam electrode in one leg and a positive mercury electrode in the other. The electrolyte – saturated cadmium sulphate solution – fills the horizontal bar of the vessel to connect the two electrodes. The e.m.f. of the cell varies very little with temperature, being given by the equation $E = 1.0186 - 0.000\,037\,(T - 293)$, where T is the thermodynamic temperature.

wet cell A type of cell, such as a car battery, in which the electrolyte is a liquid solution. *See also* Leclanché cell.

white lead *See* lead(II) carbonate hydroxide.

Williamson's continuous process A method of preparing simple ethers by dehydration of alcohols with concentrated sulphuric acid. The reaction is carried out at $140°C$ under reflux with an excess of the alcohol:
$$2ROH \rightarrow ROR + H_2O$$
The concentrated sulphuric acid both catalyses the reaction and displaces the equilibrium to the right. The product, ether, is termed 'simple', because the R groups are identical.

Williamson's synthesis A method for the preparation of mixed ethers by nucleophilic substitution. A haloalkane is refluxed with an alcoholic solution of sodium alkoxide (from sodium dissolved in alcohol):
$$R'Cl + RO^-Na^+ \rightarrow R'OR + NaCl$$
The product ether is termed 'mixed' if the alkyl groups R and R' are different. This

synthesis can produce both simple and mixed ethers.

witherite A mineral form of barium carbonate, $BaCO_3$.

Wöhler's synthesis A synthesis of urea from inorganic compounds (Friedrich Wöhler, 1828). The urea was produced by evaporating NH_4NCO (ammonium cyanate). The experiment was important as it demonstrated that 'organic' chemicals (i.e. ones produced in living organisms) could be made from inorganic starting materials.

wolfram A former alternative name for tungsten.

wood alcohol *See* methanol.

Wood's metal A fusible alloy containing 50% bismuth, 25% lead, and 12.5% tin and cadmium. Its low melting point (70°C) leads to its use in fire-protection devices.

wrought iron A low-carbon steel obtained by refining the iron produced in a blast furnace. Wrought iron is processed by heating and hammering to reduce the slag content and to ensure its even distribution. It was used to make the girders for the Eiffel tower, and is now used to make chains and fences.

Wurtz reaction A reaction for preparing alkanes by refluxing a haloalkane (RX) with sodium metal in dry ether:
$$2RX + 2Na \rightarrow RR + 2NaX$$
The reaction involves the coupling of two alkyl radicals. The *Fittig reaction* is a similar process for preparing alkyl-benzene hydrocarbons by using a mixture of halogen compounds. For example, to obtain methylbenzene:
$$C_6H_5Cl + CH_3Cl + 2Na \rightarrow C_6H_5CH_3 + 2NaCl$$
In this mixed reaction phenylbenzene $(C_6H_5C_6H_5)$ and ethane (CH_3CH_3) are also produced by side reactions.

X

xenon A colourless odourless monatomic element of the rare-gas group, and known to form compounds. It occurs in trace amounts in air. Xenon is used in electron tubes and strobe lighting. Symbol: Xe; m.p. $-111.9°C$; b.p. $-107.1°C$; d. 5.89 kg m^{-3}; p.n. 54; r.a.m. 131.3.

X-radiation An energetic form of electromagnetic radiation. The wavelength range is 10^{-11} m to 10^{-8} m. X-rays are normally produced by absorbing high-energy electrons in matter. The radiation can pass through matter to some extent (hence its use in medicine and industry for investigating internal structures). It can be detected with photographic emulsions and devices like the Geiger-Müller tube.

X-ray photons result from electronic transitions between the inner energy levels of atoms. When high-energy electrons are absorbed by matter, an X-ray line spectrum results. The structure depends on the substance and is thus used in X-ray spectroscopy. The line spectrum is always formed in conjunction with a continuous background spectrum. The minimum (cut-off) wavelength λ_0 corresponds to the maximum X-ray energy, W_{max}. This equals the maximum energy of electrons in the beam producing the X-rays. Wavelengths in the continuous spectrum above λ_0 are caused by more gradual energy loss by the electrons, in the process called *bremsstrahlung* (braking radiation).

X-ray crystallography The study of the internal structure of crystals using the technique of X-ray diffraction.

X-ray diffraction A technique used to determine crystal structure by directing X-rays at the crystals and examining the diffraction patterns produced. At certain angles of incidence a series of spots are produced on a photographic plate; these spots are caused by interaction between the X-rays and the planes of the atoms, ions, or molecules in the crystal lattice. The positions of the spots can be shown to be consistent with the Bragg equation $n\gamma = 2d\sin\theta$.

xylene *See* dimethylbenzene.

Y

ytterbium A soft malleable silvery element having two allotropes and belonging to the lanthanoid series of metals. It occurs in association with other lanthanoids. Ytterbium has been used to improve the mechanical properties of steel.
Symbol: Yb; m.p. 819°C; b.p. 1194°C; r.d. 6.97 or 6.54; p.n. 70; r.a.m. 173.04.

yttrium A silvery metallic element belonging to the second transition series. It is found in almost every lanthanoid mineral, particularly monazite. Yttrium is used in various alloys, in yttrium-aluminium garnets used in the electronics industry and as gemstones, and as a catalyst. A mixture of yttrium and europium oxides is widely used as the red phosphor on television screens.

Symbol: Y; m.p. 1523°C; b.p. 3337°C; r.d. 4.47; p.n. 39; r.a.m. 8891.

Zeisel reaction The reaction of an ether with excess concentrated hydroiodic acid. On refluxing, a mixture of iodoalkanes is formed:

ROR' + 2HI → H₂O + RI + R'I

Analysis of these gives information about the structure of the original ether.

zeolite A member of a group of hydrated aluminosilicate minerals, which occur in nature and are also manufactured for their ion-exchange and selective-absorption properties. They are used for water softening and for sugar refining. The zeolites have an open crystal structure and can be used as molecular sieves.
See also ion exchange, molecular sieve.

zero order Describing a chemical reaction in which the rate of reaction is independent of the concentration of a reactant; i.e.

$$rate = k[X]^0$$

The concentration of the reactant remains constant for a period of time although other reactants are being consumed. The hydrolysis of 2-bromo-2-methyl propane using aqueous alkali has a rate expression,

$$rate = k[2\text{-bromo-2-methyl propane}]$$

i.e. the reaction is zero order with respect to the concentration of the alkali. The rate constant for a zero reaction has the units mol dm⁻³ s⁻¹.

zero point energy The energy possessed by the atoms and molecules of a substance at absolute zero (0 K).

Ziegler process A method for the manufacture of high-density polyethene using a catalyst of titanium(IV) chloride and triethyl aluminium (Al(C₂H₅)₃) under slight pressure. The mechanism involves formation of titanium alkyls

$$TiCl_3(C_2H_5)$$

which coordinate the σ-orbitals of titanium with the π bond of ethene. The chain-

length, and hence the density, of the polymer can be controlled.

zinc A transition metal occurring naturally as the sulphide (zinc blende) and carbonate (smithsonite). It is extracted by roasting the ore in air and then reducing the oxide to the metal using carbon. Zinc is used to galvanize iron, in alloys (e.g. brass), and in dry batteries. It reacts with acids and alkalis but only corrodes on the surface in air. Zinc ions are identified in solution by forming white precipitates with sodium hydroxide or ammonia solution, both precipitates being soluble in excess reagent.
Symbol: Zn; m.p. 420°C; b.p. 908°C; r.d. 7.1; p.n. 30; r.a.m. 65.38.

zincate A salt containing the ion ZnO₂²⁻

zinc-blende structure A form of crystal structure that consists of a zinc atom surrounded by four sulphur atoms arranged tetrahedrally; each sulphur atom is similarly surrounded by four atoms of zinc. Zinc sulphide crystallizes in the cubic system. Covalent bonds of equal strength and length result in the formation of a giant molecular structure. If the zinc and sulphur atoms are replaced by carbon atoms the diamond structure is produced. Wurtzite (another form of zinc sulphide) is similar but belongs to the hexagonal crystal system.

zinc chloride (ZnCl₂) A white crystalline solid prepared by the action of dry hydrogen chloride or chlorine on heated zinc. The anhydrous product undergoes sublimation. The hydrated salt may be prepared by the addition of excess zinc to dilute hydrochloric acid and crystallization of the solution. Hydrates with 4, 3, 5/2, 3/ 2, and 1 molecule of water exist. Zinc chloride is extremely deliquescent and dissolves easily in water. It is used in dentistry, Leclanché cells, as a flux, and as a dehydrating agent in organic reactions.

zinc group A group of metallic elements in the periodic table, consisting of zinc (Zn), cadmium (Cd), and mercury (Hg). The elements all occur at the end of the three transition series and all have outer $d^{10}s^2$ configurations; Zn $[Ar]3d^{10}4s^2$, Cd $[Kr]4d^{10}5s^2$, Hg $[Xn]5d^{10}6s^2$. The elements all use only outer s-electrons in reaction and in combination with other elements unlike the coinage metals, which immediately precede them. Thus all form dipositive ions, and oxidation states higher than $+2$ are not known. In addition mercury forms the mercurous ion, sometimes written as Hg(I), but actually the ion $^+$Hg-Hg$^+$. Although the elements fall naturally at the end of the transition series their properties are more like those of main-group elements than of transition metals:
(1) The melting points of the transition elements are generally high whereas those of the zinc group are low (Zn 419.5°C, Cd 329.9°C, Hg −38.87°C).
(2) The zinc-group ions are diamagnetic and colourless whereas most transition elements have coloured ions and many paramagnetic species.
(3) The zinc group does not display the variable valency associated with transition metal ions. However the group does have the transition-like property of forming many complexes or coordination compounds, such as $[Zn(NH_3)_4]^{2+}$ and $[Hg(CN)_4]^{2-}$
Within the group, mercury is anomalous because of the Hg_2^{2+} ion mentioned above and its general resistance to reaction. Zinc and cadmium are electropositive and will react with dilute acids to release hydrogen whereas mercury will not. When zinc and cadmium are heated in air they burn to give the oxides, MO, which are stable to further strong heating. In contrast mercury does not react readily with oxygen (slow reaction at the boiling point), and mercuric oxide, HgO, decomposes to the metal and oxygen on further heating. All members of the group form dialkyls and diaryls, R_2M.

zinc oxide (ZnO) A compound prepared by the thermal decomposition of zinc nitrate or carbonate; it is a white powder when cold, yellow when hot. Zinc oxide is an amphoteric oxide, almost insoluble in water, but dissolves readily in both acids and alkalis. If mixed with powdered carbon and heated to red heat, it is reduced to the metal. Zinc oxide is used in the paint and ceramic industries. Medically it is used in zinc ointment.

zinc sulphate (ZnSO$_4$) A white crystalline solid prepared by the action of dilute sulphuric acid on either zinc oxide or zinc carbonate. On crystallization the hydrated salt is formed. The heptahydrate (ZnSO$_4$.7H$_2$O) is formed below 30°C, the hexahydrate (ZnSO$_4$.6H$_2$O) above 30°C, and the monohydrate (ZnSO$_4$.H$_2$O) at 100°C; at 450°C the salt is anhydrous. Zinc sulphate is extremely soluble in water. It is used in the textile industry.

zinc sulphide (ZnS) A compound that can be prepared by direct combination of zinc and sulphur. Alternatively it can be prepared as a white amorphous precipitate by bubbling hydrogen sulphide through an alkaline solution of a zinc salt or by the addition of ammonium sulphide to a soluble zinc salt solution. Zinc sulphide occurs naturally as the mineral blende. It dissolves in dilute acids to yield hydrogen sulphide. Impure zinc sulphide is phosphorescent. It is used as a pigment and in the coatings on luminescent screens.

zircon *See* zirconium.

zirconium A transition element that occurs in a gemstone, zircon (ZrSiO$_4$). It is used in some strong alloy steels.
Symbol: Zr; m.p. 1850°C; b.p. 4380°C; r.d. 6.5; p.n. 40; r.a.m. 91.22.

zwitterion (ampholyte ion) An ion that has both a positive and negative charge on the same species. Zwitterions occur when a molecule contains both a basic group and an acidic group; formation of the ion can be regarded as an internal acid-base reaction. For example, amino-ethanoic acid (glycine) has the formula H$_2$N.CH$_2$.

COOH. Under neutral conditions it exists as the zwitterion $^+H_3N.CH_2.COO^-$, which can be formed by transfer of a proton from the carboxyl group to the amine group. At low pH (acidic conditions) the ion $^+H_3N.CH_2.COOH$ is formed; at high pH (basic conditions) $H_2N.CH_2.COO^-$ is formed. See also amino acid.

APPENDIX

Contents

The Chemical Elements

'Relative atomic mass' in brackets is
nucleon number of most stable isotope.

Element	Symbol	Proton No.	Relative atomic mass
actinium	Ac	89	[227]
aluminium	Al	13	26.9815
americium	Am	95	[243]
antimony	Sb	51	121.75
argon	Ar	18	39.948
arsenic	As	33	74.9216
astatine	At	85	[210]
barium	Ba	56	137.34
berkelium	Bk	97	[247]
beryllium	Be	4	9.0122
bismuth	Bi	83	208.98
boron	B	5	10.81
bromine	Br	35	79.904
cadmium	Cd	48	112.40
caesium	Cs	55	132.905
calcium	Ca	20	40.08
californium	Cf	98	[251]
carbon	C	6	12.011
cerium	Ce	58	140.12
chlorine	Cl	17	35.453
chromium	Cr	24	51.996
cobalt	Co	27	58.9332
copper	Cu	29	63.546
curium	Cm	96	[247]
dysprosium	Dy	66	162.50
einsteinium	Es	99	[254]
erbium	Er	68	167.26
europium	Eu	63	151.96
fermium	Fm	100	[257]
fluorine	F	9	18.9984
francium	Fr	87	[223]
gadolinium	Gd	64	157.25
gallium	Ga	31	69.72
germanium	Ge	32	72.59
gold	Au	79	196.967
hafnium	Hf	72	178.49
helium	He	2	4.0026
holmium	Ho	67	164.930
hydrogen	H	1	1.00797
indium	In	49	114.82
iodine	I	53	126.9044
iridium	Ir	77	192.2
iron	Fe	26	55.847

The chemical elements continued

Element	Symbol	Proton No.	Relative atomic mass
krypton	Kr	36	83.80
lanthanum	La	57	138.91
lawrencium	Lr	103	[257]
lead	Pb	82	207.19
lithium	Li	3	6.939
lutetium	Lu	71	174.97
magnesium	Mg	12	24.305
manganese	Mn	25	54.938
mendelevium	Md	101	[258]
mercury	Hg	80	200.59
molybdenum	Mo	42	95.94
neodymium	Nd	60	144.24
neon	Ne	10	20.179
neptunium	Np	93	[237]
nickel	Ni	28	58.71
niobium	Nb	41	92.906
nitrogen	N	7	14.0067
nobelium	No	102	[255]
osmium	Os	76	190.2
oxygen	O	8	15.9994
palladium	Pd	46	106.4
phosphorus	P	15	30.9738
platinum	Pt	78	195.09
plutonium	Pu	94	[244]
polonium	Po	84	[209]
potassium	K	19	39.102
praseodymium	Pr	59	140.907
promethium	Pm	61	[145]
protactinium	Pa	91	[231]
radium	Ra	88	[226]
radon	Rn	86	[222]
rhenium	Re	75	186.20
rhodium	Rh	45	102.905
rubidium	Rb	37	85.47
ruthenium	Ru	44	101.07
samarium	Sm	62	150.35
scandium	Sc	21	44.956
selenium	Se	34	78.96
silicon	Si	14	28.086
silver	Ag	47	107.868
sodium	Na	11	22.9898
strontium	Sr	38	87.62
sulphur	S	16	32.064
tantalum	Ta	73	180.948
technetium	Tc	43	[97]
tellurium	Te	52	127.60
terbium	Tb	65	158.924

The chemical elements continued

Element	Symbol	Proton No.	Relative atomic mass
thallium	Tl	81	204.37
thorium	Th	90	232.038
thulium	Tm	69	168.934
tin	Sn	50	118.69
titanium	Ti	22	47.90
tungsten	W	74	183.85
uranium	U	92	238.03
vanadium	V	23	50.942
wolfram (tungsten)	W	74	183.85
xenon	Xe	54	131.30
ytterbium	Yb	70	173.04
yttrium	Y	39	88.905
zinc	Zn	30	65.37
zirconium	Zr	40	91.22

Physical Constants

Quantity	Magnitude	Unit
speed of light	$2.997\ 925 \times 10^8$	m s^{-1}
Planck constant	$6.626\ 196 \times 10^{-34}$	J s
Boltzmann constant	$1.380\ 622 \times 10^{-23}$	JK^{-1}
Avogadro constant	$6.022\ 169 \times 10^{23}$	mol^{-1}
mass of proton	$1.672\ 614 \times 10^{-27}$	kg
mass of neutron	$1.674\ 920 \times 10^{-27}$	kg
mass of electron	$9.109\ 558 \times 10^{-31}$	kg
charge of proton or electron	$\pm 1.602\ 191\ 7 \times 10^{-19}$	C
specific charge of electron	$-1.758\ 796 \times 10^{11}$	C kg^{-1}
molar volume at s.t.p.	$2.241\ 36 \times 10^{-2}$	m^3 mol^{-1}
Faraday constant	$9.648\ 670 \times 10^4$	C mol^{-1}
triple point of water	273.16	K
absolute zero	-273.15	$^\circ$C
permittivity of vacuum	$8.854\ 185\ 3 \times 10^{-12}$	F m^{-1}
permeability of vacuum	$4\pi \times 10^{-7}$	H m^{-1}
Stefan constant	$5.669\ 61 \times 10^{-8}$	W m^{-2}K^{-4}
molar gas constant	$8.314\ 34$	J mol^{-1}K^{-1}
gravitational constant	$6.673\ 2 \times 10^{-11}$	N m^2 kg^{-2}

Elementary Particles

Particle	Symbol	Charge	Mass	Spin
leptons				
electron	e^-	−1	0.511	1/2
neutrino	v	0	0	1/2
	$v\mu$	0	0	1/2
muon	μ^-	−1	105.66	1/2
baryons				
proton	p	+1	938.26	1/2
neutron	n	0	939.55	1/2
xi particle	Ξ°	0	1314.9	1/2
	Ξ^-	−1	1321.3	1/2
sigma particle	Σ^+	+1	1189.5	1/2
	Σ°	0	1192.5	1/2
	Σ^-	−1	1197.4	1/2
lambda particle	Λ	0	1115.5	1/2
omega particle	Ω^-	−1	1672.5	3/4
mesons				
kaon	κ^-	−1	493.8	0
	κ^+	+1	493.8	0
pion	π^+	+1	139.6	0
	π°	0	135	0
	π^-	−1	139.6	0
phi particle	Φ	0	1020	1
psi particle	ψ	0	3095	1
eta particle	η°	0	548.8	0

Note: it is common to measure the mass of particles in units of energy/c^2, where c_2 is the speed of light. The values above are in units of MeV/c^2. 1 MeV/$c^2 = 178 \times 10^{-30}$ kg.

The Greek Alphabet

Letters		Name
A	α	alpha
ß	β	beta
Γ	γ	gamma
Δ	δ	delta
E	ε	epsilon
Z	ζ	zeta
H	η	eta
Θ	θ	theta
I	ι	iota
K	κ	kappa
Λ	λ	lambda
M	μ	mu
N	ν	nu
Ξ	ξ	xi
O	o	omicron
Π	π	pi
P	ρ	rho
Σ	σ	sigma
T	τ	tau
Y	ν	upsilon
Φ	φ	phi
X	χ	chi
Ψ	ψ	psi
Ω	ω	omega

The Periodic Table

IA	IIA	IIIB	IVB	VB	VIB	VIIB	VIII	VIII	VIII	IB	IIB	IIIA	IVA	VA	VIA	VIIA	O
1 H																	2 He
3 Li	4 Be											5 B	6 C	7 N	8 O	9 F	10 Ne
11 Na	12 Mg											13 Al	14 Si	15 P	16 S	17 Cl	18 Ar
19 K	20 Ca	21 Sc	22 Ti	23 V	24 Cr	25 Mn	26 Fe	27 Co	28 Ni	29 Cu	30 Zn	31 Ga	32 Ge	33 As	34 Se	35 Br	36 Kr
37 Rb	38 Sr	39 Y	40 Zr	41 Nb	42 Mo	43 Tc	44 Ru	45 Rh	46 Pd	47 Ag	48 Cd	49 In	50 Sn	51 Sb	52 Te	53 I	54 Xe
55 Cs	56 Ba	57* La	72 Hf	73 Ta	74 W	75 Re	76 Os	77 Ir	78 Pt	79 Au	80 Hg	81 Tl	82 Pb	83 Bi	84 Po	85 At	86 Rn
87 Fr	88 Ra	89† Ac															

← s block →   ← Transition elements →   ← p block →

* Lanthanoids	57 La	58 Ce	59 Pr	60 Nd	61 Pm	62 Sm	63 Eu	64 Gd	65 Tb	66 Dy	67 Ho	68 Er	69 Tm	70 Yb	71 Lu
† Actinoids	89 Ac	90 Th	91 Pa	92 U	93 Np	94 Pu	95 Am	96 Cm	97 Bk	98 Cf	99 Es	100 Fm	101 Md	102 No	103 Lr